Pardo,

You are an
American hero.
I hope you enjoy the book.

V6

Matt "MeatHead"

Eagles Overhead

The History of US Air Force
Forward Air Controllers,
from the Meuse-Argonne to Mosul

Matt Dietz

Number 7 in the American Military
Studies Series

University of North Texas Press
Denton, Texas

10 9 8 7 6 5 4 3 2 1

The views expressed in this book are those of the author and do not
necessarily reflect the official policy or position of the United States
Air Force Academy, the Air Force, the Department of Defense, or
the US Government.

Permissions:
University of North Texas Press
1155 Union Circle #311336
Denton, TX 76203-5017

The paper used in this book meets the minimum requirements of the
American National Standard for Permanence of Paper for Printed Library
Materials, z39.48.1984. Binding materials have been chosen for durability.

 Library of Congress Cataloging-in-Publication Data

 Dietz, Matt, 1975– author.
Eagles overhead : the history of US Air Force forward air controllers, from
 the Meuse-Argonne to Mosul / Matt Dietz.
 American military studies ; no. 7.
 Denton, Texas : University of North Texas Press, [2023]
 Includes bibliographical references and index.
 ISBN-13 978-1-57441-890-3 (cloth)
 ISBN-13 978-1-57441-891-0 (ebook)
 1. LCSH: United States. Air Force—Tactical aviation—History. 2. Aerial
reconnaissance, American—History. 3. Close air support—History.

 UG763 .D54 2023
 358.4/50973–dc23
                                    2022051870

Eagles Overhead is Number 7 in the American Military Studies Series.

The electronic edition of this book was made possible by the support of
the Vick Family Foundation. Typeset by vPrompt eServices.

For Lauren

Contents

Acknowledgments

First and foremost, I want to thank my wife Lauren for supporting me, not only in this endeavor, but in everything for more than twenty years. Thank you for standing by me, loving me, encouraging me, and making me better every day. Thank you for all your sacrifices so I can chase my dreams. I could not have done any of this without you. Braxton and Will, thank you for being awesome, cheering me on, and for putting up with my history lessons. Hopefully this is interesting enough that you will actually read it. I love you all.

Dr. Geoffrey Wawro, Dr. Graham Cox, Dr. Alex Mendoza, Dr. Harold Tanner, Dr. Phil Haun, and Dr. John Terino—thank you for taking the time to turn a fighter pilot into an academic and, through much toil, making my writing better. Thank you, Colonel Meg Martin and the US Air Force Academy History faculty, for giving me this chance.

Thank you to the staff at the Ninth Air Force/US Air Forces Central Command (AFCENT) Historian's office, The Air Force Historical Research Agency, Penn State University Library Special Collections, the US Air Force Academy Library Special, and the Fort Worth Aviation Museum for your help and knowledge in guiding my research and making my life so much easier.

To the warriors of the Forward Air Controllers Association, the Mosquito Association, the Red River Valley Fighter Pilots Association, the OV-10 Bronco Association, the Strike Eagle Network, Friends of the Academy Library, the CAF Fighter Career Field, the OFPs, the 24 Tactical Air Support Squadron, and the Close Air Support Integration Group for adding your expertise, voice, photos, and war stories to this project. Thank you to Vicki Murphy for sharing your father's story with me—Fortune favors the Bold … Tigers! A special thanks to Hale, AAFF, Axel, Pako, Mung, Twist, Tatu, and Burt for filling in all the holes in my knowledge; your contributions are beyond my ability to repay them.

Hit My Smoke!

Introduction

On the afternoon of July 9, 1950, Air Force Lieutenants James A. Bryant and Frank G. Mitchell took off from an airfield near Daejon, Korea. The pilots flew in two Army L-17 liaison aircraft, which they had negotiated the use of from the 24th Infantry Division. The lieutenants flew into Daejon earlier in the day in Air Force L-5 Sentinel observation aircraft, but both airplanes experienced mechanical failures, forcing the pair to improvise. Despite the lack of accurate maps, they separately made their way toward the rapidly shifting front. The L-17 was a four-seat, all-metal airplane converted from a civilian design and not made for Lieutenants Bryant and Mitchell's task. By the end of the day, however, they had successfully directed ten flights of Air Force F-80 Shooting Stars in attacks against North Korean tanks and vehicles as they pressed the beleaguered US and South Korean forces. Despite the inauspicious beginning of their mission, that day, the two resourceful lieutenants became the first official US Air Force Airborne Forward Air Controllers.[1]

Charged with supplying air power from scarce US Air Force resources in the Pacific theater to the chaotic Korean battlefield, this trial mission was the brainchild of Lieutenant Colonel Stanley Latiolais, a staff officer in the Fifth Air Force operations directorate. The rapidly retreating US and South Korean forces were short on artillery support, and desperately needed airborne firepower to fill the gap. However, the jet aircraft of the day, such as the F-80, used fuel at an alarmingly high rate. Their fuel consumption increased when operating at the low altitudes required for conducting attacks against the North Korean army. Further compounding Fifth Air Force's problems was a lack of airfields on the Korean peninsula, forcing the US to fly missions from bases in Japan to battlefields in Korea. As a result, US jet fighters only had about fifteen minutes of combat time in support of beleaguered US and South Korean troops.[2]

To maximize the jet fighters' time over the battlefield, the Air Force launched Lieutenant Colonel Latiolais's Forward Air Controller experiment utilizing slower propeller-driven World War II aircraft. A Forward Air Controller or FAC (pronounced *fack*) is an aircrew that flies above the battlefield and whose primary job is to control close air support attacks of other aircraft. To accomplish the mission, the Air Force built a new unit, the 6147th Tactical Control Squadron, who used aircraft such as the L-5 observation aircraft and a modified version of the T-6 Texan for aerial reconnaissance over the front lines in search of enemy troops. Once they located the enemy, the pilots radioed an Air Force control network, which tasked American fighter bombers to the area. Simultaneously, the Forward Air Controller coordinated via another radio with ground forces, ensuring no friendly forces were nearby. When the fighter jets arrived overhead, the Forward Air Controller transmitted the target location to the jet pilots using latitude and longitude, terrain features—such as rivers or roads—and smoke rockets. When the fighters confirmed they had the targets in sight, the Forward Air Controller cleared them for bomb runs or strafe passes. Once the pilots destroyed their targets, ran out of bombs and ammunition, or were low on fuel; the team passed a battle damage assessment report over the radio to both the control network and ground forces, and the cycle began again.

Twenty-first century Air Force Forward Air Controllers still follow the methods used by Lieutenants Bryant and Mitchell in their borrowed Army airplanes. On the modern battlefield, however, they use satellite communication, onboard datalink networks, and advanced sensors. Technology enables modern Air Force Forward Air Controllers to locate targets, coordinate with ground forces, interface with command-and-control centers, and pass information to attack aircraft with a speed and precision unimagined by their predecessors. Although modern Air Force Forward Air Controllers use advanced fighter and attack aircraft, their roots grow in the mud of World War I and the first aviation mission—observation.

Despite receiving credit for creating the first official Air Force Forward Air Controllers, Lieutenant Colonel Latiolais's idea did not spring forth fully formed in 1950. Instead, the success in Korea was the product of nearly four decades of experimentation, technological advancement, and combat trial by fire. Latiolais and the pilots of the 6147th Tactical Control Squadron built on their own experiences in World War II, and the efforts of airmen since World War I, to create new procedures, organizational structures, and doctrine. Methods refined and developed in Korea were a product of fledgling ground-attack missions over the trenches of World War I and a massive air effort supporting Allied ground forces in World War II. In turn, the lessons of Korea trickled down to Forward Air Controllers operating in South Vietnam and above the Ho Chi Minh Trail. The hard lessons of Vietnam were instrumental in building permanent Air Force Forward Air Controller capabilities, which contributed to the success of Operation Desert Storm and Operation Allied Force. However, since 2001 the skills of Air Force Forward Air Controllers have remained mostly untapped. This begs the question, why has this capability been underused in modern American military operations?

The role of Air Force Forward Air Controllers in the history of aerial warfare is unmistakable, and the Forward Air Controllers' mission, in many ways, is the original air power mission. Nevertheless, the Air Force historically shunned it, because it yokes air power to land forces. The Air Force traditionally favored missions and doctrine of air superiority and strategic bombing. As a result, much of the Forward Air Controllers' mission was passed through the generations of Air Force pilots by individuals and innovators. Institutional knowledge, in the form of doctrine and tactical manuals, was often discarded as the Air Force pursued other missions. This systemic slant towards air superiority and strategic bombing is evident in almost everything the Air Force does—from doctrine to organizational structures, and aircraft acquisitions. Even the historical study and theoretical writing by air power theorists and historians tends to reflect this institutional bias and advocacy.

The bulk of air power study, writing, doctrine, and theory focus primarily on either the struggle for air superiority or strategic bombing campaigns. There are practical reasons for this emphasis. For example, as the US Air Force—and other air forces—sought independence from armies—who controlled military air services prior to World War II—the air-centric doctrine and study became desirable as a way of asserting independence from the older services, the armies and navies. Early air power theorists and operators such as Giulio Douhet, Hugh Trenchard, and Billy Mitchell envisioned air forces untethered to an army, operating against the strategic heart of the enemy, and unencumbered by the demands of land war. Independent air power as a strategic weapon became ingrained in the heart of air power doctrine and theory for good theoretic reasons, but also for political-bureaucratic reasons: to remove air forces from army control. Early doctrine sought to prove that air power must be separate from land power to achieve the strategic, war-winning effects that theorists argued only air power could deliver.[3]

Air power advocates regard the separation of air power from armies as a necessity. They argue that for air power to be most effective military leaders must allow it to take full advantage of its capabilities. To a point, they are correct. It would be absurd to wed naval power exclusively to land forces, or vice versa, or make land forces subject to every whim of air power operations. It is for this reason that air power advocates pressed for operational independence and control of air forces. However, in making their case or defending their air power turf, air forces—and the US Air Force is no exception—often turned entirely away from any tie to land forces. This attitude has resulted in a circle of doctrine and theory that nearly severs air power entirely from the land forces to achieve a wholly independent air force.

The Air Force's preference for air superiority and strategic bombing is the principal reason why analysis of the Forward Air Controller lags writing on other missions. Despite the dearth of historiography, a clear evolution and improvement in Air Force Forward Air Controller programs from World War I to Korea is apparent. However, there is no work focusing on the development of Forward Air Controllers before the Korean War, although a few books address ground-attack or close air support in early military aviation. Richard P. Hallion's *Strike from the Sky: The History of Battlefield Air Attack, 1911–1945* provides an excellent account of this period. As the title suggests, Hallion traces the evolution of ground-attack from the earliest days of aviation through the end of World War II. A few works focus on air-to-ground operations in World War II. Thomas Alexander Hughes's *Overlord: General Pete Quesada and the Triumph of Tactical Air Power in World War II*, and David Spires's *Air Power for Patton's Army: The XIX Tactical Air Command in the Second World War*, address the evolution of the Army Air Forces' command-and-control, close air support, and interdiction during World War II—all of which are critical to the development of the Air Force Forward Air Controller program in Korea. However, only Hallion gives a passing reference to the evolution of the role of Forward Air Controllers within these roots.[4]

Even though Air Force Forward Air Controllers officially began during the Korean War, only a few works focus on their role there. Accounts of F-86 Sabres in aerial combat over "MiG Alley," the area in northwestern Korea where aerial combat between US and communist air forces was most intense, dominate the writing of the era. Robert Futrell's *The United States Air Force in Korea, 1950–1953* provides a sweeping overview of the war, but offers only a small section devoted to Lieutenant Colonel Latiolais and the "Mosquito" pilots of the 6147th Tactical Control Squadron. William Y'Blood's *Down in the Weeds—Close Air Support in Korea* focuses more tightly on the close air support mission, of which the Mosquito Forward Air Controllers were a significant part. Moving in closer, W. M. Cleveland's *Mosquitos in Korea* and Major Timothy Kline's *The Airborne Forward Air Controller in the Korean War* individually examine the US Air Force's first Forward Air Controllers. Finally, Gary Lester's *Mosquitoes to Wolves: The Evolution of the Forward Air Controller* traces the transition of Air Force Forward Air Controllers from Korea to the heyday of Air Force Forward Air Controllers, Vietnam.[5]

Air Force Forward Air Controller missions exploded during the decade of American fighting in Vietnam and the Vietnam War also serves as the high-water mark for writing about them. The books range from official Air Force histories, to think tank studies, and finally memoirs spanning all eras and sub-theaters of the war. Many address the use of air power in the land war in Southeast Asia, far more than any other American conflict. Institutional histories primarily focus on the interdiction of North Vietnamese men and material, such as Bernard Nalty's *The War against the Trucks*. Other works examine the use of air power in a close air support role. Again, Nalty is a crucial contributor with *Air War over South Vietnam*. Donald Mrozek adds *Air Power and the Ground War in Vietnam* as another example of these works.[6]

Dozens of memoirs add to the official and academic histories of Forward Air Controller operations in Vietnam, far more than any other American conflict. They provide fascinating insight into the staggering number and variety of Vietnam Forward Air Controllers and their exploits. For example, writing about and by Forward Air Controllers who served in South Vietnam providing close air support to US and South Vietnamese troops is exemplified in Mike Jackson and Tara Dixon-Engel's 2004 *Naked in Da Nang*. However, their experience is entirely different from Forward Air Controllers who flew interdiction missions along the Ho Chi Minh Trail. Another example is Christopher Robbins' 1987 *The Ravens*, the foundational account of Air Force and CIA Forward Air Controllers in Laos and Cambodia. Finally, the "Fast FAC" mission in Vietnam—Forward Air Controllers who flew jet fighters instead of slower propeller-driven aircraft—offer yet another genre of the mission and historiography, exemplified by the 2006 *Bury Us Upside Down* by Rick Newman and Don Shepperd.[7]

After Vietnam, however, writing about the subject went quiet, not reappearing until 1991's Operation Desert Storm and 1999's Operation Allied Force. Yet, the explosive growth of air power writing on other themes during that decade crowded out writing on Forward Air Controllers. Of the writing specifically on Forward Air Controllers, Colonel Christopher Haave and Lieutenant Colonel Phil Haun's *A-10s Over Kosovo* represents the most significantly focused work on the subject.[8] The dual conflicts in Afghanistan and Iraq again brought land campaigns to the forefront of American military writing, and after 2001 a debate emerged in the defense community around the Air Force's ability to provide close air support to US and Coalition armies.[9] This debate included some discussion of Forward Air Controllers and their critical role in that mission. However, Air Force Forward Air Controller's contributions to any campaign since 2001 remained minimal, and by extension, so has writing about them. Outside of a handful of theses written for US military staff schools and war colleges, writing about Forward Air Controllers ends in 2001.

This walk through the historiography reveals a second question about the US Air Force Forward Air Controller mission: why is there not a complete history of the program? It seems a dedicated history of US Air Force Forward Air Controller programs is long overdue. Therefore, the primary aim of this book is to provide a critical history of US Air Force Forward Air Controllers and examine their role, status, and performance within the service's history. Thus, this book begins by examining America's initial adoption of air power and entry into World War I. It then follows Forward Air Controllers through every US Air Force air campaign from the Marne in 1918 to Mosul in 2017.

As a product of this examination, this book asks the question: why, despite the critical importance of Forward Air Controllers in the decades prior, do they rarely appear on US battlefields after 2001? Digging into the theoretical, doctrinal, institutional, and historical frameworks in the first nine chapters, I ask if the nature of air warfare changed so significantly that it left the concept and utility of the Forward Air Controller behind. Or, was the Forward Air Controller neglected since 2001 because the Air Force dislikes the capability as it clouds the service's doctrinal preferences? From these examinations, I draw conclusions about the future of Air Force Forward Air Controllers.

To tackle Air Force Forward Air Controllers' history and possible future, this book opens with the inception of military aviation in the United States prior to World War I. The first chapter traces the growth of air power from novelty to key warfighting technology through the war and into the postwar period. The next three chapters trace the creation and development of Air Force Forward Air Controllers and their mission during World War II, Korea, and Vietnam. Chapter five examines the remaking of the post-Vietnam US military and the impact of those changes on Forward Air Controllers. The Air Force's steady neglect of their mission from Operation Desert Storm, through the force reductions after the Soviet Union's collapse, and into the post 9-11 wars in Iraq and Afghanistan round out the final two chapters. Harnessing the historical examinations in the book's seven main chapters, the conclusion sketches possible futures for Air Force Forward Air Controllers.

It is important to note here that this book will focus solely on Air Force Forward Air Controllers, while mentioning the efforts of other Forward Air Controllers only in passing. This is not a slight, in fact Marine Corps Forward Air Controllers are a constant presence in Marine operations as far back as World War I. They saw heavy action in World War II, Korea, Vietnam, and Desert Storm. Indeed, Air Force Forward Air Controllers would have done well to learn from or emulate their Marine counterparts. However, the Marines' organization subjugates Marine Corps aircraft to the infantry; and Marine Corps pilots focus almost exclusively on providing support to Marines on the ground, enabling a unique working relationship between Marine air and ground units. In contrast, the US Air Force is responsible for a wide variety of missions across a much larger scope of responsibility than Marine aviation.[10] Therefore, as we will see throughout the book, a steady tension around air support to ground forces existed both within the Air Force and between the Air Force and other services, particularly after the Air Force and Army split in 1947.

Part of understanding the history and potential future for Forward Air Controllers means gaining an understanding of who they are and what they do. However, part of the problem of understanding the mission is deciphering the myriad of different terms describing Forward Air Controllers, their mission, and air power missions in general. Therefore, it is necessary to ensure readers of all backgrounds understand the terms used in this book; not just those well versed in military operations, air power, or the intricacies of Forward Air Controller operations. Further, as the military is prone to do, the terms used to describe specific military missions or capabilities changed over time; creating different lexicons across and within services depending on the era examined. The Joint Chiefs of Staff endeavored to standardize these terms, creating a standard operating language, and publishing manuals defining the terms and describing the operations thereof, but of course few people—even military people—bother to read all those manuals.

So, what is a Forward Air Controller, and what do they do? Current Joint Doctrine defines a Forward Air Controller as: "FAC(A)—Forward Air Controller (Airborne)—A specifically trained and qualified aviation officer, normally an airborne extension of the tactical air control party, who exercises control from the air of aircraft engaged in close air support of ground troops."[11] However, throughout the history of air power, Forward Air Controllers were also called Observers, Forward Observers, Airborne Forward Air Controller, Ground Forward Air Controller, Forward Air Controller Ground, Forward Air Controller Airborne, Joint Terminal Attack Controller, Tactical Air Control Party, and others. Further confusing the issue, Air Force pilots who perform this mission—both currently and in the past—most commonly use the term Forward Air Controller or FAC to describe their role as an Airborne Forward Air Controller. In contrast, Navy and Marine Aviators use FAC(A) for airborne operations and Forward Air Controller for ground-based operations. This is largely because all Navy and Marine FAC(A)s are first qualified as Forward Air Controllers on the ground before they can be qualified for the job in the air, which is not the case with the Air Force.[12]

To standardize terms, and for readability, the terms Forward Air Controller and FAC will be used here for Air Force Airborne Forward Air Controllers and defined as an Air Force officer in an aircraft, who serves as an extension of Tactical Air Control Parties, and works with ground forces to control aircraft engaged in close air support of friendly ground forces. A second term which is critical to understanding what a Forward Air Controller does is the Tactical Air Control Party (TACP). This is a team of Air Force personnel embedded with friendly ground forces designated to "provide air liaison to land forces."[13] Along with command-and-control elements, the Forward Air Controller and the Tactical Air Control Party work together to provide friendly ground forces with fire support from the air. Forward Air Controllers and specially qualified members of the Tactical Air Control Party, called Joint Terminal Attack Controllers or JTACs, are the only people qualified to control aircraft attacks in close air support situations. That means Forward Air Controllers are ultimately responsible for where, when, and how friendly aircraft attack enemy targets. They also bear the repercussions of those attacks when situations go wrong. This responsibility makes their job dangerous, challenging, and critical to protecting friendly ground forces.

A Forward Air Controller's mission must be well integrated and closely coordinated with land forces to be successful. Further, depending on the nature of the battlefield, successful missions often require overwhelming numbers of aircraft, multiple command-and-control structures, and time— things usually in short supply on the battlefield. The nature of the mission and the resources required for success historically caused the Air Force to struggle to find a place for the role of the Forward Air Controller. Most often the Air Force's difficulty with the mission was a byproduct of the types of conflicts and missions the service preferred to focus its resources on—primarily major wars and cutting-edge warfighting technology.

A 1967 RAND study of the role of Forward Air Controllers in Korea and their potential in Vietnam noted the Air Force's proclivities for more air-centric missions and focus on fighting major wars. The authors' observa-tions, which are as relevant now as they were in 1967, found the Air Force's primary interests repeatedly serve strategic capabilities and force structures designed for "maximum firepower." Further, the authors argued, the mission garnered little attention in "a service wrestling with the problems of missile forces and space stations."[14] The study feared the Air Force would turn away from Forward Air Controllers after Vietnam, as it did in the previous wars. Although the Air Force did not scrap the capability after Vietnam, as it had after previous wars, they suffered a course of steady decline, decreasing resources, and a broad neglect of the mission—except for a small cadre of pilots and planes.

After Vietnam, the Air Force briefly retained its expanded Forward Air Controller capability before allowing a steady decline in personnel, aircraft, and budget. The Air Force eliminated the capability through neglect of training and equipment as aircraft that proved successful in Vietnam were gradually retired due to age and obsolescence.[15] As the Air Force expanded and modernized in the 1980s and into the early 1990s, Forward Air Controller and close air support aircraft withered on the vine. Just as the authors of the 1967 study feared, the Air Force made no effort to fill the gaps, and the mission fell victim to an expanding and modernizing multirole fleet as defense budgets tightened in the 1990s. The resulting Air Force Forward Air Controller capabilities were unprepared to handle the workload of the post-2001 wars and, despite the clear warnings by the RAND study five decades ago, doubts about the Forward Air Controllers' future remain.[16]

To answer questions about the Forward Air Controller's possible role in future operations, we must understand where and how the Air Force applied air power on battlefields in the past, and in what circumstances they served as the conduit for that air power. Air power can be broadly applied in a variety of missions to either deter or attack an enemy or defend against an enemy attack. The focus here will be on three doctrinal applications of airpower: Strategic Attack, Counter Air, and Counter Land operations. Current Air Force doctrine defines Strategic Attack as "offensive action specifically selected to achieve national strategic objectives." Historically air forces accomplished this mission by bombing "an enemy's strategic centers of gravity, to undermine the enemy's will and ability to threaten our national security interests" and is commonly known as strategic bombing.[17] Of the Counter Air mission, Air Force doctrine states, "Counter Air is directed at enemy forces that directly or indirectly challenge control of the air" and "Counter Air operations help ensure freedom to maneuver, freedom to attack, and freedom from attack."[18] Although the missions involved in achieving this effect go far beyond killing enemy airplanes with friendly airplanes, generally, this is referred to as Air Superiority.[19]

Finally, "Counter Land operations are defined as airpower operations against enemy land force capabilities to create effects that achieve joint force commander (JFC) objectives. The aim of counter land operations is to dominate the surface environment using airpower."[20] Under the Counter Land umbrella, there are two primary air power missions: interdiction and close air support. Interdiction involves "air attacks against enemy land forces and their resources." The purpose of these attacks is to "channel enemy movement, constrain logistics, disrupt communications, or force urgent movement to put the enemy in a favorable position for friendly forces to exploit."[21] In practice, interdiction by the Air Force engages the enemy's fielded forces on portions of the battlefield beyond engagement ranges of most of the Army's weapons. From the Air Force's perspective, doctrinally speaking, interdiction is a more efficient means of using air power because it requires minimal integration with land forces, always a source of stress and friction. Further, it allows air planners to concentrate air resources at critical points on the battlefield, thereby multiplying the effects of that air power. Optimally, these operations serve to favorably shape the battlefield for land forces by engaging elements of an enemy's force days or weeks before an enemy might bring them to bear against friendly forces.[22]

By contrast, close air support (CAS) "involves using ordnance within close proximity of ground troops; that employment and the requirement for detailed integration are two characteristics that distinguish CAS from other types of air warfare."[23] This is a personal business for everyone involved because it takes place very near friendly troops. Even when things go *well*—communications work as intended, both air and land forces know the lay of the battlefield and their place on it, enemy forces are readily identified, everyone in the process understands the tactics and procedures, and the aircraft and weapons function correctly—it is a challenging and dangerous mission. When things go *badly*, the results can be tragic. Close air support requires high levels of tight interservice integration. For missions to be successful, the process often consumes immense amounts of time and personnel, again resources not usually in surplus during combat. Further, close cooperation is a two-way street. The responsibility for coordinating successful close air support operations does not fall solely on the Air Force; the Army has a role to play too. Of course, since the two services split in 1947, the Air Force had reasons for not coordinating with the Army and the Army did not always want to cooperate with the Air Force. The competition between the services often revolved around support capabilities, aircraft, and missions; but nearly as often, a literal turf war broke out over who ruled which parts of the battlefield.

INTERDICTION

Conducted beyond the FSCL

FSCL: Fire Support Coordination Line

CAS

Conducted between the FLOT and the FSCL

Enemy Forces

FLOT: Forward Line of Own Troops

Figure 1 Interdiction vs CAS

Understanding who "owns" a section of the battlefield, and what specific objectives lie within that battlefield, is critical to the coordination process. Doctrinally speaking, the Air Force and Army divide close air support from interdiction along the Fire Support Coordination Line (FSCL). Historically called the "Bomb Line," any Counter Land or Strategic Attack operations beyond that line requires minimal or no coordination between air and land components. Air power is free to conduct operations to meet the overall commander's strategy. The Forward Line of Own Troops is the line held by the most advanced portions of friendly ground forces. Close air support occurs between the Forward Line of Own Troops and the Fire Support Coordination Line, and the Army controls that area. It's the when, why, and how the Air Force or Army does the crossing of those imaginary battlefield lines that most often becomes contentious.

It stands to reason the least desirable outcome of conflict for any nation—aside from complete destruction and defeat—is forcibly subduing an enemy by ground combat. Attack by an army is destructive and costly in treasure and the blood of youth. Political will and objectives must align before a nation is willing to commit individuals to the horrors of combat. If land warfare becomes necessary, the goal of air power is to simultaneously achieve independent strategic objectives and minimize the likelihood friendly soldiers will die while covering the final ten yards before reaching their objectives. Therefore, the goal of strategic attack is to destroy the enemy's will and means of fighting at the highest levels. If this course is unsuccessful, air power strives to starve and destroy the enemy's forces beyond the reach of land-based weapons. The goal is to deny them supplies and mobility through aggressive interdiction and ultimately destroy an enemy force's capability to fight. Finally, close air support aircraft attack enemy armies in direct combat with friendly forces; destroying their defenses, their weapons, and killing enemy soldiers individually as the infantry closes the final steps to the ultimate objective.[24]

From a traditional conventional ground combat perspective, forces move across the battlefield sequentially, or nearly so. Friendly soldiers take objectives and secure their positions along the way, before the lines advance deeper into enemy territory and new objectives are taken. Air power, by contrast, works at the battlefield in the other direction, seeking to strike an enemy's ability and will to make war first. Should that fail, air power planners will begin working from strategic objectives to interdiction objectives, before reaching close air support objectives. There is, however, no requirement for sequential action. Air power may be used against targets at all points of the spectrum at any time or even simultaneously.

This division of the battlefield within and beyond the "bomb line" explains the debates between the Air Force and the Army over close air support since World War I. Matching capabilities and outcomes becomes difficult—even confrontational—when the parties involved approach problems from entirely different perspectives. For example, Marine Brigadier General Vernon E. Megee testified before Congress in 1949 and took the newly independent Air Force to task for a perceived lack of close air support in World War II. For the Marine and Army Infantry, tactical aviation and close air support are the same things. From their perspective, tactical aviation applies firepower directly and visibly against the enemy, for the benefit of the man in the mud. However, the Air Force argues interdiction has long term benefits for the infantry, even if they cannot see it. Further, it does not matter, from the Air Force perspective, if a bomber, fighter, or remotely piloted aircraft performs the task; the outcome is, arguably, the same.[25]

Forward Air Controllers enter this interservice fray when they fly over the battlefield because, while Joint Doctrine primarily defines a Forward Air Controller's role in close air support, historically Air Force Forward Air Controllers operated successfully in both interdiction and close air support missions. Ultimately, they serve to bridge the gap between inter-diction and close air support. Forward Air Controllers operate in the grey area of the battlefield, where air and land power overlap; they become connected to and beholden to *two* masters. They are the conduit through which the detailed integration of Army and Air Force planning must flow, and when done well, the critical node for success. Unfortunately, both ground and air forces shunned Forward Air Controllers at various points in history, while their mission was often complicated by the vicissitudes of defense budgets, technology, leadership, bureaucracy, and doctrine. Still, Forward Air Controllers have historically played a critical role on the battlefield and will do so in the future.

Chapter 1

1903–1939: FACs by Another Name

World War I aviation sparks the romantic notion of chivalrous warfare above the mired mass of the trenches. The phenomenon of aerial warfare made celebrities out of "Aces," men who scored five or more aerial victories. Pilots often hailed from society's wealthy elite, joined the air services straight out of Ivy League universities, polo clubs or prep schools, and relished their mythical reputations. Manfred von Richthofen, the Red Baron, had silver chalices made to commemorate each of his victories. Not to be outdone, the American pilots of the Lafayette Escadrille had a menagerie of mascots, including two lion cubs named Whiskey and Soda, and were infamous for their consumption of alcohol.[1] The reality of the air war was far different. World War I was the dawn of machine warfare. Successful assimilation of new technology and tactics to the war's reality often determined success or failure on the battlefield. Aircraft emerged as the newest technology, and, as the war ground on, nations and men adapted airplanes to the grim reality of the battlefield. By the war's end, aircraft capabilities improved dramatically, with increased performance and combat effectiveness, revealing the unique potential of aerial warfare in a wide variety of missions.

When war erupted in Europe in 1914, the United States was woefully behind in its adoption of airpower. Despite being the cradle of aviation, legal battles between American aviation pioneers, the Wright Brothers and Glenn Curtiss, hampered the development of American air power. Little technical investment by the US government and unimaginative development of technology by Army leadership further stagnated American military aviation.[2] As a result, the US significantly lagged British, French, and German air services in numbers of men and machines, aircraft technology, production capabilities, and doctrine. However, once engaged, Americans quickly adopted the aircraft and integrated it into their wartime strategy due to visionary leadership. This transition was no easy task.

The US Army did not initially embrace the Wright Brothers' successful aircraft. Much of the Army's reticence was due to Samuel P. Langley, inventor and Secretary of the Smithsonian, who failed to create a flying aircraft from a $50,000 War Department investment. In early December 1903, just two weeks before the Wright Brothers' success in North Carolina, Langley's aircraft—the *Aerodrome*—failed spectacularly in two attempts at flight. Langley's experiments took place squarely in the public eye, on the Potomac just outside of Washington, D.C., and coverage of his failure and the Army's failed investment was harsh in the national press.[3] Internally, Langley's debacle justified the military's skepticism. As a result, the War Department adopted a policy of only investing in proven technologies.[4]

For their part, the Wrights guarded details about their flights until the 1906 receipt of a patent. Due to their secrecy and obscurity, their accomplishments on North Carolina's remote wind-swept Outer Banks garnered little attention from Washington, D.C.[5] Further, Langley and others in the scientific community attempted to discredit their accomplishments. Despite the Wrights' steady improvements and expanding success in the years following the flights at Kitty Hawk, it was not until 1908 that the War Department finally agreed to trials of their machine. In the intervening years, the War Department turned its focus to dirigible airship technology. Before the turn of the century, rigid and semirigid airships gained popularity for military use in the US, France, and Germany. Dirigible airships were capable of far greater ranges and payloads compared to early aircraft. Unimpressed with the airplane, the US Army invested heavily in the investigation of airship technology before 1910.[6]

In 1908, after the Wrights successfully marketed their aircraft in France, the Army contracted with them for $100,000 to produce an aircraft and assess its feasibility for military use, yet there were still struggles. During the Army trials, on September 17, 1908, First Lieutenant Thomas E. Selfridge became the first piloted aircraft fatality in history. The aircraft, piloted by Orville Wright, with Selfridge aboard as a passenger and observer, crashed when a wing support wire broke free. Selfridge was killed, Orville injured, the aircraft destroyed, and the Army program delayed a full year. The Wrights worked to build a new airplane, and the Army weighed other options. During this time, Glenn Curtiss developed an improved aircraft, resulting in years of patent litigation between Curtiss and the Wrights. The Army hesitated and debated which aircraft to pursue. All the while, European aviators were taking to the sky, most notably in France, which heartily embraced the Wrights' flying machine.[7]

There were supporters among the doubters, however. In 1908 Lieutenant Benjamin D. Foulois arose as one of the military's leading advocates for airplanes and key pioneers in its development for military use. Foulois participated in Army experiments with dirigibles but believed aircraft would become preeminent as he observed the Wrights' trials at Fort Meyer, Maryland.[8] Foulois, born in Connecticut in 1879, enlisted in the Army during the Spanish-American War, serving in both Puerto Rico and the Philippines. After the war, he tried but failed to gain an appointment to West Point, after which he reenlisted in the Army. Assigned again to the Philippines, Foulois eventually earned a field commission.[9]

Following his commissioning, now Second Lieutenant Foulois attended the Army's Infantry and Cavalry School at Fort Leavenworth, Kansas, where he graduated near the bottom of his class. In December 1907, while at Fort Leavenworth, Foulois wrote a thesis on his vision for future military aviation, titled "The Tactical and Strategical Value of Dirigible Balloons and Aerodynamical Flying Machines."[10] In his writing, Foulois argued that the US was already behind in aviation technology due to shortsighted leadership within the US government. Further, he foretold of aerial combat, the use of aircraft by the Navy, and heretically argued that horse cavalry was already outdated and "would be used only for ceremonial purposes someday as soon as flying machines were perfected."[11]

From Leavenworth, Foulois joined the Signal Corps, which was responsible for Army aviation, and joined the investigations of the Wrights' aircraft. In October 1909, he and the Army's two other pilot trainees, Lieutenants Frank Lahm and Frederick Humphries, were assigned to College Park, Maryland, for further flight training. However, before his arrival, Lahm and Humphries damaged the aircraft, preventing any flights until the Wrights arrived from Dayton to repair it. That same month Lahm and Humphries received assignments out of the Signal Corps, "leaving Foulois and the Wright aircraft to constitute the Army's entire heavier-than-air flying force."[12]

In December 1910 Foulois, the aircraft, and a cadre of enlisted men moved to Fort Sam Houston near San Antonio, Texas. His commander told him, "Your orders are simple Lieutenant. You are to evaluate the airplane. Just take plenty of spare parts—and teach yourself how to fly."[13] During 1911, despite funding shortages, Foulois made significant progress in evaluating the aircraft and learning how to fly. The advancements made at Fort Sam Houston included such revolutionary ideas as replacing the airplane's skids with wheels to make landings and takeoffs safer.[14] By 1912 Congress appropriated funds for the purchase of five new aircraft; three Wright Bs and two Curtiss aircraft. Through 1912 and into 1913, Foulois continued operations at Fort Sam Houston. The Army opened a second training school in Dayton, Ohio—with winter training in Savannah, Georgia. Still, the service purchased few additional aircraft and made little progress towards integrating aviation into Army tactics or strategy. On March 5, 1913, in response to rising tensions along the US-Mexican border, the First Aero Squadron, the Army's first operational combat squadron, entered service at Texas City, Texas.[15]

Despite the progress, the United States was dramatically behind in the development of military aviation. "In 1914, just prior to the outbreak of World War I, the United States stood eleventh in total funds allocated for military aviation—well below such world powers as Greece and Bulgaria."[16] At the onset of the war, Congress allocated funds to increase the size of the

Army's aviation branch to 60 officers and 260 enlisted men. By comparison, the French l' Aéronautique Militaire had 3,500 men assigned. The British Royal Flying Corps (RFC) had 2,073.[17] On the Western Front in August of 1914, France had approximately 136 frontline combat aircraft, Great Britain 36, and Germany 180.[18] Even the woefully unprepared and outdated Austro-Hungarian Army had a larger combat air force than the United States in 1914.[19] Organizationally, America was behind Europe as well. The world's first air service was the French air service, l' Aéronautique Militaire, founded in 1910. That same year the Imperial German Air Service was born as well. The British Royal Flying Corps began service in the spring of 1912.[20] By comparison, the US Army Air Service was born a subordinate branch to the Army Signal Corps. The Air Service only achieved temporary independence from the Signal Corps in May of 1918 and remained under Army control from 1919 until 1947.

European military aviation rapidly accelerated after 1914 despite very humble beginnings. Aerial combat began in 1911 when Italian pilot Giulio Gavotti dropped hand-grenade-sized "bombs" over the rail of his cockpit into an Ottoman troop encampment in Libya. During the Mexican Revolution, two American pilots, each flying for a different Mexican faction, fired revolvers from their cockpits at each other, becoming the first pilots to engage in air-to-air combat. Aircraft functioned as artillery observers and scouts during the First and Second Balkan Wars. Airplanes also proved invaluable in the earliest days of World War I, despite their rudimentary technology. Reconnaissance and artillery spotting were vital in several decisive battles of 1914. On September 3, a French aircraft spotted a gap between the German First and Second Army, enabling French and British forces to maneuver into defensive positions along the Marne and halt the German advance. On the Eastern Front, German airmen enabled their Eighth Army to find, envelop, and defeat the Russian First and Second Armies at Tannenberg. On October 5, 1914, French pilot Joseph Frantz scored what is considered to be the first aerial victory with a front-mounted 8-millimeter Hotchkiss machine gun on his Voison aircraft.[21] Bombing attacks by aircraft also increased during 1914 as both the French and German air forces built dedicated bombers. The British and German navies both embraced aviation, building the world's first aircraft carriers, the HMS *Ark Royal* and SMS *Glyndwr.* The British Navy Flying Service was particularly aggressive, conducting the first shipborne air attack of targets on land and the first night bombing raid.[22] European aircraft rapidly advanced in speed, altitude, range, and payload early in the war and their air forces dramatically expanded too.

Meanwhile, on March 15, 1916, as the British and German air services were developing state-of-the-art fighter aircraft and tactics in preparation for the deadly aerial battle of the Somme, the US 1st Aero Squadron arrived by train in dusty Columbus, New Mexico. The entirety of the Army's only operational squadron "included ten trucks, eight airplanes, six motorcycles, eleven officers, eighty-two enlisted men, an aviation mechanic, and two Hospital Corpsmen."[23] The Army dispatched the squadron to New Mexico as part of General John Pershing's punitive expedition against Pancho Villa. Led by now Captain Benjamin Foulois, they served as scouts and a communication link for Pershing's column as it moved into Mexico. The squadron flew Curtiss JN-3 Biplanes, a training aircraft so underpowered they were unable to carry a machine gun when loaded with a pilot, an observer, and fuel.[24] Further hindering operations, the JN-3s were not able to fly above ten thousand feet, substandard considering many contemporary European aircraft were able to operate above fifteen thousand feet. The aircraft also did not handle the rugged terrain and high wind of the American southwest well, making it difficult for the squadron to maintain contact with Pershing's force. By the end of April, six of the eight aircraft crashed, and the other two were not serviceable. Foulois repeatedly wrote his superiors for improved aircraft, better parts, and increased supplies to no avail.[25]

Replacement aircraft, more powerful Curtiss R-2s, were slow in arriving. When the squadron did receive the new airplanes, the correct parts for reassembly were not included in the shipment, forcing Foulois's maintenance team to cobble them together with the remains of the old aircraft. The 1st soldiered on trying to fix the problems, but a lack of spare parts and poor aircraft unsuited for the operating conditions continued to plague them. When Pershing's expedition withdrew from Mexico in February of 1917, the 1st Aero Squadron's contributions to the overall effort proved minimal.

Adding insult to injury, Army quartermasters charged Foulois with misappropriating funds for purchasing local fuel and parts during the campaign, a course he pursued due to the poor support the Army provided for his command. Foulois defended every purchase to the penny, and the Army dropped the charges.[26]

When America entered World War I only three months later in April 1917, the US Army's aviation branch was completely unprepared. From acquisition of the first Wright flyer in 1909, through America's declaration of war in 1917, the Army purchased only 224 aircraft. Of these, only 55 remained in service, none of which were combat-capable.[27] In the spring of 1917, Major William "Billy" Mitchell traveled to France to observe the French l' Aéronautique Militaire and the Royal Flying Corps. Ironically, Mitchell had been a critic of the pursuit of Army aviation, arguing that airplanes were useful only in a reconnaissance role. However, after the outbreak of World War I, Mitchell became an outspoken advocate for airpower.[28]

Mitchell witnessed frontline combat from both the ground and the air. What he saw convinced him of the urgent necessity of embracing airplanes on the battlefield. Mitchell envisioned an American air service ready for combat in the fall of 1917, utilizing French aircraft designs built in the US, and flown by pilots trained at French flight schools in Europe. His recommendations went unheeded in America.[29] Mitchell remained in France, continuing to report what he saw and advocating increased US aviation investment. When America entered the war, Mitchell joined Pershing's staff as the aviation attaché officer and was promoted to Lieutenant Colonel. When the first American combat division deployed to France in July of 1917, the American air force in Europe consisted of one Nieuport aircraft and one pilot—Mitchell.[30]

Even if the American Air Service was not yet in the fight, American aviators were participating in the war as volunteers in other air forces. When the war began, neutrality laws forced any American who swore an oath to any other nation to forfeit their US citizenship. To circumvent this, some volunteers joined the Red Cross, others the French Foreign Legion, and some chose to forfeit their American citizenship to join the war. Those who volunteered with the French Foreign Legion pledged allegiance to the Legion, retaining their citizenship through a technicality.[31] Despite the restrictions, more than five hundred Americans joined the British, French, and Italian air services. Most of the American volunteers, just over three hundred, joined the RFC, while 269 served with the French. By far the most famous of these volunteer groups was the Lafayette Escadrille.[32]

The Lafayette Escadrille was the brainchild of Norman Prince, Elliot Cowdin, and William Thaw, who came to France in 1914 to join the war. Through their persistence and the help of a committee of Americans in Paris, the men lobbied their way first into cockpits and then into establishing an American Squadron, or Escadrille. Their efforts paid off on April 17, 1916, with a dinner commemorated the official beginning of l' Aéronautique Militaire No. 124, l' Escadrille Américan, better known as the Lafayette Escadrille. The Escadrille rapidly formed and moved to the front, flying Nieuport 17 aircraft. American pilots and a handful of French officers made up the initial cadre of the Lafayette Escadrille. The Americans made their first patrol on May 13, 1916, and on May 18, Kiffen Rockwell scored the Escadrille's first victory.[33] Most Americans who served in the Escadrille transferred to US units after America entered the war and, by July 1918, the Lafayette Escadrille's aircraft and remaining men joined the American 103rd Aero Squadron.[34]

While American pilots participated in aerial combat on the Western Front with the British and French air services, their most significant contribution was attention in the American press. Americans began romanticizing aviation, generating increased attention from American political and military leaders.[35] Soon, "high officials and oracles of the press began to tell the world that America would win the war with an air force which ... would blacken the skies of Germany with tens of thousands of planes."[36] In June 1917, two months after America declared war on Germany, the Aircraft Production Board Announced ten thousand pilots would be flying American airplanes by year's end. Chief of the Signal Corps, General George Squire, testified to the House Military Affairs Committee that twelve thousand aircraft and twenty-four thousand engines would be produced by 1918.[37] As a result of the positive press and cooperation generated by Americans abroad, the United States agreed with France to produce twenty thousand aircraft, five thousand pilots, and fifty thousand mechanics in America for deployment alongside l' Aéronautique Militaire.

Despite the grand intentions, three problems faced the Americans before flying a single combat sortie. The first problem was a resurgent enemy. In response to the American public announcements of an overwhelming air force, the Germans ramped up production and adopted new tactics garnered from the battles of 1916 and 1917.[38] They reorganized their units and built several new aircraft.[39] The Imperial German Air Service regained the upper hand with The Red Baron and his Flying Circus dominating the headlines and the aerial victories count.[40] The Germans further innovated during 1917 and into 1918 by experimenting with dedicated ground-attack aircraft, which worked in concert with artillery and infantry. These new German tactics supported offensives in early 1918, enabling them to make territory gains unseen on the Western Front since August 1914. The new tactics also ushered in a new era of combined arms warfare, hinting at the Nazi Blitzkrieg of 1939.[41]

The second problem facing the Americans was aircraft production. In June 1917, Major Raynal C. Bolling traveled to Europe to examine which aircraft types the US industry should produce for combat. After reviewing Allied aircraft, Bolling recommended a mix of British bombers and observation aircraft, French fighters, and Italian night bombers. However, Washington politics, foreign licensing difficulties, a shortage of raw materials, and few aircraft factories hampered production.[42] The June 1917 production targets exceeded the goals budgeted by the Congressional Special Committee on Production in April 1917. However, American aircraft production was not on track to meet either goal. The piecemeal production caused consternation between the United States and the other Allied powers before any American units arrived in France. The production capability gap was so wide by the end of the war that less than one-third of US squadrons flew American-produced aircraft.[43]

Slow aircraft production was also opening rifts within the American Air Service leadership and contributed to the third problem faced by the Americans—which strategic and operational objectives the yet-to-deploy

Air Service should pursue first. The problem stemmed from an ongoing feud between Mitchell and Foulois. Promoted from Major to General in July 1917, Foulois arrived in France in November and took over as the American Expeditionary Force (AEF) Chief of the Air Service. Mitchell, who had been in theater for months, wanted to focus the American effort on putting aircraft and pilots into combat as soon as possible. Foulois wanted training and logistics systems in place to ensure American success once units moved into the frontlines. Further deepening the rift was the choice of leadership within the Air Service. Foulois brought with him a large staff, many of them non-pilots, to address the logistics of operating a combat deployed air force. He also followed a more political and pragmatic approach, promoting officers to the Air Service staff with connections within the Army and experience with the Army bureaucracy.[44]

Mitchell openly criticized the course Foulois pursued to Pershing and other American staff officers. Mitchell also disliked Foulois's choice of staff, his emphasis on logistics over operations, and the lack of available aircraft for the ever-increasing numbers of American pilots. Further, Mitchell doubted Foulois's flying experience. This final criticism, while not wholly unfounded, was quite hypocritical. Although Foulois was one of the Army's first pilots and led the first combat deployment, he had not flown since the end of the Pershing Expedition. Instead, Foulois sat in Washington, addressing the shortfalls in American aviation, and testifying before Congress about the lack of preparedness in the American air service. In contrast, Mitchell only learned to fly after the outbreak of World War I and was not a rated Army pilot until after American entry into the war. In fact, he only started flying in France as part of his attaché duties.[45]

Although Mitchell took tremendous exception to non-flying Army officers in charge of aviation matters, Foulois had a strong motivation for his logistical focus. After his experience on the Mexican-American border, Foulois believed a securely established infrastructure was necessary to sustain combat operations in France. According to Foulois, Mitchell failed in his duties to build the necessary logistics framework to receive Air Service personnel and support flying operations. Foulois fired back at Mitchell's criticism, writing in his memoirs, "I could find no evidence of solid accomplishment that I could attribute to his efforts. To say I was disappointed would be much too kind a statement. I was furious at his gross incompetence and reported my fury to General Pershing."[46]

Fuming at Mitchell's neglect of his duties, Foulois removed Mitchell from his position as attaché, assigning him instead as the Commander of Air Service components, 1st Army. Through the winter of 1918 Mitchell and Foulois continued their feud. Mitchell's public criticism forced Foulois to reorganize the Air Service leadership. Through Pershing, he promoted Major General Mason Patrick to Chief of the AEF Air Service, Foulois took over Mitchell's job, and then demoted Mitchell to 1st Brigade Air Service command. Foulois sought to take advantage of both his and Mitchell's tactical knowledge and "practical experience with planes and pilots—not as a manager of construction projects."[47] Although Mitchell continued to chafe at Foulois's leadership, and the Army Air Service in general, the two embraced a common ideal for future US air operations.

In order to gain combat experience during its first few months of operations, the Air Service operated under French tactical control in the quiet Toul sector.[48] The 1st Aero Squadron became the first US unit to be assigned to the front in April 1918. On April 14, the 94th Aero Squadron scored the first aerial victories for the US as Second Lieutenant Alan Winslow downed an Albatross D-V, and First Lieutenant Douglas Campbell scored a Pfalz D-III.[49] From Toul, the Americans moved into the Chateau-Thierry sector in early June to support the Allied effort to blunt the German Spring Offensive and support the Allied counterattack.[50] The Americans added the first observation and bomber aircraft to their frontline units. The 91st Aero Squadron began observation missions over the front on June 6. On June 12, the 96th Aero Squadron conducted the first US aerial bombardment mission against rail yards at Dommary-Baroncourt.[51]

During these early operations, American squadrons worked diligently to perfect their craft. Observation squadrons overcame a lack of radio and photographic equipment to provide support to ground forces. Pursuit and Observation squadrons worked on their formation tactics and aerial gunnery techniques critical for success in combat. Pursuit squadrons also benefited from the injection of experienced pilots from the Lafayette Escadrille.

Unfortunately, bomber operations did not fare as well. In early July, the 96th Aero Squadron suffered the loss of an entire flight on a bombing raid due to low clouds and rain. This severely crippled American bombardment operations with pilots and planes already in short supply.[52]

During the summer, Mitchell commanded the American 1st Air Brigade in the sector. The American units were subordinate to the French Sixth Army, which created friction between Mitchell and his subordinate commanders, Mitchell and the French, Mitchell and American Army commanders, and Mitchell and Foulois. However, for the early July attack on Vaux, the American 1st Observation Group was able to support the French and American infantry with ninety-six artillery-spotting and reconnaissance missions. The 1st Pursuit Group's aircraft provided air cover against the German Jasta Groups 1 and 3. Jasta Group 1 had a formidable reputation and was commanded by Manfred von Richthofen until his death in April 1918.[53] Although these first independent American efforts counted as minor actions in the course of the war, "they had achieved command of the air against a numerically superior and more experienced opponent...and proved they could conduct modern combined-arms warfare."[54] With the increasing proficiency and tactical success, Foulois elected to return Mitchell to the post of First Army Aviation Service Commander on July 25, despite the continued friction between Mitchell and most of his colleagues.[55] Foulois persuaded Pershing to make the change, which placed Mitchell in charge of planning and leading American air operations for the upcoming fall Allied offensives. Foulois privately worried, "Mitchell would either end up the war a hero if I were right, or in a straitjacket if I were not."[56]

In August 1918, in preparation for the St. Mihiel offensive, Mitchell took command of 1,481 Allied airplanes—the largest air force under a single command on the Western Front.[57] American observation aircraft began working at the St. Mihiel salient, conducting reconnaissance of German positions and assisting in the registration of artillery. Bad weather hampered artillery observation during the offensive, but the earlier preparations proved invaluable to the infantry attacks. Allied pursuit planes conducted aggressive patrol operations, strafed German infantry columns, and attacked observation balloons to prevent the detection of the increased troop concentrations. German and American pursuit aircraft engaged in pitched battles in September as the German Army withdrew under pressure from the Allied offensive. American bombing missions were less successful as bomber formations pushed deeper behind German lines, due to poor weather and increased German air cover. Despite the bombers' continued struggle, American observation and pursuit units proved to be equal to the challenge of combat over the Western Front during St. Mihiel. Further, the unified command structure relieved much of the tension between Mitchell and other commanders.[58]

The American attack in the Meuse-Argonne quickly followed the St. Mihiel offensive, beginning the final offensive across the Western front. The British and French withdrew aircraft placed under Mitchell's command to support their respective armies during the campaign—the Americans were operating on their own now. American Observation Squadrons began reconnaissance and artillery preparations for the opening of the ground assault, but poor weather made locating German artillery batteries difficult. Communications equipment also continued to be a weak point, forcing pilots to use handwritten messages dropped from their cockpits to ground forces. This inefficiency resulted in missed opportunities to cut off retreating German infantry.[59]

The need to improve observation techniques was critical to the AEF. The earliest use of aircraft in war was for observation. When patrolling the front, observation aircraft tracked enemy troop movements and mapped trench lines. The first air power mission proved to be the most critical mission during the Great War. Accurate and effective artillery fire decided the course of battles during the war.[60] Observation aircraft allowed heavy artillery to improve accuracy and spotted enemy artillery positions, enabling critical counterbattery fire. When it became evident that a symbiotic relationship between aviation and artillery developed, aircraft began attacking enemy observation aircraft. From that point, combatants built dedicated pursuit aircraft to protect friendly observation aircraft and attack the enemy's patrols.[61]

Early observers communicated by dropping notes and sending Morse Code via smoke generating machines or light signals. The development of

cameras specifically for aerial photography in 1915 and an improvement
in radio technology allowed an increased level of specialization between
observation sorties and ground units. By the Battle of the Somme, improved
radio technology allowed observation aircraft to communicate directly with
artillery and infantry units via Morse Code and voice. As the AEF prepared
for the Meuse-Argonne offensive in 1918, observation units worked to
improve their craft, which led to the development of the first Forward Air
Controller tactics.[62]

Shortly after the American entry into the war, the War Department
began translating l' Aéronautique Militaire aviation manuals, including
material taken from the Germans, for artillery and infantry liaison and
communication. These manuals served as the basis for American observation tactics. They spelled out infantry, artillery, and observation aircraft
radio procedures. The manuals also standardized smoke signals, message
drops, and a series of ground-based signals utilizing colored panels.[63]
Pilots and observers used maps with a grid system, similar to the modern
Military Grid Reference System, to direct artillery attacks and coordinate
with infantry units.[64]

The Air Corps began hybridizing these tactics to meet specific operational needs as the AEF went into combat. Mitchell's orders to the First
Army's aviation units ahead of the Meuse-Argonne offensive emphasized
the importance of air-to-ground support missions. He wrote, "Greater
importance will be given to Infantry Liaison work—the airplanes should
bring in direct information to the Infantry (to the Divisions, regiments, and
even battalions on the first line)—a greater share, too, should be given to
the attack of troops on land (rearguards, nests of resistance, reinforcements
or reserves)."[65]

Further, Mitchell recounted in his diary both the effectiveness of
German ground-attack missions in support of infantry assaults and the
damage to morale the attacks imposed: "the effect produced by enemy
pursuit planes firing on our troops with machine guns and small bombs
is decidedly worth the development of this type of offensive aviation."[66]
To meet Mitchell's orders, American observation pilots used a new
technique for directly supporting infantry during the Meuse-Argonne

campaign. Combining French observation techniques with German attack aviation methods, the Americans created the first forward air controller missions. Called Cavalry Reconnaissance Patrols, the formations flew low over the battlefield, closely following the artillery barrage and the first wave of infantry. As they orbited over the front, these proto-Forward Air Controllers located German strong points and other obstacles for the infantry. Pilots strafed these positions to neutralize them, directed other planes in their flight to attack, coordinated artillery strikes, or notified advancing troops of the dangers.[67]

Lieutenant Henry Dwight of Brookline, Massachusetts, flew Cavalry Reconnaissance patrols in the waning days of the war. Dwight did not arrive at the front until October 1918, despite enlisting in April 1917. He completed his pilot training in Fort Worth, Texas, in March 1918 and observation pilot training in France in August before joining the 12th Aero Squadron. Dwight's daily writings as a young Air Corps pilot reflect the thrill, frustrations, and dangers of combat aviation. In less than a month of frontline experience, he flew in the cold and dreary weather plaguing the Western Front in the last month of the War. German fighter planes and antiaircraft fire covered their Army as it retreated from France, offering stiff resistance as Lieutenant Dwight and his squadron flew reconnaissance, escort, and ground-attack missions.[68] On November 2, Dwight was "called at 6 for an early infantry mission. Wilson and Over in the mission plane, Stenak and I in protection … We had a fine time with the doughboys in their shell holes."[69] Two days later, he wrote, "We left Remicourt on a mission and put in 2½ hours at the front on infantry … our doughboys are still going ahead."[70] Exhausted from the strain of two weeks of relentless flying and three straight days of two-and-a-half-hour Cavalry Reconnaissance missions, Dwight wrote, "After a long sleep, felt better. Good meal served in our floorless barracks [helped]."[71]

As the saga of Henry Dwight reveals, American aviation was slow to arrive on the Western Front. Aircraft and pilot production never met expectations. While the US managed to produce fifteen thousand aircraft and nearly five thousand pilots, few made it to combat. Internal struggles between Foulois, Mitchell, and other Army leaders led to disorganization and delays. As of November 11, 1918, serving in frontline combat units, there were "45 American squadrons and 767 pilots, 481 observers, 23 aerial gunners and the compliment of soldiers ... equipped with 740 airplanes ... twelve of these squadrons were equipped with American built airplanes."[72] Comparatively, the US Air Service was inconsequential: "at the time of the Armistice, the French air services had 90,000 officers and men and 11,000 aircraft; the British 291,200 men and 22,000 airplanes; the Germans 80,000 men and 9,000 aircraft; and the Italians 100,000 officers and men."[73] Pershing and Foulois jealously guarded American independence of operations, which led to further delays in American deployment. However, the Americans learned quickly and proved to be innovative once they were in combat.[74]

Despite the importance of observation and the improvisation of the Cavalry Reconnaissance Patrols, Mitchell pressed for more combat effective American bombers and pursuit missions. The Americans conducted several coordinated bomber-pursuit offensives, and two-thirds of American aerial victories during the Meuse-Argonne offensives came in bomber-pursuit combined offensive operations. The zenith of the American bomber effort was a 353-aircraft raid on October 9 against the cantonment area between La Wavrille and Damvillers. The bombers dropped thirty-two tons of munitions while the fighter force scored twelve aerial victories—all with the loss of only one American aircraft.[75] Although these bomber missions did little to support the AEF effort and were minuscule in scope compared to bomber operations of World War II, these raids reflected Mitchell's vision for the future of air power, a vision that grew to dominate the Air Corps between the World Wars. The hard-learned lessons of combined arms techniques, like the Cavalry Reconnaissance Patrols, were viewed as insignificant in warfare.[76]

After the war, Mitchell and Foulois began a campaign for an independent air service in the United States. Although Mitchell received the lion's share of the press, Foulois pressed equally as hard. He tirelessly worked in the halls of Congress and testified several times in committee to bring about a US Air Force.[77] Foulois was the first US air power theorist, advocate, and practitioner. He was among the first qualified pilots in the Army Air Corps, led its first combat effort during the expedition against Pancho Villa, and led the air component of the American Expeditionary Force during World War I.[78] Yet, Mitchell and Foulois each represented two divergent ideas within American air power theory. Famously, Mitchell relentlessly advocated the ideas of an independent air arm and strategic bombing. "Mitchell and other Army flyers held that the airplane, being able to bypass ground and sea forces and go directly to important targets, represented an entirely new kind of warfare."[79] Foulois, in contrast, represented a more nuanced approach of combined arms, with air power often subjugated to land forces, but still a staunch independent air force advocate.[80] "What he was after was the reorganization of America's defense structure, so that military aviation might grow to realize its offensive potential, becoming second in importance only to the 'queen of battle.'"[81]

In 1919 Mitchell became the deputy chief of the Air Service, conducted bombing demonstrations against captured German warships in 1921 and 1923, and in 1925 published *Winged Defense: The Development and Possibilities of Modern Air Power—Economic and Military.*[82] Mitchell's outspoken nature famously earned him a court-martial in 1925. Foulois also suffered for his advocacy and leadership in aviation. He became the Chief of the Air Corps in 1931 and endeavored to turn around the fractured and divisive command.[83] In 1934, President Franklin Roosevelt abruptly canceled the US government airmail contracts with the airlines and handed the job to Foulois and the Air Corps. The Air Corps was woefully underfunded and poorly equipped to handle the mission, despite Foulois's confidence. The results were tragic— Air Corps aircraft were not capable of flying at night nor in the miserable winter weather frequent along the air mail routes. The Air Corps matched its previous annual fatality rate in the first two months of 1934 and was forced to reduce air mail service to less than half of the commercial routes. Critics harangued the Air Corps and Foulois for this failure, but the Air Corps developed crucial night and weather flying experience, which it eventually harnessed to train future pilots.[84] The Air Mail struggles, coupled with scandal over Air Corps equipment acquisitions and continued animosity within the War Department over Foulois's drive for Air Corps independence, led to Foulois's retirement by the end of 1935.[85]

Critically, the Air Force took steps toward independence and developed key theoretical ideas during the period of Mitchell's and Foulois's leadership, despite their ignominious departures. The Air Service achieved independence from the Signal Corps in 1920 and became the more independent Air Corps in 1926. Under Foulois's leadership, the Air Corps established its own General Headquarters (GHQ), Air Force. From a theory and doctrine perspective, the 1926 formation of the Air Corps Tactical School (ACTS) was the most significant accomplishment during his tenure. In 1931, the school moved from Langley Field, Virginia, to Maxwell Field, Alabama. Between Mitchell and Foulois, Billy Mitchell had been the most outspoken advocate and prolific air power writer. However, under Foulois as the Chief of the Air Corps, ACTS developed critical doctrine and theory, which directly impacted the Air Corps' procurement, training, and employment through the end of World War II, and laid the foundation for all future US Air Force theory and doctrine.[86]

In the 1920s and 1930s, many of Mitchell's and Foulois's contemporaries took the ideas the two pioneers first articulated and developed them into the doctrinal and theoretical frameworks of US air power. The torchbearers of the Mitchell-Foulois legacy were men like "Hap" Arnold, Ira Eaker, Clair Chennault, Jimmy Doolittle, Carl Spaatz, Haywood Hansell, Pete Quesada, Hoyt Vandenburg, Nathan Twining, and Curtis LeMay. Many of them were World War I veterans or participated in the groundbreaking aviation milestones of the 1920s and 1930s. They would lead the Army Air Forces during World War II, and the independent Air Force after 1947. Their vision of air power shaped the Air Force for generations. Critically, most of these men were Air Corps Tactical School instructors or graduates in the interwar period.[87]

Aside from the doctrinal developments, in the period between World War I and World War II aviation technology boomed. Aircraft became faster and flew further while carrying heavier payloads. The biplanes made of wood and canvas, with their bulky struts and wires, were replaced by aluminum monocoque designs with cantilevered wings.[88] The Air Corps experimented with improved instruments for flying in bad weather, air-to-air refueling, and transoceanic flight. Mitchell, Foulois, and others tirelessly advocated for increased readiness by American aviation. Independence was their goal, but fearing the US might be caught unprepared for war, the Air Corps also wanted to continue to improve upon the lessons of the Great War. Because of the overall decline of military budgets and personnel, the Air Corps' budget actually increased as a percentage of the Army's total budget. However, the role of the forward air controller, embodied in the observation and cavalry reconnaissance missions, became lost in the interwar advances as the Air Corps focused its miniscule budget allocation on improving bombers.[89]

The need to improve attack aircraft survivability and capability after the war overshadowed the small steps the Air Corps made toward a FAC mission. Cavalry reconnaissance patrols and ground-attack missions suffered high loss rates due to the fragility of World War I aircraft.[90] Airplanes constructed from wood and canvas operating close to the ground and over the front were quickly shot down. The need to protect attack aircraft from ground fire even caught the attention of Mitchell: "low flying is becoming more and more a necessity for many purposes, the question of protection against fire from the ground assumes more and more importance. This is particularly true in considering the development of attack squadrons."[91] Attempts to reinforce attack aircraft with heavier wood or metal panels made underpowered World War I aircraft slower and less maneuverable, increasing their vulnerability to enemy fire. During attack experiments in the 1920s and 1930s by the 3rd Attack Group, the Air Corps increasingly adopted twin-engine all-metal aircraft as attack aircraft. Developed similarly to light and medium bombers of the era, attack aircraft moved away from the observation mission in favor of bomber-based interdiction and attack missions further behind enemy lines.[92]

Observation aircraft also moved away from its frontline support mission. Photo reconnaissance aircraft, similar in design to attack aircraft and bombers, began to replace the photographic and intelligence portions of the observation mission during the 1920s and 1930s. With increased speed, payload, and range, these aircraft also began to conduct missions beyond the front lines.[93] The longer-range missions increasingly collected target intelligence for bombing missions, rather than supporting infantry and artillery. As Rebecca Hancock Cameron wrote about the training of the era, "reconnaissance units trained for long-range surveillance flights in order to search out and determine the nature of the target, the best route of approach, the location of antiaircraft and enemy airfields, and the type of bombs to be carried."[94]

Technological developments and doctrinal changes gradually pushed the Air Corps' focus away from supporting troops at the front, shifting instead to interdiction and strategic bombardment. Doctrine documents, the lessons of the Air Corps Tactical School, and the operational capabilities of combat squadrons traced the path the Air Corps traveled. In 1923 Billy Mitchell, then the Assistant Chief of the Air Service, conducted a comprehensive inspection of units across the country, testing unit readiness and combat capabilities. Among the units inspected were the 1st Pursuit Group at Selfridge Field, Michigan, the 12th Observation Squadron at Fort Bliss, Texas, and the 1st Attack Group at Kelly Field, Texas. Just five years after the end of World War I, the divergence of the attack and observation groups from their original mission supporting ground forces at the front is evident. Mitchell reported the 12th Observation squadron conducted no training in conjunction with the cavalry unit also located at Fort Bliss. Further, the squadron's aircraft were woefully neglected, with only one flyable aircraft. While the 12th foundered, the observation squadrons of the 1st Pursuit group only conducted recon-naissance missions to support simulated airfield bombing missions.[95]

Mitchell also found the squadrons of the 1st Attack Group lacking in equipment and knowledge of their missions. For their inspection, he had the Group conduct attacks against a simulated truck convoy in a bombing range outside of Laredo, Texas. While Mitchell reported the results of the exercise to be satisfactory, he found the lack of knowledge among the aircrew troubling. Mitchell wrote a short paper on the nature of the attack mission for the pilots of the group and included a copy in the inspection report.[96] In the report, Mitchell summarizes the mission of attack aviation as follows: "Attack aviation is that branch of an air force which is organized trained and equipped to attack any military object on the ground or water. Its principal targets are those which are moving along roads and lines of communication." Later in the paper, he spells out the means of using attack aircraft. "In application of attack aviation the target must be found first by surveillance aviation; next, the air cleared and attack aviation brought up,

protected by pursuit aviation; and third, the direct attack by attack aviation."
Mitchell goes as far as instructing attack pilots to land their aircraft and
fight from entrenched ground positions. He makes no mention of integration
with ground units by supporting ground force attacks or responding to enemy
attacks of friendly positions.[97]

Just a few years later, lectures at the Air Corps Tactical School separated
attack and observation aviation even further from their ground support roots.
Haywood Hansell, who went on to be one of the writers of American air war
plans in preparation for World War II and command bomber forces in the
Pacific, lectured in the 1930s about the roles and missions of the Air Force.
He specifically divided observation into two classes, one supporting the Army
and one supporting air forces. Further, he relegated air force observation to
supporting bombing missions: "Observation, and remember we are speaking
of air force observation—will be required to make deep penetrations ...
Increased range of bombardment will require detailed information of deep
lying objectives."[98] The accompanying map problems for this lecture given
in the 1937–1938 term also gave little thought to coordination with ground
forces, applying a formulaic approach to air power application instead. "Map
Problem 2, Immediate Support of Ground Forces—The Strategic Defensive"
called only for airpower to interdict railroads and roads beyond the front, and
enemy forces as they moved to the front. The problem included simulated
intelligence reports of key rail bridges and junctions, the air forces assigned
to imaginary commands, and critical tasks to complete the exercise. The solution, provided after the students completed the assignment, suggested attacking railroads and communication centers, then columns of enemy troops on
the roads and in reserve. Air force capabilities and distance from the front
dictated attack priorities. There is no mention in the exercise, or the solution,
that air force should coordinate with ground commanders to establish priorities, nor are there any discussions of attacking enemy forces engaged with
friendly forces.[99]

As the Air Service and then the Air Corps moved toward independence,
it drifted away from its ground support roots. In 1935 the Air Corps created
the General Headquarters Air Force, establishing a separate air combatant
command designed to consolidate American air power under a single

commander. In a speech after World War II, Hansell cited the MacArthur-Pratt agreement, which created the GHQ Air Force and summarized the way air power advocates desired to use the Air Force in the future. He said the GHQ Air Force should primarily be strategic, used for "long range reconnaissance, for interdicting enemy reconnaissance, for demolition of import installations, and for interdiction of enemy movements."[100]

In the 1930s, the Air Corps increasingly pressed strategic bombardment as the primary air power mission. The service developed long-range bombers during the decade, culminating in the production of the B-17 Flying Fortress. Several key technological leaps in the 1930s, coupled with Air Corps experiments and doctrine developments within ACTS, began a definitive shift within the Air Corps to develop the ideals of strategic bombing into the high-altitude daylight precision bombing doctrine of World War II. First was the development of all-metal, multiengine monoplane bombers such as the Boeing Y1B-9—an experimental aircraft that never entered production—and the Martin B-10. These aircraft were significantly faster, and carried heavier payloads over further distances, than their wooden biplane ancestors. These bombers served as the starting point for the development of the famed World War II bombers like the B-17, B-24 Liberator, and the B-29 Superfortress.

The second technological development was the Norden Bombsight.[101] Developed for Air Corps bombers beginning in 1929, an improved version of the Norden arrived for Air Corps testing and fielding in 1935, the same year the first B-17 bomber arrived for testing.[102] "The arrival of the B-17 and the Norden bombsight had finally made material the Air Corps' dream of a long-range bomber that could, in-turn, make the ACTS theory of high-altitude daylight precision bombing a reality."[103] Throughout the 1930s, the Air Corps Tactical School faculty produced a series of lectures that preached the gospel of strategic bombardment and aided in the creation of Air War Plans Division-1 (AWPD-1). This was the Air Corps' first independent war plan, comparable to the war plans of the Army and the Navy directed at America's most likely foes. AWPD-1 offered the prospect of defeating Germany and Japan with an air campaign alone.[104] "Central to the plan was an independent strategic bombing campaign using high-altitude precision bombing."[105]

Air War Plans Division-1, later updated to Air War Plans Division-42, became the backbone of the Combined Bomber Offensive in Europe and the US bombing campaign in the Pacific. The primary difference between the plans was not only the order and nature of the bombing targets but the ACTS' hopeful expectations of an independent air arm. "AWPD-1 assumed that an air offensive might eliminate the necessity for a subsequent ground invasion, whereas AWPD-42 looked toward the establishment of an air ascendency necessary to subsequent surface operations."[106]

Fortunately, the steady work of air power advocates, such as Foulois, Mitchell, and General H. H. "Hap" Arnold convinced Congress and the Roosevelt administration to increase the Air Corps budget and readiness. With war looming in Europe, the president asked for a $500-million Air Corps expansion in 1938. Congress responded with $300 million and expanded the size of the Air Corps to twenty-four combat groups. By the spring of 1940, Roosevelt called for the production of fifty thousand airplanes.[107] When the United States entered World War II in 1941, the Air Corps had two decades of preparation and over two years of advanced aircraft production. However, air power proponents neglected the development of tactical and close air support aviation, leaving the early Forward Air Controllers—the Cavalry Reconnaissance mission—forgotten to history. Instead, Air Corps planners and leaders embraced Strategic Bombing as the principal air power contribution to winning a war.[108]

Chapter 2

World War II: Foundations of FAC Operations

C ompared to the First World War, US air power was far more prepared for World War II. Before entering the war in 1941, the Roosevelt administration set in motion the production of a massive, modern air force with the approval of a force expansion to fifty-four groups.[1] The planned force centered around the strategic bombardment concepts developed by the Air Corps Tactical School and spelled out in AWPD-1. However, the Army Air Force invested far less in tactical capabilities, organizations, and doctrine—particularly in close air support operations.[2] As a result, the Army Air Forces' capability to provide close air support during the war developed organically out of the necessity of combat. Airmen and soldiers created Forward Air Controller capabilities ad hoc, using resources at their disposal in both the European and Pacific Theaters. By the end of the war, these on-the-spot creations evolved—along with command-and-control, organizational structures, doctrine, and tactics—into an effective joint force capability able to provide planned and on-call close air support for US ground forces directed by Forward Air Controllers and Tactical Air Control Parties.

The development of modern Air Force Forward Air Controllers began with the steady evolution of tactics and organizational structures during the

Allied offensives in North Africa, Sicily, and Italy. During Operations Torch, Husky, Avalanche and Shingle in 1942-1943, the Twelfth Air Force—the US Army Air Forces primarily responsible for air operations in the North African and Mediterranean theaters—steadily adapted to the conditions of the battlefield and ground commanders' demands for better air support. Each of the early offensives added a critical piece of the puzzle and enabled airmen and soldiers to adopt, test, and perfect new techniques and tactics for close air support and Forward Air Controller operations.[3]

The Twelfth Air Force, created in August 1942 under the command of Brigadier General Jimmy Doolittle, was responsible for providing US air power in North Africa. The Twelfth Air Force was organized and equipped as a spinoff of the Eighth Air Force with six bomb groups, three fighter groups, and one transport group. The Twelfth was further subdivided into XII Bomber Command, tasked with conducting bombing raids on German infrastructure in the Mediterranean Theater; XII Fighter Command, which was responsible for providing air superiority and bomber escort; and finally, XII Air Support Command. The XII Air Support Command was a hybrid organization consisting of both fighter and bomber groups tasked with providing support to US ground forces after they came ashore in North Africa.[4]

Operation Torch began November 8, 1942. Attempting to seize a foothold in North Africa, the Allies came ashore at three widely separated landing zones in Algeria and Morocco, on both the Mediterranean and Atlantic coasts. Allied commanders divided the landing zones into Eastern, Western, and Central Task Forces. The British Eastern Air Command supported the Eastern Task Force. XII Fighter and Bomber Command supported the Central Task Force, while XII Air Support Command was responsible for the Western Task Force. Finally, British and US Navy gunfire initially supported the Torch landings, with the air forces taking over the responsibility once the Allied forces were ashore.[5]

Field Manuals (FM) 1-10, 1-15, and 31-35, Army Air Forces doctrine in use in November 1942, did not adequately address the intricacies of coordinating air support amid the complex problems presented by combined Allied air, ground, and maritime operations in North Africa. *FM 31-35* dictated the primary air-ground coordination doctrine.[6] As envisioned by the manual, the Air Support Command was the primary air power entity responsible for providing air support—in the form of both interdiction and close air support—to an Army unit. Doctrinally, each army command had an Air Support Command under its control, with the subordinate corps and divisions supported by Air Support Parties. *FM 31-35* dictated that ground units channel their air support requests through the Air Support Parties, to the Air Support Command before receiving approval from the army commander.

Ultimately the system for air support spelled out in *FM 31-35* proved to be overly cumbersome and unadaptable to the realities of combat. The system relied on a static battlefield for the successful coordination of air attacks and the army commander maintaining awareness of the capabilities of every air unit under his control.[7] Instead, Allied chains of command and areas of responsibilities overlapped and conflicted, hampering the ability of air power to adequately assist the land campaign as it unfolded in the winter of 1942 and 1943. The arrangement of the air commands created a mismatch between aircraft allocated to the various US air task forces and their assigned tasks. For example, the heavy bombers of XII Bomber Command were designed for attacking large industrial complexes—which were nonexistent in German-controlled North Africa—and were ill-suited to provide direct support to ground forces. The fighters of XII Air Support Command struggled to provide air superiority and prevent Luftwaffe attacks on Allied troops, because the German fighters outclassed American planes early in the war. Unfortunately, the fighters were also poorly suited for ground attack.[8]

The distances of the North African theater also hampered Twelfth Air Force operations, particularly after Allied ground forces moved east into Tunisia. Operating from fields in Algeria, the aircrews faced a four-hundred-mile or more round-trip flight to the front. The distances left little fuel for supporting Allied ground forces and hampered coordination between air and ground units. The extended flight times delayed air response to support requests and increased aircraft mechanical issues. As Allied ground forces occupied airfields in Eastern Algeria and Tunisia, the Twelfth Air Force was able to move aircraft closer to the fighting, but it was a slow process. As Torch forces moved eastward across Algeria and into Tunisia, the air forces increasingly overlapped responsibilities with the British Desert

Air Force, under the independent control of General Bernard Montgomery, moving west from Egypt.[9]

Further complicating the Army Air Forces' effort was a lack of systems and institutional knowledge in place to use the doctrine and techniques spelled out in *FM 31-35*. Although the Army conducted several large exercises across the southern United States in 1941, aside from liaison units, it did not use air and land power together in large complex maneuvers before the war.[10] Compounding the struggles, while *FM 31-35* spelled out the air support structure, which should have been in use during Operation Torch, few of these structures existed. Allied ground forces suffered air attacks from the Luftwaffe and lacked air support when faced with aggressive combined arms attacks by the Afrika Korps because of the lack of command-and-control architecture, little practical experience in joint operations, overlapping areas of responsibility, and the suboptimum mix of aircraft within each task force. Both air and land commanders were displeased with the situation. Ground commanders accused the Army Air Forces of leaving them unprotected, while Allied air commanders felt unnecessarily restricted. Ground commanders at all levels wanted air power to focus on targets and threats closer to Allied forces on the front lines, while air commanders wanted to attack Axis ground and air forces deeper in Tunisia before they could be brought to bear at the front.[11]

When the Afrika Korps counterattacked the American II Corps in February 1943 at Kasserine Pass, the dysfunctional system and poor weather exposed the severe flaws in American air support. German air attacks spooked Allied ground forces and led to them firing on Allied aircraft attempting to provide support. The situation became so poorly orchestrated that "B-17's attempted to bomb the famous pass, but got lost and struck an Arab village more than *100 miles* from the battle area."[12] The German attack overran airfields near Allied front lines, forcing the air forces back to fields in Algeria and pilots to again execute missions from near the maximum range of their aircraft.[13]

The muddled effort led to the reorganization of the air forces in North Africa at the conclusion of Torch. General Dwight Eisenhower placed US Army Air Forces General Carl Spaatz in command of all Allied air forces in North Africa, creating the new Northwest African Air Command in February 1943. This new command was a subordinate organization of the newly created Mediterranean Air Command led by British Air Marshall Sir Arthur Tedder. The reorganization brought all Allied air power in theater under a single commander, a first step in correcting the issues facing the US Army Air Force. Allied tactical air power—including the fighter, attack, and light bombers of XII Air Support Command—fell under the command of Air Marshall Sir Arthur Coningham. A more streamlined process improved joint air and ground planning and gave Allied ground commanders better access to air support during combat.[14]

As Allied forces prepared to jump from North Africa to Sicily, improvements in command-and-control systems strove to synchronize air efforts and improve the Allied air-to-ground support campaign. In preparation for Operation Husky, planners placed Tactical Air Control personnel aboard ships supporting the landings. These controllers used radio and radar equipment to direct flights of fighters and bombers to attack Axis forces in the landing areas, interdict Axis supply routes and reinforcements, and attack Axis naval and air forces during Husky's preassault phase. Tactical Air Controllers also moved ashore with the Husky landing force, working in conjunction with ship-based controllers to direct missions as the Allied Fifteenth Army Group moved north from the landing areas. Despite the improved organization, planning, and preassault air attacks, close air support missions still did not meet ground force commanders' expectations. XII Air Support Command flew no close air support missions in the first two days of Husky. Aircraft flying from fields in Algeria and Tunisia still required thirty minutes or more to arrive over Sicilian battlefields, limiting combat time. Radio and radar equipment often failed as it was not waterproof, nor was it suited for rapid movement across rugged terrain. Further, the mountainous terrain of Sicily hampered coordination by blocking radio signals between ground forces, Tactical Air Controllers, and aircraft. Finally, there was no ability for Allied ground forces engaged in combat with Axis forces to communicate directly with aircraft tasked to provide support.[15]

Further compounding the problems in the first stages of Allied Mediterranean operations was the fact that few aircraft available to the Twelfth Air

Force were suited for close air support missions. XII Air Support Command flew a hodgepodge of fighter and light bomber aircraft, such as the P-40 Warhawk, P-38 Lightning, P-39 Airacobra, A-36 Apache, B-25 Mitchell, A-20 Havoc, and A-26 Invader. Although P-40s were flown from the deck of the USS *Ranger* to support the Torch landings, both the P-40 and P-39 were poorly suited for air-to-ground operations. Both had a limited range, and when carrying additional fuel tanks to extend their range, they could only carry one or two 250-pound bombs. Making matters worse, the pilots rarely trained in ground-attack missions and the aircraft were far outmatched in aerial combat by their German counterparts. In contrast, the P-38 was a twin-engine long-range fighter that could carry two 500-pound bombs, but because it performed well at high altitude, had a long range, and proved more capable against Luftwaffe fighters in the air, the aircraft primarily served as an air superiority fighter during the early stages of the Mediterranean campaign. The B-25, A-20, and A-26 were all similarly designed twin-engine bombers. Both the A-20 and A-26 had a complement of forward-firing cannons fixed in their nose, which worked well against Axis armor and mechanized forces. Although excellent in an interdiction role, the bombers were not well suited for supporting friendly troops engaged in close combat with Axis forces because of the inaccuracy of World War II bombers.[16]

The one aircraft in Twelfth Air Force's inventory well suited for the close air support mission was the A-36. The A-36 was a ground-attack version of the more famous American fighter, the P-51 Mustang. Powered by a lower horsepower Allison engine, versus the P-51's Rolls Royce Merlin, and outfitted with dive brakes, the A-36 could carry two 500-pound bombs and sported six .50 machine guns. The dive brakes enabled a pilot to literally brake during his dive to slow his airspeed, allowing for steeper, more accurate dive-bombing attacks. Unfortunately, lack of coordination and communication between Allied ground and air forces led to A-36s executing multiple attacks on friendly positions during Operation Husky, including three attacks on General Omar Bradley's headquarters, giving the aircraft a bad reputation among Allied armies. The P-47 Thunderbolt eventually replaced the A-36 as the Army Air Forces' primary ground-attack fighter.[17]

The Forward Air Controllers of World War II did not develop from tactical aviation communities such as fighters or bombers, but instead from the
observation aircraft of the Army and the Air Force. As the Army Air Forces
developed bomber, reconnaissance, attack, and fighter aviation doctrinally divergent from close air support of ground forces during the interwar period, the Army developed observation and liaison aircraft to fill the
traditional observation void. In 1941, in preparation for US Army maneuver exercises, the Army Air Forces contracted with Piper, Taylorcraft, and
Aeronca aircraft companies to provide small observation aircraft for evaluation during the maneuvers. The aircraft proved invaluable in artillery spotting, reconnaissance, delivery of small amounts of cargo and messages, and
casualty evacuation. The airplanes earned the nickname "Grasshoppers"
due to their ability to operate out of unprepared grass fields, their drab
green paint schemes, and the Grasshopper logo on the civilian pilots' shirts
during the initial operational tests.[18]

The utility of the aircraft during the 1941 maneuvers resulted in
contracts with the three test companies, plus Stinson and Interstate.
Depending on the production company, the airplanes were designated
the L-2 through L-6. Grasshopper aircraft were powered by small,
rugged engines that produced between fifty and ninety horsepower and
the airplanes were simply constructed from welded metal tubing, wood,
and canvas. Grasshoppers had two seats in tandem—one seat in front
of the other. The pilot occupied the front seat and an observer or cargo
flew in the back. The Stinson-manufactured L-5s were the largest and
most powerful of the Grasshopper aircraft, with the later models powered
by 185 horsepower engines. With a cruise speed of 115 miles per hour,
a range of nearly four hundred miles, and a payload of six hundred pounds
with full fuel, the L-5 was faster by thirty miles per hour with a range
nearly double, and carried nearly one hundred pounds more payload,
than any of the other Grasshoppers. In November of 1942, the Army Air
Forces activated the 115th Liaison Squadron at Laurel Field, Mississippi,
as the first Air Force observation unit to fly the "Grasshopper" aircraft,
with Army artillery units operating the airplanes soon after the trials in
1941. Grasshoppers served in every operational theater, proving infinitely

adaptable (and the L-5s served in all branches of the military through the Korean War).[19]

The Grasshoppers faced their first combat test on November 9, 1942. Launched from the aircraft carrier USS *Ranger* sixty miles offshore, a flight of three Army aircraft was tasked to support the landings of Operation Torch. Unfortunately, the flight, led by Captain Ford Allcorn, faced antiaircraft fire from Allied guns as they came ashore, and all three aircraft fell victim to friendly fire. Luckily, the three pilots survived.[20] In late November, the Army Grasshoppers began an artillery observation school at Sidi-bel-Abbes, Algeria, and assisted II Corps as Allied forces drove across Tunisia in March and April of 1943.[21] Due to their limited flight range, both the Army and Air Force Grasshoppers launched from ships in support of Allied amphibious operations. Beginning with the Husky landings in Sicily, the Grasshoppers used temporary wooden or steel mesh runways constructed over the decks of Navy LSTs.

The "mini-aircraft carriers" launched the Grasshoppers from the decks to provide naval gunfire and artillery spotting for the landings. Once the ground forces secured a beachhead, the Grasshoppers came ashore and used the beach, fields, or roads for runways.[22] A second unique launch and landing system known as the "Brodie System" suspended the small aircraft in a harness. "A pilot could take off or land with the aircraft hooked to a trolley that ran along a cable. On landing, the trolley provided braking for a smooth stop. The cable and trolley could be rigged on very short jungle fields, or even on ships."[23] Grasshoppers continued utilizing ship-based launches during Allied landing operations in the Mediterranean, the invasion of France, and operations in the Pacific.

Operation Avalanche, the Allied invasion of Italy, began on September 9, 1943, less than one month after the US Seventh and British Eighth Armies completed Operation Husky with the capture of the Sicilian city of Messina. Like the Torch landings, Avalanche involved three widely separated operations areas. A smaller landing, code-named Baytown, consisting of British and Canadian forces comingjust across from Messina on the "toe" of the Italian mainland on September 3. The Baytown force moved northward from their landing areas to link up with the main Fifth Army force, commanded by US General Mark Clark, which landed just south of Salerno. A third, smaller force, came ashore in the "heel" of Italy, near the town of Taranto, and maneuvered north to link up with the other Allied forces.[24]

The Allied air forces in the Mediterranean Theater tripled in size in the year after Operation Torch. The air component also made great strides to fix the shortfalls exposed during Torch. The release in July 1943 of a new doctrine document, *FM 100-20: Command and Employment of Airpower*, tried correcting some of these issues but turned out to be highly controversial, particularly with Army ground commanders.[25] *FM 100-20* placed air power and air commanders equal with land power and commanders. The manual also established the priorities of tactical air power, listing them in order as follows: first, gain air superiority; second, interdict enemy troops and supplies and prevent them from moving in or around the theater; and finally, provide close air support to assist ground forces in gaining their objectives. While land commanders worried this was a move by the Army Air Forces towards independence and spelled the end of air support to ground forces, the new doctrine sought to remove the conflicting and overlapping areas of responsibility that remained within the organization.[26]

Reorganization of command structures after the Kasserine debacle, coupled with realigned air power priorities, led to a better delineation of tasks for air component commanders preparing for Operation Avalanche. The Strategic Air Force and Coastal Air Force commands supplied strategic attack and air superiority missions inland from the landing areas and north up the Italian peninsula. As during Torch and Husky, the XII Air Support Command operated an amalgamated force of fighters and bombers assigned the dual mission of providing support for the ground forces and ensuring Allied air superiority over the landing areas. All three commands conducted preinvasion attacks, and both the Coastal Air Force and the XII Air Support Command provided aerial cover for Allied naval forces as they moved to the assault areas.[27]

To improve coordination between the XII Air Support Command and Fifth Army, the XII's commander, Major General Edwin House, established a command center aboard the USS *Ancon*, which he eventually moved ashore and colocated with General Clark's headquarters.

The command center processed air support requests and supported Navy gunfire through Grasshopper observations. A second initiative begun by House and Clark was the introduction of airmen down to the division level. These Air Liaison Officers advised ground maneuver forces about the best use of air power and assisted units in coordinating air support requests. (The Air Force still uses Air Liaison Officers today to advise and assist Army units with coordination of air power.) The lack of reliable communication gear, the rugged Italian terrain, and the distances between Allied airfields in Sicily and the landing beaches in Salerno and Taranto continued to hamper communications and slow air support response times, in spite of the improved command structures, injection of liaison officers, and better cooperation between senior air and land commanders.[28]

XII Air Support Command response times improved as Italian airfields fell under Allied control, and Allied fighter units were able to move to the Italian mainland. However, some ground commanders began to bypass the air support request system and instead passed their requests directly to nearby fighter units. This direct coordination skipped the official command-and-control system but allowed for the two units to directly coordinate before the aircraft were airborne. It also enabled them to communicate directly once the fighters were overhead the front.[29]

The incremental changes to Army Air Forces doctrine and Allied command structures steadily enabled better close air support for Allied forces. Nevertheless, the lack of adequate communications between ground forces and aircraft inhibited the final and most critical phase of the process—the delivery of weapons against enemy forces directly engaged with friendly forces. Providing close air support to Allied ground forces became increasingly critical as Operation Avalanche slowed to a grinding slog during the winter of 1943 and 1944. Weather and terrain, combined with stiffening German resistance, slowed the Allied advance south of Rome. Between September and October 1943, the Fifth Army advanced northward about one hundred miles, but between November 1943 and January 1944, the Allies only gained a few dozen miles against the well-defended positions near Cassino. In January, under political pressure to speed the advance, Clark initiated Operation Shingle and landed VI Corps sixty miles behind the main Axis line and just thirty miles south of Rome. However, after a month, VI Corps remained stalled just inland from the landing beaches, with the remainder of the Fifth Army still stuck in the central Italian mountains.[30]

During the spring of 1944, the last pieces fell into place, enabling true Forward Air Controller operations and providing much-needed support to the stalled Allied advance. The first step was the use of pilots at the front to help coordinate air attacks, a process dubbed "Rover Joe." The second step was a command-and-control network called "Pineapple," which allowed more responsive air attacks closer to friendly troops. The final development married Rover Joe, Pineapple, and the Grasshoppers into a system called "Horsefly."[31]

Rover Joe was the American derivative of a British concept used in previous Mediterranean campaigns nicknamed "Rover David" and "Rover Paddy." The Rovers featured a two- or three-man team consisting of pilots, radio operators, and drivers. Equipped with a jeep, maps of the area, and a radio capable of communicating directly with aircraft and Tactical Air Controllers, Rover Joes patrolled the front both in their jeeps and on foot, assisting Allied ground forces in the coordination of close air support missions. Usually chosen from fighter groups near the sector of the front where they patrolled, Rover Joes often controlled aircraft from their own unit. Rover Joe was the first use of Air Force Tactical Air Control Parties, or TACPs, to control close air support missions. Since the introduction of Rover Joe, Air Force TACPs have been the primary attack controllers of Air Force close air support missions.[32]

The Pineapple System used Tactical Air Controllers, Rover Joe, reconnaissance aircraft, and fighter aircraft to attack enemy forces operating in areas between the Bomb Line and friendly forces. The Bomb Line was the closest distance to friendly forces Allied aircraft could attack enemy positions without coordinating with the ground forces. The Allies created them to prevent air attacks on friendly forces, as had happened in North Africa and Sicily. However, due to poor communications, the area between the Bomb Line and friendly troops became a haven from air attacks for the enemy. To solve this problem, flights of fighters and tactical reconnaissance aircraft patrolled the area between friendly troops and the Bomb Line. If they located enemy forces the airmen radioed the Pineapple controllers or Rover Joe, who in turn coordinated with Fifth Army command posts to ensure no friendly troops were in the area. If the area was clear, Allied aircraft could attack the enemy positions. This system enabled pilots to strike targets of opportunity below the Bomb Line and tactical reconnaissance missions to request immediate strikes when they discovered Axis forces. Eventually, VI Corps used Pineapple with scheduled intervals of armed fighter patrols to strike Axis transportation and armor below the bomb line.[33]

In April 1944, the Fifth Army finally began to break out along the western end of the main front. The units making the breakout moved quickly along the relatively flat coastal planes south of the VI Corps areas around Anzio. The rapid movement of Allied ground forces north taxed the air component's ability to keep pace with the rapidly changing operations. In May 1944, the XII Air Support Command, now renamed XII Tactical Air Command, expanded the Pineapple system in an attempt to maintain support for the rapidly shifting front of Operation Diadem. During the breakout, by chance, an L-5 Grasshopper landed near Captain William Davidson's Rover Joe position in search of fuel. A short discussion between the unknown pilot and Davidson sparked an experiment between the 1st Armored division and the 324th Fighter Group. The two units began using a combination of Pineapple, Rover Joe, and radio-equipped L-5s to find enemy positions, direct fighter aircraft to the area, and provide control from the L-5 to the fighters. The new system, named Horsefly, represented the foundational moment for Air Force Forward Air Controllers.[34] Though the names, technology, and systems for supplying air support to ground forces changed over time, the triad of the Tactical Air Controller, embodied by Pineapple; Tactical Air Control Party, embodied by Rover Joe; and Forward Air Controller, embodied in the Grasshopper, remained constant in the decades since 1944.

Using L-5s as Horseflies for the experiments was the most logical choice. The aircraft's more powerful engine enabled it to carry the heavier payload of the two-man Horsefly crew plus the SCR-522 radios used to communicate with Rover Joes and fighter aircraft. Other Grasshopper aircraft, which had lower power engines and no electrical system, required modifications of wind-powered generators and reconfigured cargo areas to accommodate the radios. The radio equipment alone weighed ninety pounds and required nearly two cubic feet of cargo space. Crystal frequency controls allowed the operator to use up to four preset VHF frequencies—a revolution at the time—but the complicated parts left the radio fragile and vulnerable to damage. The size, weight, power requirements, and fragility of radio sets continuously bedeviled communication between ground forces and aircraft during WWII.[35]

The Horsefly missions began with frontline units transmitting targets and air support requests to their division headquarters. The division headquarters vetted and prioritized targets with the assistance of their Air Liaison Officers. The division team radioed targets to the Pineapple controllers, who either contacted the Horsefly operations to dispatch an aircraft to the area or radioed a nearby Horsefly already airborne. During the experiment between the 1st Armored division and the 324th Fighter Group, planned flights of four fighters were assigned Horsefly missions at one-hour intervals. As the system evolved and spread to other commands, fighter and Horsefly missions were used on-call for Allied forces engaged with enemy forces, or as targets of opportunity arose. Once the Horsefly was in the target area, the pilot and observer worked with Rover Joe and Pineapple to confirm the target types and locations.

Fighter flights checked in with the Pineapple controllers, who directed them to the Horsefly's target area with geographic reference points—such as towns or rivers—and coordinates. Horsefly aircraft had their upper wing surfaces painted corresponding to the aircraft's callsign. For example, "Horsefly Red" had red wings, which made it easier for strike pilots to distinguish them from other Grasshoppers or Horsefly aircraft in the area. Once in contact, the Horseflies escorted the fighters to the area and described the types of targets and their locations using geographic reference points and coordinates as well.

Horseflies ensured the fighters attacked the correct targets. If the Forward Air Controller was not confident the fighters were going to strike the correct target, or if the target was no longer in the area, the fighters moved on to a backup target. One enthusiastic L-5 pilot mounted bazookas under the wings of his airplane, using them to attack targets or mark them for attacking fighters. Other Horsefly and Grasshopper pilots adopted this modification. Eventually, wing-mounted rockets replaced the bazookas, and Air Force Forward Air Controllers have used rockets as one of their primary tools of the trade ever since.[36]

The 1st Armor and 324th's Horsefly experiment lasted for two weeks in June 1944, utilizing several different techniques attempting to refine the procedures. Fighter pilots from the 324th took over flying the L-5s from 1st Armor's usual artillery spotters because they were more familiar with the fighter operations. The familiarity with capabilities and procedures enabled the team to get weapons delivered onto the targets faster than their artillery-spotting counterparts. The slow speeds of the L-5s made leading the faster fighters to the target area difficult. Foreshadowing the Fast Forward Air Controller missions of Vietnam, the experiments even tried P-51 reconnaissance aircraft as Horseflies. In the end, however, the Grasshoppers remained the Horsefly aircraft for the remainder of the war. 1st Armor's thoughts about the experiment were positive but tentative, concluding, "It is doubtful that 'Horsefly' technique will expand into an operation covering anything wider than a division front due to the complexities of control installations." Further, 1st Armor estimated "a month of intensive training would be necessary to insure [sic] success of an operation."[37]

In contrast, the 324th was enthusiastic about the development and transmitted the best practices to XII Tactical Air Command headquarters, who passed the techniques on to other commands. XII Tactical Air Command continued to use Horseflies during Operation Anvil/Dragoon in support of Seventh Army's drive from Southern France into Germany.[38] Most critically, the Horsefly reports were available to Ninth Air Force's IX and XIX Tactical Air Commands in time for the Allied breakout from Normandy. Ninth's Tactical Air Commands continued to refine Horsefly operations and pioneered advances in air-to-ground doctrine and tactics, which shaped the future of Army-Air Force cooperation and Forward Air Controller operations. "This deft cooperation paved the way for allied victory in Western Europe and today remains a classic example of air-ground effectiveness.

It forever highlighted the importance of air-ground commanders working closely together on the battlefield."[39]

Brigadier General Elwood "Pete" Quesada stands out as the leading proponent of air-ground integration among Army Air Forces leaders during World War II. Quesada was one of the few voices pressing for better cooperation during a time when most air power advocates supported strategic bombardment and strict independence from ground forces.[40] Commander of XII Fighter Command during operations in North Africa, he was promoted to command IX Fighter Command and IX Tactical Air Command in preparation for Operation Overlord. Quesada's counterpart at XIX Tactical Air Command was Brigadier General Otto P. "Opie" Weyland. Like Quesada, Opie Weyland was a staunch supporter of air-to-ground integration.[41] In preparation for Overlord, General Dwight Eisenhower aligned Allied air and ground components to streamline cooperation. Ninth Air Force and its subordinates IX, XIX, and XXIX Tactical Air Commands aligned with Twelfth Army Group and its respective subordinates First, Third, and Ninth Armies.[42] Additionally, drawing on the combat experience of its leadership, Ninth Air Force conducted training lectures and tabletop exercises on air-to-ground integration prior to Overlord. Harnessing Twelfth Air Force's experiences in the Mediterranean, the training focused on new Air Liaison Officers. The Ninth's new Air Liaison Officers gained valuable insight on understanding the ground commanders' scheme of maneuver, integrating into ground command staffs, and interpreting air power for ground commanders.[43]

Aligning their planning with the doctrine of *FM 100-20*, Allied air staffs prepared a three-phase air campaign designed to achieve air superiority over northern France, isolate German armies on the battlefield, and then destroy Axis ground forces. For all practical purposes, the first mission of the Eighth Air Force in August of 1942 began Phase One of the Overlord air plan, establishing air superiority. The American component of the Allied Bomber Offensive targeted German industrial capacity and repeatedly aimed at Luftwaffe production and petroleum infrastructure necessary to sustain German air power. In the spring of 1944, VIII and IX Bomber Commands began to focus their attention on German frontline airfields to reduce the Luftwaffe's capability to resist the Allied invasion. The introduction of long-range fighter

escort missions and fighter sweeps by VIII Fighter Command steadily ground
down the Luftwaffe's fighter inventory.[44]

In contrast to the Fighter Groups of Ninth Air Force, VIII Fighter
Command had, in modern parlance, an air dominance mission. Its organization, training, and equipment were focused exclusively on aerial combat.
While the fighter groups of VIII Fighter Command conducted some
ground-attack missions, only 7 percent of all Eighth Air Force ordinance
supported tactical or close air support targets. Lieutenant Joe Fulton, who
flew P-51s in the 353rd Fighter Group, represents a typical Eighth Air
Force fighter pilot. Between October of 1944 and April of 1945, Fulton
flew 61 combat missions. During those missions, he ferried one aircraft
to France from England, aborted one mission due to mechanical failure,
and flew five strafing missions. The fifty-four remaining missions were
all bomber escort or fighter sweep missions. During his tour Fulton earned
three aerial victories—one ME-109, one FW-190, and one JU-88—but
Fulton dropped zero bombs.[45]

With Allied air power controlling most of northern Europe, Phase II,
isolating the battlefield, came next. "The heaviest and most critical responsibilities assigned to any single air organization fell upon the Ninth Air
Force."[46] In preparation for the invasion, the Allied air forces dropped nearly
100,000 tons of bombs against targets in and around the landing areas.
Fighter and bomber groups of Ninth Air Force attacked railyards and railroad
junctions, troop concentrations, road networks, bridges, and gun batteries.
76,200 tons of ordinance were delivered against transportation targets, while
23,094 tons of ordinance targeted German artillery batteries. The logistical
requirements to support the air assault of the Ninth Air Force alone taxed the
Allied transportation network already strained by the preinvasion build up.
Further compounding the logistics effort, Allied air planners carefully distributed preinvasion air attacks to balance the needs of striking all the necessary
targets against revealing the location of the Overlord landing area—dropping
three bombs on targets away from Normandy for every one dropped in or
near the landing beaches.[47]

For the pilots of Ninth Air Force, little changed between June 5 and
June 6, 1944. As Allied armies landed on the beaches of Normandy, Allied

air forces continued their assault of German defense networks, troops, and transportation systems. Isolating the battlefield remained the primary objective of the air attacks, but the importance of the mission increased with friendly forces vulnerable to counterattacks in the landing areas. The Allies' dominion of the sky allowed ground-attack missions by fighter and bomber aircraft, along with Overlord landing operations, to proceed unmolested by the Luftwaffe. IX Fighter command flew 2,300 sorties in twenty hours in support of landing operations on June 5 and 6.[48] Due to the pace of operations, and the variety of targets hit, the transition between Phase Two and Phase Three of the Allied air plan happened organically, rather than by command decision. However, IX and XIX conducted initial close air support operations in support of the landings without the assistance of Rover Joes. XII Tactical Air Command and Fifth Army were still experimenting with the Rover Joe, Pineapple, and Horsefly concepts as the Allies came ashore in Normandy.[49]

Command-and-control networks, communications, and integration between Ninth's Tactical Air Commands and their Army counterparts benefited immensely from the Mediterranean trials of Twelfth Air Force. In addition to the air-land command alignment structure, each of Ninth's tactical components benefited from an extensive and improved radar network. Complementing the radar networks were robust communications with redundant radio, telephone, and messaging systems. This network tapped into Air Liaison Officers dispatched to battalion and higher Army headquarters, allowing IX's and XIX's Tactical Air Control Centers to receive and respond to requests for air support much faster than previous ground operations. To further decrease response times, Ninth's fighter groups fed aircraft to the front at regular intervals, just as the 324th did during the Horsefly experiments. Finally, to maintain smooth coordination, decrease response time, and increase aircraft time over the combat zone, fighter squadrons from Ninth Air Force moved into France. On June 8, General Quesada landed an aircraft on a hastily constructed landing strip in Normandy to join the VII Corps command post with General Bradley.[50]

Quesada's move to Normandy was the tip of a massive effort to bring Ninth Air Force to France. By the end of June, IX Engineer Command built nine all-weather airfields and was in the process of building seven more. Further, Ninth Air Force fighter groups flew more than thirteen thousand ground-attack missions in June alone.[51] The effort of the Ninth Air Force to provide support to Allied forces in Normandy was summarized as follows: "From 6 June to 24 July ... 13 fighter-bomber groups and one reconnaissance group had crossed the channel, and a highly efficient and effective radar control system had been established on the beach head."[52]

Not all air power efforts were as successful, however, particularly as Allied advances slowed in the dense hedgerows of northern France. Allied commanders hatched a plan known as Operation Cobra, an armored spearhead scheduled for July 21, to ignite an Allied breakout from the hedgerows of Normandy. In support of this operation, air planners conceived a massive bombardment of German defenses around the town of St. Lo. This bombardment included the use of heavy and medium bombers from all Allied bomber commands to augment the tactical air power of Ninth Air Force. Senior Allied air commanders, including General Carl Spaatz, commander of US Strategic Bombing Forces, and Deputy Supreme Commander Sir Arthur Tedder, heavily influenced the air operation.[53]

Bad weather delayed Cobra until July 24. However, weather on the 24th again hindered operations. The operation called for IX Tactical Air Command to begin the attack in a designated two-by-eight-kilometer box concentrated on the 2nd Panzer Division east of St. Lo. Only half of the six fighter groups from IX Tactical Air Force made it to their targets, and most of those reported unsuccessful attacks. With the strategic bombers already approaching the target area, commanders decided to allow the attacks to continue. The lead bomber formation did not release its weapons because cloud cover obscured the target; the next group made three attempts before releasing their bombs; and most of the third wave dropped weapons. Fortunately, commanders canceled the subsequent bomber attacks, but the results of the previous waves were disastrous. Bombs fell short into friendly lines, a group of fighters attacked Allied ammunition dumps, and one bomber inadvertently attacked one of the Ninth Air Force fields in Normandy.[54]

On the 25th, Allied air forces made a second attempt. This time, Allied ground forces pulled back from the front lines to create a safety buffer.

IX Tactical Air Command fighter groups again were slated to begin the operation and were assigned targets in sections of the target area nearest friendly lines. Also, General Bradley wanted the heavy bomber attacks conducted parallel to the Allied lines, but bomber planners routed the bomber stream perpendicular to the front. As the bombs began to fall, it became apparent that things were again not going as planned. The conflagration of dust and smoke from explosions along the front obscured smoke bombs dropped to mark the target area. As a result, some Allied bombers again delivered their weapons onto friendly forces, and although the US units under friendly attack could talk to the fighters of IX Fighter Command, no communication link existed between ground forces and bombers. Errant bombs killed 101 Allied troops, including Lieutenant General Leslie McNair, and wounded 463. Though the attacks by Allied strategic bombers were accurate by contemporary standards, the damage to the reputation of Allied air power was on par with that of Operation Husky.[55]

Fortunately, the damage done to the German forces exceeded the damage done to the confidence in air power built by Quesada and Weyland. For nearly three hours, the Germans endured attacks by 1,900 heavy and medium bombers along with 550 fighter attacks. The melee destroyed armor and mechanized forces along with communication infrastructure and command posts, tearing a hole in the German lines for Allied armor to attack through.[56] For the next month, Allied forces, pivoting around the town of Argentan, moved the front more than one hundred miles. Besides providing the critical break in the front at St. Lo, close air support and interdiction missions by IX and XIX fighters secured the Allied armies' flank and continued to break German formations in advance of friendly forces.[57]

To lead the Allied armies across France, Ninth Air Force improved the Rover and Horsefly tactics created in Italy. IX and XIX Tactical Air Commands adopted a technique called "armored column cover." Essentially Rover Joe in a tank, the tactic put pilots with armor units at the leading edge of the front. Equipped with the same SCR-522 radio found in Allied fighters and Horsefly aircraft, the Rover Joe Tactical Air Control Parties directed close air support strikes from the relative security of Allied tanks. Air planners paired them with Horsefly aircraft and flights of four fighters scheduled at one-hour intervals for close air support missions. When coupled with Ninth Air Force P-47 fighters, the tactic proved devastating to German forces.[58]

The P-47 Thunderbolt, perhaps more than any other fighter in World War II, was ideally suited for the close air support and ground-attack missions. The largest Army Air Forces fighter of the war, the aircraft operated initially as a long-range bomber escort. However, the aircraft's ability to carry a massive bomb load, rugged construction, and eight .50 machine guns made it a formidable ground-attack platform. Ninth Air Force pilots, operating from temporary fields in France, were able to carry nearly every bomb in the Army Air Forces inventory between one hundred and two thousand pounds. For close air support missions, the P-47 usually carried two 500-pound bombs and two racks of high-explosive rockets, allowing pilots to destroy Axis troops, trucks, trains, tanks, aircraft, and buildings.[59]

Participating in the Ninth Air Force efforts from D-Day to VE-Day was Lieutenant Quentin Aanenson, a P-47 pilot in the 391st Fighter Squadron of the 366th Fighter Group. His story is a testament to the effort by the Ninth Air Force supporting Allied armies in Western Europe, and the rugged P-47 in ground-attack and close air support missions. Aanenson completed flight training in the US just in time to join Ninth Air Force for D-Day. He flew his first combat mission on June 4 as part of a flight hastily assembled from new arrivals, unclaimed aircraft, and just enough veteran pilots to lead the group into combat. The 391st and Lieutenant Aanenson moved to St. Pierre du Mont, France, on June 17, 1944, as the Allied hold in Normandy expanded. Flying from newly constructed airfields, Aanenson joined a flight of four P-47s from the 391st, which led the attacks at St. Lo at the opening of Operation Cobra. On August 3, he scrambled on a close air support mission to support VII Corps during the German counterattack near Mortain. During the mission, 20 mm flak heavily damaged his P-47. With his engine and cockpit on fire, Aanenson tried to bail out of the crippled aircraft, but the canopy refused to open, damaged by German flak. He managed to get the fires out and nurse the crippled aircraft back to the 391st's base. Struggling to maintain control of his damaged aircraft, Lieutenant Aanenson landed almost one hundred miles an hour faster than the P-47's normal landing speed. The aircraft careened off the runway, and in the crash, he was knocked unconscious, dislocated his shoulder, and broke three ribs. After a few days of recovery in London, he was back flying combat missions on August 10.[60]

As the advancing armies moved south and east from Normandy, the Allied advance picked up momentum. The Allies trapped portions of three German armies in a salient between the towns of Failase and Argentan, where Allied air power decimated the retreating Germans.[61] Aanenson described a similar scene of destruction further east, along the Seine river.

> It was late August 1944, and Patton's Armored Divisions were in a mad dash to the Seine River, trying to catch the rapidly retreating Germans before they could escape. I was flying in a flight of four Thunderbolts patrolling the Seine to do everything we could to prevent their crossing.
>
> Up to this time most of the Germans had been crossing at night to escape our attacks, but on this particular day—with Patton's tanks rapidly approaching them—the Germans were forced into trying to cross during the daytime. It was late afternoon near the town of les Andelys when we suddenly spotted them. What happened during the next 10 minutes will stay fixed in my memory as long as I live.
>
> The German troops were crowded on barges, in small boats, just anything that would float. We caught the barges in midstream, and the killing began. I was the third plane in the attack, and when I pulled in on the target, a terrible sight met my eyes. Men were desperately trying to get off the barges into the water, where large numbers of men were already fighting to make it to shore. My eight .50 caliber machine guns fired a hundred rounds a second into this hell. As the last P-47 pulled off the target, the first plane was making its second strafing pass, and the deadly process continued. In about three passes we had used up our ammunition, so we pulled up and circled this cauldron of death.[62]

In November 1944, the 391st moved to an airfield near Asche, Belgium, supporting First Army. When the German counteroffensive in the Ardennes began in December, Lieutenant Aanenson joined VII Corps as a Rover Joe. With his radio team, Horsefly aircraft, and tank or jeep, he supported close air support missions for the divisions of VII Corps as they fought the Germans in the Battle of the Bulge. By March, VII Corps was preparing to cross the Roer River near the western German city of Duren. Taking up a position in the tower of the eight-hundred-year-old Merode Castle, the recently promoted Captain Aanenson directed close air support missions supporting the 8th Infantry Division's crossing. Soon his position in the castle came under attack by German artillery fire. During the barrage, one of the Rover Joe radio operators was killed by a German shell so close to Annenson, the dead man's blood splattered his uniform, radios, and maps. Despite the carnage, he continued to direct attacks against German positions. By the end of the day, he controlled twenty-four squadrons in attacks around the town and across the river. Aanenson remained the VII Corps air liaison and Rover Joe until the end of March 1945, when VII Corps crossed the Rhine at Cologne.[63]

In seven months of air combat on the Western Front, Aanenson flew seventy-one combat missions: sixty in the P-47, eleven in the P-51, and three in the P-38s. He served as Rover Joe and air liaison for four months during the heaviest ground fighting on the Western Front, directing hundreds of close air support missions for VII Corps soldiers. Aanenson received his orders home at the end of April and was back in the United States just before VE-Day. Meanwhile, the 391st continued to support the Allied advance into Germany, ending the war at an airfield near Muenster.[64]

While the air war in the European Theater began with an emphasis on strategic bombing by the Eighth Air Force and transitioned to Ninth Air Force's tactical airpower providing integrated support to Allied armies, in the Pacific the process went the other way. The nature of the war in the Pacific required close cooperation between naval, ground, and air forces to gain footholds in the vast theater. Steadily grinding from the periphery, the campaigns slowly brought the heart of Japan within striking distance of long-range B-29 bombers. It was not until 1945, after three years of relentless fighting, that air power finally began its devastating strategic bombardment of Japan, culminating in the nuclear bombings of Hiroshima and Nagasaki.[65]

Between the Japanese attack on Pearl Harbor and the check of the Japanese advance in the Solomon Islands and New Guinea in August 1942, the Army Air Forces in the Pacific did little more than support a fighting retreat. The Hawaiian Air Force was effectively obliterated on the ground by the Pearl Harbor attacks. Afterward, only 79 of the command's

231 aircraft were operational; the attack destroyed 64 completely, the rest were severely damaged and unflyable. The bulk of Army Air Forces in the Pacific belonged to the Far East Air Force and were widely scattered across the Western Pacific. The command operated a hodgepodge of bombers and older fighters in five bomb groups, three fighter groups, two transport groups, and one photo squadron. This was a force roughly equivalent to Twelfth Air Force at the start of Operation Torch but spread across the entirety of the Pacific west of Midway Island, a theater roughly twenty times as large as the North African landing operations.[66]

In March 1942, the Allies reorganized the Pacific theater, along with the numbered Air Forces there. The newly created Eleventh, Seventh, and Thirteenth Air Forces took responsibility for the North, Central, and South Pacific Areas. Fifth Air Force, formerly the Far East Air Force, integrated into the Southwest Pacific Area, under the command of General Douglas MacArthur. Tenth Air Force and Fourteenth Air Force, added later, took responsibility for the China and India-Burma theaters, respectively. Finally, the Twentieth Air Force took responsibility for strategic bomber operations in the Pacific Theater after April 1944. Several factors shaped air operations in all areas of the Pacific Theater. First, Navy operations dominated the theater, driving a level of joint operations far exceeding the European Theater. Second, distances in the Pacific Theater shaped the nature of the air campaign. Both bomber and fighter aircraft continually operated at the maximum extent of their range. The distances, rather than the targets or desired bombing effects, most often determined the type of aircraft used in operations. Finally, the Europe First Allied strategy left the Pacific Theater constantly short on aircraft, supplies, and personnel. Airmen fought the bulk of the war with aircraft less capable than their counterparts in Europe and on a shoestring budget of aircrew, ammunition, and spare parts.[67]

The South and Southwest Pacific Theaters saw the most integrated air-to-ground operations. Beginning with the Allied attack on Guadalcanal and New Guinea in August and September 1942, operations in the South Pacific developed a rhythm. Weeks before beach assaults began, Navy and Air Force fighter sweeps of Japanese-held areas gained local air superiority for the next phases. Fighter sweeps and bombing missions targeted aircraft in the air and on the ground. Bombers and attack aircraft bombed Japanese command-and-control networks, communications, radar facilities, and support infrastructure. Once plans for landing and ground operations were complete, fighter and bomber attacks focused on Japanese troops, critical strong points, and artillery positions to shape the battlefield for the landings. When the landings began, a mix of naval gunfire, Navy and Marine Corps carrier-based aircraft, and Army Air Forces land-based fighter, attack, and bomber aircraft provided fire support to ground forces. During ground operations, Army Air Forces usually focused on maintaining air superiority over the landing areas and interdicting Japanese reinforcements inland. The final air power missions were Air Force and Navy long-range maritime interdiction patrols focused on attacking resources bound for mainland Japan or personnel and supplies intended for the widespread Japanese outposts.[68]

Close air support developed along a similar path in the Pacific as in Europe and North Africa. During the 1942 operations on Guadalcanal and in New Guinea, the air support request and command-and-control systems followed the dictates of *FM 31-35*. Just as in Operation Torch, poor coordination and lack of communication hindered close air support missions. By 1943 all three services began utilizing ship-based Tactical Air Controllers. Air support parties akin to the Rover Joes in Europe, called SAPs or K-Ration Teams, went ashore beginning with landings in June 1943 at the Kiriwinan Islands.[69] By December 1943 an air support system similar to the Pineapple and Rover Joes developed.[70] However, "there was no uniformity. A party at this stage of the war … might be attached to a division, brigade, or regimental headquarters" and "no less than seven different types of ground parties controlled strikes during these southern [Pacific] campaigns."[71]

Fifth Air Force commander, General George Kenney, applied some order to the system by standardizing procedures and staffing Air Liaison Officers to Army divisions. Kenney, a staunch advocate of independent air power under a single air commander, came to the Pacific in August 1942 to turn around the command. His personality was large enough to stand against the weight of MacArthur, which allowed Kenney and Fifth Air Force to develop an integrated and capable fighting force. Besides streamlining and standardizing the air support request system in the Southwest Pacific, Kenney and Fifth Air

Force developed new tactics and weapons to address the unique problems of the Pacific. First, they developed a skip bombing technique for attacking Japanese ships. They also developed a "parafrag" bomb for use against Japanese airfields and troops. Delivered from V Tactical Air Command's A-20s and A-26s at low altitude, the bombs used a small parachute to slow their descent. This technique allowed the aircraft to attack from lower altitudes, which increased their accuracy while allowing the aircraft time to escape the explosions, and this proved particularly useful in the steep terrain and tropical weather conditions of the South Pacific.[72]

Despite the joint nature of the operations in the Pacific, control of airstrikes usually belonged to the ground force making the request. Therefore, Marine Tactical Air Control Parties directed flights of Marine attack aviation in support of embattled Marines ashore. For US Army units, fire support most often came from artillery ashore, such as mortars. Army Air Forces aircraft providing close air support were most often controlled by Army Air Forces Tactical Air Control Parties in concert with the ship-based Tactical Air Controllers. Although many of the same air-ground support systems developed in the Pacific as in Europe, weather, terrain, and distance conspired to hamper the final steps in developing true Forward Air Controller operations in the Pacific. The jungle vegetation on many of the islands of the South Pacific hampered the ability for pilots to identify friendly and enemy positions from the air. Without Tactical Air Control Parties to confirm friendly locations, pilots hesitated to use bombs close to friendly positions. Long flight distances required to support operations prevented fighter aircraft from remaining over the target area for extended periods, further hampering their ability to discern the situation on the ground.

The conditions of the Pacific Theater of Operations should have created an ideal scenario for the use of Forward Air Controllers, but instead Fifth and Thirteenth Air Force relied on close cooperation during the planning phase, along with a robust air liaison and Tactical Air Control network to provide close air support. Ashore or afloat controllers directed fighters and bombers assigned to support ground forces to their target areas. Once the fighters were inland, Rover Joes used their radios, colored smoke, and colored canvas panels to identify friendly and enemy positions for them. The Tactical Air

Control Parties oriented the panels into arrows and other shapes, just as in World War I, to communicate their message in the absence of reliable radio communication. Smoke shells, delivered by artillery or grenades, proved to be the best marker to point pilots towards their targets, and the variety of smoke colors allowed Tactical Air Control Parties to differentiate between enemy and friendly positions.[73] Although the Australians developed a Forward Air Controller capability during operations in New Guinea and the US Marine Corps built a similar capability, none of the Pacific Air Force commands broadly implemented the Horsefly or air-tank team tactics.[74] Grasshopper pilots only sporadically assisted fighter airplanes finding targets.[75]

Instead, beginning with the May 1943 battle for Attu, Grasshoppers and other aircraft only occasionally adopted a FAC-like role. During the fights on Attu and the Marshall Islands, the Grasshoppers and other aircraft served as "Air Coordinators," an airborne radio relay between ground parties and Tactical Air Control Centers onboard assault force ships. The following January, on Kwajalein, Marine and Air Force Grasshoppers occasionally coordinated to lead fighter and bomber aircraft to target areas but provided only minimal direction for attacks. The rough terrain and jungle conditions of the South and Southwest Pacific hampered ground and air controllers' radio communication, making reliable communications difficult. Often— even with the help of Grasshoppers, ground controllers, and smoke marking enemy positions—attack pilots often made multiple passes over an area to confirm friendly and enemy positions. Despite the difficult conditions, ground commanders generally retained direct control of air attacks. Grasshopper pilots who managed to maintain radio contact with ground forces merely served as a radio relay between the Tactical Air Controllers and the flights of aircraft providing close air support.[76]

The US Army, under the command of General MacArthur, bore the brunt of fighting during the longer duration campaigns of the Southwest Pacific, including the nearly year-long effort to retake the Philippines in 1944 and 1945. However, the lack of air resources available to the Fifth Air Force limited the amount of air support to ground operations. Keeping with the pattern developed in the previous year of operations, General Kenney focused the command's efforts in securing the task force's flanks and sea lanes against

Japanese attack. Also, General Kenney sent most bomber missions ahead to attack targets on the islands slated for the next invasion. In short, the focus of Kenney's small force was not close air support.[77]

As Sixth Army moved northward from Leyte in October 1944, to Mindoro, and eventually Luzon in January 1945, maintaining communications between Tactical Air Control Centers onboard ships and Tactical Air Control Parties ashore became increasingly difficult. Further, when Tactical Air Control Centers moved ashore, the terrain and jungle environment blocked radio signals and caused equipment to malfunction, the same as ashore Tactical Air Controllers experienced during the campaigns in Sicily and Southern Italy. Due to the distances from their bases, B-24 Liberator heavy bombers were put into service as Air Coordinators, with Grasshoppers filling the role after ground forces secured airfields on each island, but Tactical Air Control Parties on the ground still maintained control of air attacks.[78] As the final operations began on Luzon in January 1945, radio-equipped Grasshoppers finally began to direct close air support attacks. As a result, in late January 1945 a handful of Grasshopper pilots flying radio-equipped L-5s began providing control for airstrikes, covering for Rover Joes unable to contact the close air support aircraft. Some of these pilots even experimented with the use of wing-mounted rockets, just as happened in Italy, but oddly none used them in combat to direct attacks like the Horseflies in Europe.[79]

While operational focus and size of the force in the Southwest Pacific blunted the broader development of the Horsefly techniques, the distances and naval-centric operations in the Central Pacific also prevented their use. In the Seventh Air Force's Central Pacific area of responsibility, the distances practically negated the use of land-based fighter aircraft in a close air support or Forward Air Controller role for most of the war. The Central Pacific magnified the problems faced during the earliest phases of Operation Torch. The first use of the command's land-based fighters required a creative solution. Like the P-40s launched from the USS *Ranger* during Torch, P-47s from the 318th Fighter Group launched from the aircraft carrier USS *Manila Bay* to support the June and July 1944 invasions of Guam, Saipan, and Tinian. Finally, a small contingent of Seventh Air Force's fighter aircraft joined the war. Seizure of airfields on the islands allowed the remainder of the group's P-47s to finally join the fight as well.

To control the growing number of close air support aircraft, the fighting on Guam also saw the use of the Air Coordinator. The Central Pacific Air Coordinator most often flew US Navy torpedo bombers, which usually had a three-man crew and additional space for radios to help coordinate attacks. The Navy pilots also used strafe runs and 100-pound bombs to mark targets for both Navy and Army fighters to attack. Even though the Seventh Air Force P-47s finally contributed on Guam, the bulk of the close air support came from the Navy or Marines.[80] Army ground-based fighters again saw action during the April 1945 invasion of Okinawa and a joint command was established to coordinate air support efforts during the campaign. The air operations in support of the invasion of Okinawa were much larger than previous air operations in the Central Pacific. Tenth Army created a new Tactical Air Force command to control and organize the larger air operations. A Marine Corps officer commanded the new organization, which the Allies "charged with providing air defense, troop support, and photographic reconnaissance." The Tactical Air Force "was eventually to be composed of 23 Army and 16 Marine squadrons, but Marine aircraft were in the majority during the height of the fighting on Okinawa."[81]

However, beyond creation of the new joint Tactical Air Force, the battle for Okinawa did not create any new tactics or control structures. In fact, close air support proved to be difficult during the three-month-long battle. The Japanese Army used extensive defensive entrenchments and the island's cave complexes, shielding them from air attack. The result was a brutal and bloody hand-to-hand fight for US soldiers and marines. As the battle moved close to Japan, the Army Air Forces shifted focus from ground support missions to the strategic bombardment of the Japanese mainland.[82]

The captured airfields on Saipan and Tinian brought Tokyo in range of the Air Force's B-29 bombers. In April 1944, the Twentieth Air Force took over B-29 operations in the Pacific. Simultaneously, with the bulk of the close air support supplied by the Navy and Marine Corps, the Seventh Air Force's fighters based on Okinawa transitioned from ground support to bomber escort missions. Bomber operations steadily increased during 1945, transitioning from high-altitude daylight attacks against industrial targets to low-altitude night incendiary attacks against cities during the spring of 1945. Then, in the early morning hours of August 6, 1945, the B-29 *Enola Gay* took off from Tinian to make history by delivering the world's first atomic weapon in combat.[83]

The methods, organizations, and doctrine of Army Air Forces support to combined operations in the Pacific mirrored those of Europe and North Africa. General Kenney's and General MacArthur's personalities and relationship also shaped tactical airpower development in the Southwest Pacific, where the bulk of the Army-Air Force coordination happened. The command relationship between MacArthur and Kenney mirrored Quesada's and Weyland's relationship with their Army counterparts. These positive relationships drove innovation and cooperation in both Europe and the Pacific. Through trial-and-error tactics, training, equipment, organizations, and doctrine evolved to meet the reality of the battlefield. However, the environment, vast distances, and shoestring budgets of the Pacific theater ultimately stunted widespread Forward Air Controller operations. Although the Grasshopper pilots provided critical radio relay networks, they only provided attack control sporadically. There was no theater wide implementation of the innovations seen in Europe, like air-tank teams and Horsefly. By late 1944 and early 1945, when these developments were at their peak in Europe, the operational reality of the Pacific theater shifted the Air Force's focus away from air-to-ground missions, and instead strategic bombardment of the Japanese homeland became the main effort.[84]

Steady innovation produced results for soldiers on the ground, and both the Army and the Air Force wanted to keep up the cooperation after the war. Close air support made up 33 percent of all missions flown by the three tactical air commands of the Ninth Air Force, and their contribution to victory was undeniable.[85] Ninth Air Force's postwar evaluation called for an examination of doctrine and organizations, suggesting they "be reviewed and revised on the basis of practical air-ground experience in [Europe] and other combat theaters."[86] The report focused on the shortfalls of *FM 31-35*, command-and-control systems, and training. Ninth's staff emphasized the need for close cooperation provided by air liaisons and sought to expand the tactics and methods such as the air-tank teams. Further, the analysis advocated training for bomber crews in close air support along with improved systems for the use of bombers in close air support and interdiction missions.[87]

Interest in the Forward Air Controller mission continued as well. In August 1945, the 162nd Liaison Squadron conducted an expansive test of Horsefly capabilities and tactics. "Seven aircraft and liaison pilots were assigned for test on the following phases: Rocket Firing; Bombing; Night Navigation; supply Dropping; Horsefly; Vulnerability to Enemy Fire and Miscellaneous Phase."[88] The test concluded that Horseflies were capable in the Forward Air Controller role. Utilizing rockets and radios to point out targets to P-51 strike aircraft, the test replicated the combat experience of the Italian and Western European campaigns. Further, the tests found the Horsefly reasonably effective at delivering bombs, giving the diminutive aircraft an outsized punch. Unfortunately, the tests also found the small aircraft vulnerable to antiaircraft fire, even at night, due to the increased use of radar-directed antiaircraft guns. Like the Ninth Air Force report, the test board recommended exploring the concept further, including the use of larger and faster aircraft.[89]

In March 1946, the Army Air Forces established Tactical Air Command (TAC), Strategic Air Command (SAC), and the Air Defense Command (ADC) as the three Air Force warfighting commands. The first TAC commander was Pete Quesada, unquestionably the best choice for the position. Quesada and TAC sought to expand and improve all facets of tactical aviation, and air-to-ground coordination was a high priority. On September 18, 1947, the National Security Act of 1947 created the US Air Force (USAF) as an independent service, part of the broader reorganization that also created the Department of Defense (DOD).[90] General Carl Spaatz became the first Chief of Staff, but by April 1948 General Hoyt Vandenburg took over. Vandenburg and Quesada had a turbulent relationship history back to when the pair worked together in Europe. In an era of demobilization and shrinking budgets, Spaatz, and later Vandenburg, invested heavily in long-range bombers and nuclear weapons technology. As a result, SAC received approximately 60 percent of the Air Force procurement and flying hour budget.[91] Soon after ascending to Chief,

Vandenburg downgraded TAC's status, subordinating it along with ADC to the newly formed Continental Air Command, and stripped the command of men and aircraft. The decision sought to increase nuclear-capable bomber units within the Air Force while remaining within the Congressionally mandated size limit of the Air Force. Quesada applied for retirement, which Vandenburg denied. Eventually, in October 1951, Vandenburg relented and allowed Quesada to retire. Tactical Air Command remained sidelined, along with further developments in air-to-ground coordination, until December 1950, six months after the outbreak of the Korean War.[92]

Chapter 3

Korea to Vietnam: Codifying the FAC Mission

From the moment the United States and the Soviet Union divided the Korean peninsula after World War II, skirmishes between the two Koreas steadily simmered along the 38th Parallel. On the morning of June 25, 1950, infantrymen of the North Korean Peoples' Army, supported by T-34 tanks, assaulted across the line in earnest. For the next month North Korean attacks drove the unprepared, undermanned, and poorly equipped US and South Korean forces south, inflicting heavy casualties, and eventually surrounding the combined force in a forty by sixty-mile perimeter around the South Korean port city of Busan. The North Korean leader, Kim Il Sung, expected to drive the defenders into the sea, but the perimeter held, tenuously, for two months. Finally, on September 15, General Douglas MacArthur orchestrated a successful US amphibious landing at Inchon. Combined with a breakout en masse from Busan, the offensive cut off and decimated North Korean forces, then pushed north, nearly to the Chinese border. This development prompted the Chinese to intervene in November 1950. By January 1951 the combined communist forces again pushed United Nations forces out of Seoul and some 80 miles south of the 38th parallel. However, the advance stalled again under the weight of sustained air attack and repeated UN counter attacks. UN forces managed to fight their way back to the 38th parallel by the end of July 1951, where the war stagnated for the next two years.[1]

While the war on the ground in Korea featured World War II vintage equipment and stagnated into trench warfare reminiscent of 1918, the war in the air pitted cutting-edge frontline Soviet and American swept-wing fighters engaged in the first jet versus jet aerial combat. Communist air forces, consisting primarily of Soviet combat units, engaged in a defensive strategy focused against UN heavy bombers and fighter bombers.[2] American air planners initially executed a bombing campaign reminiscent of World War II, primarily targeting North Korean industrial infrastructure. After the intervention of Chinese forces, however, the focus of the air campaign shifted to interdiction. The mountainous terrain and dearth of major roads and railroads led planners of the Far East Air Force to adopt a course of action resembling Operation Strangle, the Allied interdiction campaign in Italy during World War II. For the duration of the war, UN airpower concentrated most sorties against communist supply and troop movements, road and rail bridges, and transportation choke points in North Korea.[3] Although interdiction missions consumed the bulk of USAF efforts, air planners recognized the need to provide close air support to UN forces from the beginning of the war. Recreating Forward Air Controller capabilities in the Horsefly mold was central to this effort.[4]

When the war began in June 1950, the responsibility for the air war in Korea fell on the Far East Air Force. Mirroring the state of American Army units in Korea, the Far East's subordinates—the Fifth, Twentieth, and Thirteenth Air Forces—were inadequately staffed and poorly equipped for the challenge.[5] A lack of bases on the Korean peninsula made the already difficult situation even more challenging. Aircraft supporting the rapidly retreating US and South Korean forces were forced to fly from US bases in Japan, Okinawa, or Guam. The Eighth and Forty-Ninth Fighter Bomber Groups, flying F-80 Shooting Stars, bore the brunt of the close air support missions during the first few weeks of the war. While the Eighth made a two-hundred-mile round trip from Itazuke Air Base near Nagasaki at the southern end of the Japanese mainland, the Forty-Ninth flew from Misawa Air Base at the far north end of the island—a one-way distance of five hundred miles. Distances, weather, poor communication with ground forces, and the F-80's high fuel consumption combined to severely limit the Air Force's efforts.[6]

While the introduction of jet aircraft into the newly minted Air Force had advantages, war on the Korean peninsula quickly highlighted the shortcomings of jet aircraft in the close air support and forward air control roles. Forward Air Controllers must fly slow enough to assess the terrain and disposition of friendly and enemy ground forces, but still fast enough to avoid enemy ground fire. Close air support operations rely on the ability of Forward Air Controllers and attack aircraft to remain in a target area for extended periods. As the North Koreans pushed south, they forced the US and South Korean artillery to relocate or abandon equipment continually, decreasing the ability of artillery to provide firepower support to the beleaguered infantry. Aging World War II vintage F-51 Mustang and F-82 Twin Mustang fighters, along with B-26 Marauder and B-29 Superfortress bombers, were called upon to fill the gap.[7] Still, airpower was critically needed to provide close air support. "With an urgency demanded by a critical ground situation, the Fifth Air Force put together the first comprehensive Forward Air Controller program, a complete system, operating over the entire battle area during all daylight hours."[8]

Although the Fifth Air Force lacked the men and equipment, the infant US Air Force was not wholly unprepared for the situation. Despite significant doctrinal and organizational hurdles after 1948, leaders within the defunct Tactical Air Command, like Pete Quesada, continued to press for improved air-ground integration. Three exercises conducted between the spring of 1949 and the eve of the Korean War—Tarheel, Swarmer, and Portrex—maintained some of the progress made during World War II in the close air support and Forward Air Controller missions. The exercises also revealed how much both the Air Force and Army needed to learn about smoothly integrating air and land forces. Communication and radar capabilities continued to hamper air-ground cooperation during these exercises, just as it had during World War II. Neither service correctly prioritized the personnel and equipment placement for Tactical Air Control Parties and Tactical Air Control Centers during sealift and airlift to the exercise locations, resulting in misplaced equipment and disjointed or significantly reduced operations.

Further, both Army and Air Force units neglected training with Forward Air Controllers, Tactical Air Control Parties, and the Control Centers in the years before the exercises. As a result, pilots, controllers, planners, and ground commanders were unfamiliar with tactics and procedures.[9]

Several positives, however, came out of the exercises. These exercises established the capabilities of jet aircraft to provide firepower in a close air support situation. Despite the limitations of jets in this role, the exercises proved them to be more accurate in weapons delivery when compared to propeller-driven aircraft. Without a huge propeller spinning on the front, jets had better forward visibility, fewer adverse handling characteristics, and the ability to mount guns along the central axis of the airplane. Like the Grasshopper experiments of August 1945, the exercises also confirmed the need for a more survivable Forward Air Controller aircraft with expanded communication capabilities. Finally, the exercises allowed for the Army and the Air Force to refine command-and-control procedures. Air Force planners, Tactical Air Control Centers, Tactical Air Control Parties, and Air Liaisons were able to integrate with the Army's new Fire Support Coordination Center. Just as the Air Force worked to streamline its air support request procedures during World War II, these exercises allowed the Army to do the same before Korea. The Fire Support Coordination Center served as the central clearinghouse and prioritization center for air support requests within a division. Despite the frustrations, the struggles in peacetime training provided critical lessons that the Air Force put into practice in Korea.[10]

With the lessons of World War II and the recent exercises in mind, the Far East and Fifth Air Force staffs moved quickly to shore up the rapidly deteriorating situation in Korea during July 1950. First, the Air Force dispatched Tactical Air Control Parties to provide air-to-ground coordination for the 24th Infantry Division, the first American infantry division injected into Korea after the North Korean attack. A Joint Operations Center opened at Itazuke on July 4 to prioritize air support requests, and on July 6, a forward element of the Joint Operations Center moved to the Eighth Army Headquarters in Daejon, Korea. In a scene reminiscent of Operations in North Africa or the Pacific in 1943, the long lines of communication and flight distances between the Korean Peninsula and Japan slowed aircraft response times. The time delay, poor weather, and rugged Korean terrain further muddled the picture for pilots tasked to provide air support. As a result, pilots went to the wrong areas, could not contact their assigned Tactical Air Control Parties, were not up to date on the ground situation, or could not find the targets assigned to them. To improve the response of airpower and help stop a potential disaster, Fifth Air Force introduced a plan constructed by Lieutenant Colonel Stanley Latiolais to use Forward Air Controllers on the battlefield.[11]

When the Joint Operations Center moved to Daejon, a T-6 Texan aircraft also flew to Daejon to provide an airborne platform to assist in sorting out the mess on the battlefield, but there was insufficient staff to man both the Eighth Army's Air Liaison office and fly the aircraft. So, on July 9, additional personnel, including Lieutenants Bryant and Mitchell, went to Daejon as the core of a new squadron. Although Bryant and Mitchell brought L-5s to serve as the first Forward Air Controllers, Latiolais's plan called for the use of the T-6 to serve as the new Forward Air Controller aircraft. The Texan was a World War II training aircraft used to transition pilots into combat fighters. Powered by a large radial engine, the modified T-6s could carry nearly the same weapons payload as a World War II fighter. Although not as fast as a World War II fighter, its higher speeds and all-metal construction made the T-6 more survivable on the battlefield than the L-5, but the aircraft was slow enough to provide the controller with a good view of the terrain. Most importantly, the aircraft had ample electrical power and space for the additional radios necessary for the Forward Air Controller mission. Since it was a training aircraft, the T-6 also had a second seat used by an observer who assisted in locating targets.[12]

Latiolais's idea was part of the lessons of World War II still alive within the Air Force's corporate knowledge, and a product of his, and others, first-hand experiences in the previous war. From January to October 1944 he had commanded the Ninth Air Force's 493rd Fighter Squadron. He flew P-47s on air-to-ground support missions in northern France—including attacks under the direction of the armored column cover's Rover Joe and Grasshoppers. Working with Latiolais on the new Forward Air Controller mission was Lieutenant Colonel John Murphy. Like Latiolais, Murphy commanded a P-47 squadron in Ninth Air Force. He went on to command a fighter group and flew 139 combat missions during World War II. Murphy and Latiolais flew F-80s together at the 49th Fighter Bomber Group at Misawa Air Base, Japan, before joining Fifth Air Force. When the Korean War broke out, Murphy was serving as the Fifth Air Force Director of Operations and was Latiolais's boss. The pair sold the Forward Air Controller idea to the Fifth Air Force Commander, General Earl E. Partridge. Partridge embraced the idea and sent Murphy to run the new Joint Operations Center at Daejon, charging him with the task of building the Tactical Air Control network for the Korean War.[13]

Partridge understood the level of effort required to provide air support to ground forces in combat. Before World War II he flew in the 3rd Attack Group experimenting with new ground-attack aircraft, weapons, and tactics. During the war, Partridge served as the Northwest African Air Force Operations Officer, taking over the position just before Operation Husky, and he oversaw the efforts to improve Allied air-to-ground cooperation and capabilities in the Mediterranean. Partridge's superior in 1950, the Far East Air Force Commander, was the critical cog in securing the necessary personnel and equipment for the fledgling Forward Air Controller experiment. Fortunately, the new Far East Air Force Vice Commander of Operations was General "Opie" Weyland, who implemented a host of air-to-ground innovations for General Patton's Third Army as commander of XIX Tactical Air Command.[14] Weyland's blessing of the project secured the support of the most senior air officers in theater, and the core of a new unit emerged with the "specific purpose of tactical reconnaissance and tactical control in close coordination with friendly ground units."[15]

Initially saddled with a lengthy moniker, Operations Section of the Joint Operations Center, the Forward Air Controllers reported to Lieutenant Colonel Murphy. The Operations Section supported the Eighth Army with three missions, which, even today, constitute the core of a Forward Air Controller's responsibilities. First, Forward Air Controllers conducted tactical reconnaissance of the front line. Since June 25, the situation on the Korean Peninsula was confused and chaotic, with the US and South Korean forces rapidly falling back. As a result, the Forward Air Controllers' primary job was assisting Eighth Army Headquarters in getting an accurate picture of the front. The second Forward Air Controller mission was controlling close air support airstrikes. As the front collapsed southward toward Busan, airstrikes provided ground forces sufficient relief from the North Korean attack to establish and hold a defensive perimeter. Finally, the Forward Air Controllers of the Operations Section controlled interdiction strikes near the front.[16]

Due to the dynamic battlefield, Air Force pilots attacking near the front struggled to discern the ground situation. Often, in the time between flight briefing at bases in Japan and target attack along the front in Korea, "the situation changed before the pilot arrived in the target area ... unfortunately, however, it was practically impossible [for pilots] to change once the prebriefed plan was carried into execution."[17] The situation was so bad that the Air Force was taking a beating in the press.[18] To alleviate the confusion, formations of fighters checked in with Forward Air Controllers, who remained in constant radio contact with the Joint Operations Center and maintained a critical awareness of the shifting front. The Forward Air Controller updated attacking pilots with changes to the situation and assigned new targets if needed. "Gratifying results were immediately forthcoming, with the initiation of the developmental stage (phase) of this new method of tactical air control, and the resulting increase in effective utilization of available fighter bombers."[19]

Lieutenants Bryant's and Mitchell's first missions on July 9 were an auspicious start, even though a North Korean Yak-3 fighter attacked Lieutenant Bryant, and none of the twenty F-80 pilots the pair controlled were aware of the new Forward Air Controller procedures. The next day, the first T-6 mission took place, piloted by Lieutenant Harrold Morris. In a FAC role, the T-6 used an ARC-3 radio for air-to-air and air-to-ground communication. With eight preselected radio channels available to the Forward Air Controllers, the ARC-3 doubled the number of radio channels used by the World War II era SCR-522s. Ironically, the radio on board Morris's T-6 failed. Still, he was able to use visual signals to direct a flight of Royal Australian Air Force F-51s—also a first for the war—in an attack against a column of enemy tanks. On July 11, Major Merrill Carlton took over as the unit's commander. Two other pilots also joined on the 11th, bringing the personnel total to seven pilots and three maintenance technicians. On July 13, in what would be the first of many moves in the seesaw Korean War, the Joint Operations Center and the Forward Air Controllers of the Operations Section were forced to evacuate Daejon for the K2 airstrip outside of the town of Daegu, just ahead of the North Korean advance. All of Operation Section's equipment and personnel fit into just one truck, minus those who flew out the handful of airplanes.[20]

Much like the Horsefly experiments in World War II, the Operations Section did not yet have an official designation as an Air Force squadron, nor did it have an official parent organization. When the unit moved to K2, the 51st Fighter Group provided for their logistical needs. Later in July, the newly arrived 6132nd Tactical Air Control Group took over the section's administrative and logistics needs. None of these arrangements were satisfactory, and the group of misfits scrounged for everything from sleeping quarters to aircraft parts. The one permanent thing the Operations Section did manage was a name. On July 15, the Joint Operations Center assigned Forward Air Controller missions the callsign "Mosquito," and the name stuck. The unit kept the Mosquito callsign until the Air Force deactivated them in 1956. The first mission of the day flew as Mosquito Able, the second Mosquito Baker, and so on in sequence to the last flight of the day, Mosquito Howe—seven flights in all.[21]

Along with a callsign, the section had plenty of missions to accomplish. Each morning Major Carlton drove from K2 to the Joint Operations Center in Daegu for a daily situation update. Major Carlton and the intelligence personnel briefed the Mosquito Able pilot who, upon returning from his mission, passed intelligence updates via phone back to the Joint Operations Center, and then gave a face-to-face update to both the Operation Section's intelligence personnel and the Mosquito Baker pilot. This debriefing and briefing procedure repeated at the start and end of each mission until the Mosquito Howe pilot landed after dark. Although the Mosquitoes provided critical Forward Air Controller capabilities for close air support and attack missions, "as Mosquito pilots brought in the most current and reliable information, [Joint Operations Center] intelligence increasingly depended on the Mosquito for the latest front-line dispositions."[22]

Between July 9 and July 31, the group grew to twenty-four Mosquito pilots and four observers who logged 670 hours of flight time on 269 missions.

Other than a running list of personnel and the number of missions flown, the group kept few official records. The only record of the Operation Section's accomplishments covered the eight days between July 23 and July 31. The unit history reported that the Mosquitoes directed the destruction of "27 tanks, 16 trucks, 127 other vehicles, including jeeps and animal drawn carts, three artillery pieces, three bridges, three fuel dumps, a marshalling yard and a large building." Further, fighters under the direction of the Mosquitoes claimed "38 tanks, 3 trucks, 107 other vehicles, and 14 artillery pieces" damaged.[23] Although the Mosquitoes lost three aircraft and faced constant threat of enemy air and ground attack, the unit lost no pilots or observers in the first month of operations.[24]

Given the amount of destruction visited upon North Korean forces by Mosquito pilots, it is obvious why the enemy advance stalled and why the group began grabbing the attention of senior American commanders and the American press. The Fifth Air Force Commander, General Earl Partridge, flew on a Mosquito mission as a pilot, and later Eighth Army Commander General Walton Walker flew along as an observer. General Jimmy Doolittle, and Associated Press reporter and Pulitzer Prize winner Hal Boyle, also accompanied the Mosquitoes on missions.[25] Despite the successes, and the attention, many within the Operations Section felt that the Mosquitoes were underused.[26]

To capitalize on the unit's growing success, and because of its continuously growing size, the Operations Section became the 6147th Tactical Control Squadron (Airborne) on August 1, 1950, with Major Carlton as its first commander. Originally slated to be the 6147th Tactical Liaison Squadron, Major Carlton convinced Fifth Air Force Headquarters to change Liaison to Control to better delineate the squadron's mission, and also add the phrase (Airborne) to differentiate it from the ground-based tactical control squadrons. The squadron maintained its intelligence, close air support, and control of interdiction missions but added the responsibility for developing Forward Air Controller operating procedures and tactics. The first change the Mosquitoes made to their operations was assigning each pilot to a specific sector of the front and associating them with the Tactical Air Control Party patrolling that area. This enabled the pilots to become intimately familiar with their sector, allowing them to recognize changes in the enemy disposition instantly, and, by developing an understanding of the friendly plans and maneuvers, build trust between the pilots and the ground forces in their area.[27] To delineate their affiliation, Mosquito pilots adopted the Tactical Air Control Party in their sector as part of their callsign. For example, the Mosquito pilot who flew in the 25th Infantry Division's area worked with the Tactical Air Control Party "Vaudeville" and flew with the callsign "Mosquito Vaudeville."[28]

During August the Mosquitoes grew exponentially in every way possible. By the end of the month, the 6147th had twenty-seven T-6 aircraft, fifty-five pilots, and forty-four observers on its roster. The planes and aircrew logged 1032 sorties and 2777 hours, achieving a high-water mark of forty-one missions on August 20, 1950, and again on August 21. Seventeen pilots reached the fifty-mission mark during the month, but, unfortunately, the Mosquitoes suffered their first members killed in action: Second Lieutenant Ernest Reeves and his observer Master Sergeant Herschel Bushman did not return from their mission on August 20, 1950. Initially listed as missing, the Mosquitoes discovered later they died when their aircraft went down near Uiryong. Two days later, pilot Second Lieutenant Robert McCormick and his observer Second Lieutenant Charles Wenzl died when their T-6 crashed while working with the Tactical Air Control Party Vaudeville near the town of Masan.[29] Despite the losses, the Mosquitoes continued to grow and accumulate a growing list of enemy equipment damaged and destroyed.[30]

As part of their regular Forward Air Controller missions, Mosquitoes gathered an immense amount of intelligence on enemy force dispositions. Occasionally, Mosquito pilots conducted missions further behind the front, specifically to collect information on North Korean movements. Intelligence personnel posted the most up-to-date information on a massive map displaying the entire front situated in the squadron's intelligence section. As Mosquito pilots returned from missions, they used the map as a briefing tool for Mosquitoes headed out for the next mission. More importantly, the Mosquitoes passed this information to the Fifth Air Force and Eighth Army headquarters. The squadron history boasted that "it can never be denied that the Mosquito was the primary source of Fifth Air Force intelligence."[31]

Mosquito reports to the Fifth Air Force and Eighth Army brought in the most current locations of enemy troops and supply depots, the best areas to interdict vehicle convoys, the latest enemy trends in camouflage techniques, and evaluations of "friendly air and ground action against the enemy in both tactics and type of ammunition used."[32]

Locating camouflaged vehicles and equipment was one of the Mosquitoes' specialties and was no easy feat. Finding hidden tanks, trucks, troops, and supply stashes required a high level of skill and familiarity with the terrain. Because the Mosquitoes worked the same areas day after day, they were able to spot small variations in the scene below. Lieutenant John Planinac recounted in the Mosquitos's history that "the 'camouflage' of branches and leaves were of a different hue from the green of the surrounding foliage."[33] Decades later, in the jungles of Southeast Asia and the hills of the Balkans, the uncanny sense of these minor changes in shades of green and brown, slight disturbances of dirt, and things which just seemed out of place made Forward Air Controllers masters of finding and targeting the enemy.

By the end of August 1950 the squadron was finally large enough to address essential administrative functions. They constructed better living quarters, and a chow hall served meals twenty-four hours a day to keep pace with the growing operational tempo. Major Carlton also established a rotation policy relieving the Mosquito pilots from an endless cycle of dangerous frontline combat. He set the rotation requirement at fifty combat missions; once a pilot or observer obtained that number, they could return to the US or rejoin their original combat squadron in Korea. This new policy enabled the first two Forward Air Controllers of the Mosquito program, Lieutenant Frank Mitchell and Lieutenant James Bryant, with fifty-five and fifty-six missions respectively since July 9, to rotate out of the 6147th.[34] Several Mosquitoes logged far more than the fifty-mission minimum, and Lieutenant Ken "Scotty" Wilson logged the most, flying 232 missions as a Mosquito observer. Scotty, a member of the Royal Scots Regiment of the British Army, was also one of a handful of UN soldiers who flew with the Mosquitoes, primarily as observers.[35]

With the squadron steadily adding new Forward Air Controllers, Lieutenant Planinac created a method for training new arrivals. The training program also served as a means of standardizing and codifying the Mosquito tactics, techniques, and procedures. As part of the training program, the squadron created the first Standard Operating Procedures (SOP) in August 1950. Over the next three months, the squadron produced eleven SOPs, covering the mission of the squadron, training procedures, preflight briefings, aircraft inspection procedures, pilot and observer responsibilities, and most importantly, Forward Air Controller procedures and tactics. The Forward Air Controller procedures spelled out how to coordinate with the command-and-control systems in Korea, how to prioritize targets, and how to direct different types of fire from aircraft and artillery. The most critical part of any Forward Air Controller's job is providing the attack pilot with a "talk on" to the desired target. This process begins with the Forward Air Controller passing the attacking aircraft the target's coordinates and description, such as tanks or artillery, and then the Forward Air Controller coordinates a rendezvous point and altitude using major features in the area to describe the scene. The SOP states that "if possible, the Mosquito will select an obvious feature in the vicinity as a reference point and will direct the fighter from there."[36] If needed, the Forward Air Controller might fire rockets in the target area to assist the attacking pilot in locating the target.[37]

The rockets most commonly used by Air Force Forward Air Controllers since the Horsefly experiments of World War II are 2.75" diameter folding fin rockets. About four feet long, and weighing about twenty pounds each, aircraft carry these rockets in a wing or fuselage mounted pod, most commonly holding seven rockets each. However, on both the L-5 in World War II and the T-6 in Korea, the rockets were mounted directly to the airplane's wings in racks of three or five. They have small stabilizing fins that fold up when loaded on the aircraft and pop out when fired. While the rockets can carry various explosives, in the Forward Air Controller mission they usually contain a small white phosphorus charge. Still used by modern Air Force Forward Air Controllers, they are commonly called "Willie Petes" because of the WP abbreviation for the rockets' white phosphorus charge. Upon impacting the ground, the Willie Pete explodes and produces a distinctive bright white smoke plume about fifty feet tall or more, depending on terrain and weather conditions. This smoke is easily seen from the air, even from many miles away, and serves as a quick method for Forward Air Controllers to point out or confirm a target for an attacking aircraft.

Lieutenant John Thompson authored a report on Forward Air Controller tactics as part of the growing wealth of knowledge in the 6147th's SOPs. Among the topics covered by his report were the necessity of rockets, smoke grenades, and flares for marking targets in different terrain and lighting conditions. The rockets were more accurate as markers than smoke grenades but had the downside of "creating the impression the Mosquito was an attack aircraft. It follows that the man on the ground being fired at with rockets reasons that he might just as well shoot back."[38] Smoke grenades worked well against entrenched enemy positions, which offered little above the ground for a Forward Air Controller to aim a rocket at but were easily seen from directly above. Like observers and Cavalry Reconnaissance pilots of WWI, Mosquitoes dropped the smoke from the T-6 cockpit, which had a sliding canopy the pilot or observer could open in flight. The smoke grenades also came in an assortment of colors, allowing Forward Air Controllers to differentiate friendly and enemy positions, or multiple enemy positions, with different colors.[39]

In addition to rockets, the Mosquitoes also used illumination flares for night close air support missions. Illumination flares revealed enemy movements, which increasingly happened at night as the war ground on, thanks to American air superiority and the constant attacks directed by the Mosquitoes. Night operations offered extreme danger in the poorly charted mountains of Korea but gave a prowling Mosquito the cover of darkness. Further, muzzle flashes were easily spotted from several miles away at night, allowing observers to pinpoint the position of a gun from a safe distance. The Mosquitoes marked targets with flare guns or directed UN artillery to fire illumination flares into a target area. This allowed night fighters or bombers to locate and attack enemy positions despite the darkness.[40] Lieutenant Thompson's report declared, "No Mosquito, regardless of day or night schedule, should leave for his area without a full supply of all three; rockets, grenades, and flares."[41]

Over the next year, the Mosquito enterprise continued to grow. The 6147th moved up and down the peninsula with the tide of the war. In December 1950 they pressed north to K-16 airfield just outside of Seoul, only to be pushed south again by March 1951 to the K-37 Airfield near Daegu.[42] On April 25, 1951, significant changes were made to the Mosquito's operation. The new 6147th Tactical Control Group formed at K-6 Airfield, about forty miles south of Seoul. Again, Carlton—now a Lieutenant Colonel—became the group's commander. Five new squadrons joined the group: the 6147th Air Base Squadron, 6147th Maintenance and Supply Squadron, and 6148th, 6149th, and 6150th Tactical Control Squadrons. The Air Base Squadron was responsible for operating the day-to-day living conditions and upkeep of base facilities at K-6. The Maintenance and Supply Squadron kept the new group's T-6s, other aircraft, and vehicles operating. The 6148th and 6149th became new combat Mosquitoes, joining the pilots of the 6147th, which now operated as the headquarters for the group. Finally, the 6150th joined the group, created for the specific purpose of training new Forward Air Controller pilots, observers, and Tactical Air Control Parties.[43]

Although the 6148th and 6149th maintained small training sections, these primarily functioned to transition new pilots from the 6150th into frontline combat missions. The 6150th used the significantly expanded 6147th operating procedures as the basis for its training development. The new system allowed the chronically personnel-deficient Mosquitoes to train new pilots quickly and consistently. Pilots and observers rotated into the Mosquitoes from stateside and from Far East Air Force combat fighter squadrons, flew fifty missions, and then returned to their squadrons. Even the popular comic "Terry and the Pirates" rotated the main characters, pilots Terry and Charles, through duty with the Mosquitoes. With the addition of the 6150th, the group kept a steady flow of trained replacement Mosquitoes moving into the new frontline squadrons.[44]

The 6150th also trained new Air Force Tactical Air Control Parties. Like the Rover Joes of World War II, the team consisted of a radio operator and a pilot who served as the attack controller. After training, the new controllers joined the 502nd Tactical Control Group, which operated the Tactical Air Control Centers and oversaw the Tactical Air Control Parties. Between World War II and Korea, the 502nd was the only group of its kind.[45] The controllers of the 502nd embedded into their assigned ground unit, usually a division, and assisted in the planning, coordination, and control of

Air Force support to the division. In combat on the front lines of Korea, the Mosquito Forward Air Controllers worked as an arm of the Tactical Air Control Party by providing a bird's eye view of the battlefield and controlling airstrikes in their assigned sector. The team worked together to provide the Army forces the best airborne firepower possible. In training, pairing the Forward Air Controllers with the Tactical Air Control Parties in the 6150th allowed them to build both the skills and the teamwork needed for successful operations in combat.

The initiation of a training program in a combat zone was one of the Mosquitos's critical innovations in creating a permanent Air Force Forward Air Controller program. It was a model the Air Service used in World War I to train all combat aviators, but ironically it was never used again except in creating Forward Air Controllers in Korea and Vietnam. The second innovation the Mosquitoes developed involved utilizing other aircraft as radio relays to expand the Tactical Air Control System. When the Mosquitoes began operations during July 1950, the rugged Korean terrain and distances between the front, the Joint Operations Center, and fighter bases in Japan hampered a low flying Forward Air Controller's ability to coordinate between all the parties involved. Since they still had obsolete L-5 aircraft on hand, the Mosquitoes put them to use. Placing them back from the front and away from the danger of enemy guns, the L-5s replicated the Air Coordinators in the Pacific theater during World War II. They served as a radio relay to headquarters and operations centers, augmenting the Forward Air Controller's critical communication web. The limits of the L-5 and its SCR-522 radios soon revealed this system worked only as a temporary fix. In early 1951 the 6147th instead began using a C-47 Skytrain cargo aircraft with a bank of radios and an experienced Forward Air Controller situated in the cargo area. The aircraft began flying under the callsign "Mosquito Mellow."

Linked to the Tactical Air Control Centers and the Mosquitoes via radio, Mosquito Mellow also tapped into the 6147th's operation center, dubbed "Mosquito Control." Mosquito Control harnessed the vast collection of intelligence gathered by Mosquitoes and telephone links to the Joint Operations Center, fighter bases in Korea and Japan, Tactical Air Control Centers,

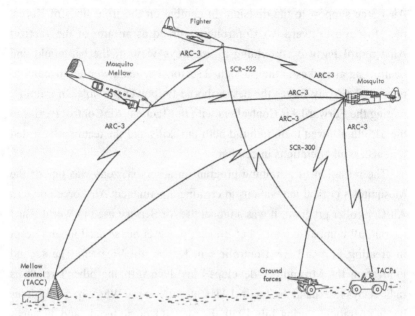

Figure 2 The Mosquitoes' control network. Drawing courtesy of the US Air Force.

and Fifth Air Force and Eighth Army headquarters to deliver airpower across the entire Korean theater. The Air Force used the Mosquitoes' innovation of airborne command-and-control—linked to ground controllers and headquarters—again in Vietnam. Further, the airborne command center became a model used by militaries all over the globe for modern battlefield command-and-control.[46]

The final, most important innovation the Mosquitoes made in the creation of Air Force Forward Air Controllers was the Standard Operating Procedures. From creation of the first SOP in August 1950 to the Mosquitoes disbandment in 1956, the 6147th made a concerted effort to standardize and codify the how and why of Forward Air Controller operations. The innovations and procedures use by Mosquito FACs, Mosquito Mellow, and Mosquito Control permeated future Air Force doctrine and tactical manuals from Vietnam to modern Air Force doctrine and tactics manuals. Lieutenant Colonel Carlton rightly deserves credit as the father of modern Air Force Forward Air Controllers.

He shepherded the 6147th from a small ragtag collection of pilots and aircraft to a paradigm-shifting theater-wide air power system critical to UN success in Korea. Just as the lessons of Grasshopper, Rover Joe, and Horsefly lived in the collective knowledge of the Air Force through people like Latiolais and Murphy, the lessons of the Mosquito survived into Vietnam and beyond, thanks to Carlton's herculean effort.

Not all the Mosquitoes' contributions were as serious—they also invented songs about their endeavor. Soldiering and singing are practically universal across history and culture and are uniquely tied together beyond the pomp of the drums and bugles of marching and maneuver. Soldiers sing songs to pass the time, lament their condition, assuage their fear, and get revenge on overbearing superiors. The songs usually take their cue from military songs, religious music, or popular music. Sometimes serious and sorrowful, they are just as often hilarious and bawdy. When men took to the sky in combat, their songs adapted as well, and a deep tradition of pilots and song remains in the Air Force today.

The Mosquitoes added to their Forward Air Controller legacy by creating their own songs. They even created a Mosquito songbook, adding it as an appendix to their official unit history in November 1950. The songbook expanded over the course of the war, but one song included in the official Mosquito history best captures the Mosquito Forward Air Controller mission. Sung to the Bing Crosby tune "Galway Bay," it goes:

If you ever come across the sea from Japan,
Then maybe at the closing of your day,
You will watch the old Mosquito leading fighters,
While flack and ground fire come from every way.

Oh they don't have any guns or bombs to hit with
But they are feared by North Koreans all,
For they know that when they see the old Mosquito,
Soon hell breaks loose and death begins to fall.

Oh they say the T-6 wasn't made for combat,
And I am here to tell you that is so,
But I'll fly another mission in the morning
Although I'm not convinced I ought to go.

All ye pilots gather round to hear my story,
So not talk of jets and Baker 29's
For this evening we'll observe a minute's silence,
In memory of Mosquitoes on the line.[47]

In a trifecta of finales, Lieutenant Chester Brown of the 6149th flew his fiftieth mission, and the last Mosquito mission, on July 27, 1953—the last day of the Korean War. Of the milestone, he reported, "I am very proud to have the honor to be the last one to fly ... Now I can go home to my wife and children." In his time with the Mosquitoes, Brown controlled sixty close air support missions, including a strike by four Navy fighters on an enemy cave complex. When the weapons exploded in the cave, he recounted, "the mountain literally blew its top."[48]

Close air support, thanks to the success of the Mosquitoes, represented just over 20 percent of the total combat missions in Korea and accounted for more sorties than the more famous air superiority missions.[49] The Mosquitoes controlled some 90 percent of all close air support and inter-diction airstrikes along the front. The 6147th Tactical Air Control Group flew more than forty thousand sorties in their three years of combat.[50] The rapid response and improvisation in support of the Army demonstrated that the Air Force could operate as an independent entity while still providing critical close air support to the Army. The Air Force emphasized the forward air control mission as well as the integration of Army and Air Force operations. However, the Air Force saw the Mosquitoes as a temporary endeavor. Despite the success, the Army and Air Force continued to feud over command, control, and allocation of close air support missions.[51] The Mosquitoes's equipment was obsolete by the end of the war. Sadly, there was no dedicated training plan for Tactical Air Command to integrate or expand the Forward Air Controller mission, despite the command's rebirth in 1950.[52]

The Forward Air Controller mission died again in Korea when the 6147th Tactical Air Control Group closed its door permanently in 1956. Instead, as Tactical Air Command and the Air Force moved toward nuclear bombers, supersonic interceptors, and away from the conventional ground-attack and close air support missions, the responsibility of close air support integration returned to Fort Bragg, North Carolina. This is where it lived before the Korean War when in September 1950 the Air Force established the Air Ground Operations School. The school was chartered to train Air Liaison Officers and Tactical Air Control Parties, and in the decade after the Korean War it formed the backbone and model of Air Force air-ground integration. Air Liaison Officers trained at the Air Ground Operations School before joining Army combat divisions to serve as the division's airpower conduit.

These Air Liaison Officers, along with the radio operators of the Tactical Air Control Parties, belonged to the Tactical Air Control Groups, such as the 502nd in Korea or the 507th at Fort Bragg. However, the pilots who rounded out the Tactical Air Control Party and served as a division's primary attack controller remained assigned to Air Force fighter squadrons. Therefore, whenever a division required air support, either in exercises or combat, pilots who previously attended the Air Ground Operations School were drawn from a fighter unit and sent to integrate into the division's Tactical Air Control Parties. With sufficient notice, pilots were sent to the Air Ground Operations School en route to joining the division. Not surprisingly, this created friction between the two services and between the pilots and Tactical Air Control Parties they augmented. The weeklong course was not enough for pilots with no previous knowledge of Army operations or air-ground integration to develop the necessary skills to successfully coordinate the complex missions. As the Ninth Air Force reported in 1954, the course "did not prepare them for their duties," leaving the Tactical Air Control Party teams unprepared to successfully integrate air and land forces.[53]

For a brief period in 1953, Ninth Air Force experimented with an entire squadron consisting of Tactical Air Control Party trained pilots to augment the Tactical Air Control Groups. The experiment was short-lived, and the Ninth Air Force shuttered it by year's end due to personnel shortages in operational fighter squadrons and a lack of volunteers. Ironically, the Ninth Air Force Commander at the time, Lieutenant General Edward Timberlake, was the Vice Commander of Fifth Air Force during the Korean War. He saw firsthand how effective the Forward Air Controller and Tactical Air Control Party team performed during the war. Still, Timberlake made no effort to revitalize the Forward Air Controller enterprise.[54]

The Air Force seemed conflicted about its air-to-ground missions and the need to integrate with the Army. Unit histories of the Ninth Air Force in the 1950s are replete with complaints about lack of adequate air-to-ground training ranges and opportunities for realistic training. Paradoxically, the command also complained about the outcome of exercises with the Army when given the chance to train with actual ground forces on large ranges. For example, in 1954 Ninth Air Force units participated in several maneuver exercises with the Army in various scales, but few proved to be a positive experience for the airmen involved. The command complained smaller-scale exercises "provided little of training value for either the Air Force or the Army" and the small exercises "distort the contribution of tactical air power."[55] In contrast, units participating in larger exercises were complex, with too many participants for adequate individual training opportunities. During several large exercises that year, the command reported units were "unable to accomplish the planned gunnery training because the ranges were not made available."[56]

Despite its repeated criticism of Air Force air-to-ground support, both before and after the Korean War, the Army also showed little interest in maintaining close air-ground integration. During 1954 the Ninth Air Force's Joint Air Ground Instruction Team toured Army units, briefing them on the most current air-to-ground procedures and tactics. The Ninth Air Force reported during briefings that "nothing was accomplished" but "three demonstrations for the 11th Airborne Division at Fort Campbell ... were more successful."[57] The improvement, however, came only after a significant reduction of the presentation and an overhaul that used catchy florescent symbols highlighted on maps by blacklights. Instead, the Army too turned inward, seeking to restructure and modernize its forces and doctrine in the post-Korean War decade.[58]

For the rest of the decade, interest and support for interservice cooperation and training waned and became increasingly hostile as the two services squabbled over doctrinal and core mission issues, such as airlift and close air support. During a large joint force exercise named LOGEX 58, for example, the Army used helicopters as aerial transport. As a result, the Air Force withdrew support from the exercise, citing Tactical Air Command guidance, which directed "units not to participate in any planning or execution of a field or command post exercise that violates accepted documented doctrine."[59] Further pushing the Air Force and Army apart was the expansion of nuclear doctrine and the rise of Strategic Air Command. The race for nuclear capabilities drove both Tactical Air Command and the Army to increasingly embrace the nuclear mission. While the Army adopted General Maxwell D. Taylor's ideal for a nuclear battlefield Army in the Pentomic Divisions, the Air Force directed Tactical Air Command to increase nuclear readiness, "including development of plans to provide mobile atomic strike forces for use in tactical air operations in any area of the world."[60]

In March 1962 a joint Army-Air Force exercise called Bristle Cone took place at Fort Irwin, California—just five months after the US Air Force began training South Vietnamese pilots in Operation Farm Gate. During what should have been an excellent training and integration opportunity, Bristle Cone's broad objectives tested both services' airlift, close air support, and command-and-control planning and execution. Among the exercise goals was a test of the Army's new Army Air Ground System (AAGS). This was the Army's mirror of the Air Force's Tactical Air Control System (TACS). In theory the two systems were to work together to optimize air power. The Army used AAGS to coordinate and prioritize air support requests generated by maneuver units in the field and funneled through division level Fire Support Coordination Centers. TACS, in response, used the Air Support Operations Center (ASOC) to match those requests with available air power. The Air Force replaced the Joint Operations Center from Korea with the Air Support Operations Center to convey the entity's purpose more accurately. These systems were designed to work in concert during the planning and execution of operations across an entire theater.[61]

However, Bristle Cone revealed significant shortfalls in TACS-AAGS cooperation. Sadly, the issues replicated, nearly word for word, the struggles the Army and Air Force faced in the Tarheel-Swarmer-Portrex series of exercises in 1949-1950, and during the opening weeks of the Korean War. Tactical Air Control Parties could not coordinate with the Air Support Operations Center or with the controllers trying to distribute airpower to them. Coordination between the Air Support Operations Center and the

Air Liaison Officers was lacking and difficult. Tactical Air Control Parties and the ground units struggled to identify targets for Air Force fighters to attack, and the fighters could not see the targets the Tactical Air Control Parties did find. Fighters roamed the battlefield searching for their assigned Tactical Air Control Parties. Headquarters for both the Army and the Air Force lacked timely and accurate tactical intelligence. The two services struggled to allocate their aviation resources to the correct missions, a problem complicated by the Army's use of both fixed-wing aircraft and helicopters. Finally, Air Force personnel lacked adequate training, and most did not attend the Air Force Air Ground Operations School. Both services walked away from Bristle Cone unhappy with the results. If either service bothered to assimilate the lessons of the Mosquitoes into their air-to-ground doctrine, these frustrations may have been avoided.[62]

Although they were not the first, the Mosquitoes of the Korean War were genuinely the birth of the modern Airborne Forward Air Controllers. The injection of the Mosquito Forward Air Controllers onto the battlefield and their dedicated effort to codify training, tactics, and procedures solved many of the issues of Army-Air Force cooperation. Unfortunately, few of the lessons learned filtered into Air Force doctrine; just six years after the group's deactivation revealed that neither the Army nor the Air Force kept the lessons of the Korean War alive. Doctrine, organizations, and personnel dedicated to the Forward Air Controller missions went neglected. "In fact, the stature of the airborne Forward Air Controller in Air Force Doctrine and attitudes never changed from 1946 to 1966."[63] Neither the Army nor the Air Force put forth the effort to maintain the air-to-ground coordination machinery. In the years after the Korean War, both the Army and Tactical Air Command suffered at the feet of Strategic Air Command's nuclear force. Cooperation between the services waned, driven by budget and personnel shortages, and exacerbated by doctrinal differences.[64] At the World-Wide Fighter Symposium held at Maxwell Air Force Base in July 1954, former Ninth Air Force commander Major General Homer Sanders opined about the Air Force's neglect of tactical aviation in favor of nuclear bombers. General Sanders experienced firsthand the Air Corps' neglect of tactical aviation before World War II and Korea and rightly surmised the Air Force was repeating the same mistakes in the wake of the hard-learned lessons of Korea.[65] Vietnam proved him correct.

Chapter 4

War in Indochina:
The FAC Comes of Age

ar in Vietnam did not erupt in an unexpected attack as it had in
World War II and Korea. Instead, US presidents from Harry
Truman to Lyndon Johnson gradually fed military and financial aid into
Southeast Asia, first to assist the French, and then to prop up the Republic of
Vietnam. In 1949, in response to the Chinese Communist victory, National
Security Council 48/2 recommended allocation of $75 million to combat
the spread of communism in Asia. As a result, the Truman administration
allocated an initial injection of $15 million and 128 members of the Military
Assistance Advisory Group, or MAAG, to aid the French struggle against
the Viet Minh in 1950.

By 1960 assistance to South Vietnam grew to more than $190 million
administered by 685 officers and enlisted personnel. By the time Congress
approved the Gulf of Tonkin Resolution in August 1964, the US was already
deeply committed to the war in Vietnam. The day the Gulf of Tonkin
Resolution passed, August 5, 1964, the MAAG became Military Assistance
Command Vietnam (MACV) and steadily increased its Air Force personnel
and expanded their training and advisory mission to the South Vietnamese
Air Force. But three years prior to the Gulf of Tonkin Resolution, US Air

Force advisors were already aiding the South Vietnamese Air Force, under the code named "Farm Gate."[1]

Farm Gate was technically Detachment 2 of the 4400th Combat Crew Training Squadron, which began in April 1961 at Hurlburt Field, Florida. The new squadron had its origins in President Kennedy's Flexible Response doctrine. Flexible Response, which replaced Dwight Eisenhower's nuclear-based New Look doctrine, emphasized special forces, clandestine operations, and support to groups and governments fighting communists around the world. The squadron was charged with providing training to foreign air forces, primarily in airlift and close air support missions. The 4400th's Detachment 1 operated in Mali, while its Detachment 3 operated in Panama. By 1962 the 4400th expanded, becoming the 4400th Combat Crew Training Group, and by 1963 transitioned to the 1st Air Commando Wing, adopting the name of Wingate's Raiders from World War II. By year's end, the airmen of Jungle Jim operated in fourteen countries in Europe, Central and South America, Africa, the Middle East, Southeast Asia, and the Pacific.[2]

The pilots of Farm Gate had three missions. First, they were to advise and assist the Republic of Vietnam's Air Force in building a combat-capable force. The training primarily focused on providing air support to the Army of the Republic of Vietnam. The second Farm Gate mission mirrored the first and focused on educating the South Vietnamese in the use of air power to support ground operations. Finally, the Farm Gate team assisted the South Vietnamese in building a command-and-control system that facilitated timely air support to their ground forces. To facilitate the third mission, Americans assisted the South Vietnamese in the creation of a Theater Air Control System modeled on contemporary Air Force doctrine—from Tactical Air Control Centers down to Air Liaison Officers and Tactical Air Control Parties. Ironically, Farm Gate included the training and use of South Vietnamese Forward Air Controllers, despite the US Air Force's neglect of the mission, as proven by the Bristle Cone exercise.[3]

The decision to use Forward Air Controllers in Vietnam stemmed from several factors. Under the rules of engagement, airstrikes in South Vietnam required the direction of a Tactical Air Control Party or Forward Air Controller. However, the terrain and environment of South Vietnam conspired against the sole use of ground-based attack controllers. The terrain of South Vietnam varied from the jungled mountains of the Central Highlands to the flooded rice paddies of the coastal plains and the Mekong Delta. Though only about forty miles wide from the western border to the eastern shore at its narrowest point, the country stretched more than seven hundred miles from its border with North Vietnam to its southern tip. Both US and South Vietnamese headquarters in Saigon, in the flat southern plains, struggled to maintain reliable communications with Vietnamese units and their American advisors operating in the field. Use of Forward Air Controllers and airborne control aircraft modeled on Mosquito Mellow enabled Tactical Air Control Centers and headquarters staffs to retain critical communication links with far-flung units. Further, the triple canopy jungles and small farm plots broken by densely wooded areas limited the sightlines of any soldier engaged by an enemy. The jungle also prevented pilots, particularly jet pilots, from readily spotting targets from the air. As a result, a Forward Air Controller's ability to orbit the battlefield and discern the variations in the terrain below, spot muzzle flashes through the trees, and mark targets with smoke rockets, proved to be a necessity for conducting close air support airstrikes.[4]

Farm Gate operations were highly classified when they began, and the members of the unit stripped the rank and service information from their uniforms before arriving in Vietnam. The first five American Forward Air Controllers and radio operators arrived at Bien Hoa Air Base, just north of Saigon, in November 1961. Sent on a ninety-day temporary assignment, the pilots had no aircraft to fly initially and instead served as both Air Liaison Officers and Tactical Air Control Parties for the South Vietnamese Army. In January 1962, a Farm Gate Forward Air Controller became the first American to direct an airstrike in Vietnam. The mission featured an American pilot flying an L-19 aircraft, acting as the Forward Air Controller with a South Vietnamese observer. The combined use of American pilots and South Vietnamese observers was common during Farm Gate operations because the rules of engagement required a Vietnamese observer or Forward Air Controller to be on the scene when American attack aircraft supported South Vietnamese forces.[5]

Farm Gate attack pilots flew a motley mix of aging World War II and Korean-era aircraft, including the T-28 Trojan, A-1 Skyraider, A-26 Invader, and C-47 Skytrain. In the role of light fighter and attack aircraft, Farm Gate used the T-28 and the A-1. The A-1 Skyraider entered service with the Navy at the end of World War II as a dive bomber. It remained in service with the Navy and Marine Corps through the end of the Korean War, providing close air support for the Marines. In 1962 the A-1 joined the Farm Gate team as a close air support, Forward Air Controller, and rescue aircraft—roles it performed for both the US Air Force and South Vietnamese Air Force through the end of the war. A massive and rugged airplane, the Skyraider carried up to eight-thousand pounds of ordinance. The airplane's capability and durability, coupled with the A-1 pilot's dedication to the mission, earned it a devoted following among Vietnam veterans. The T-28 replaced the T-6 as the Air Force's primary training aircraft after World War II. Like the T-6, a large radial engine powered the T-28. It also had a two-seat cockpit and carried a similar load to a World War II fighter, but at slightly slower speeds.[6]

The Douglas A-26 Invader was a World War II attack bomber used extensively in interdiction missions by the Fifth Air Force in the Southwest Pacific and Ninth Air Force in Europe. The A-26 also served during Korea as a night-attack aircraft, interdicting communist troops, supply convoys, and trains. The A-26 reprised that role in Vietnam, seeing action beyond the Farm Gate program. It supported US special forces throughout Indochina and conducted night interdiction along the Ho Chi Minh Trail. Just as in Korea, aging C-47s served as airborne command-and-control aircraft for Farm Gate operations. Truly a fly-by-night operation, the first Farm Gate pilots and maintenance technicians worked together in primitive conditions in the Vietnamese heat and humidity to refuel aircraft, load bombs, and repair battle damage taken during missions.[7] The first losses of the program came "in February 1962, a Farm Gate C-47 on a leaflet drop mission in the highlands near Bao Loc was shot down, killing the six airmen, two soldiers, and one Vietnamese crewman on board."[8]

South Vietnamese Air Force and Farm Gate Forward Air Controllers usually flew the O-1 Birddog. Initially developed for all services of the

US military in 1949 as the L-19, the O-1 derived from the civilian four-seat Cessna 170. An all-metal two-seat aircraft resembling the L-5, the Birddog used short dirt fields or roads for runways like its Grasshopper ancestors of World War II. Although the O-1 carried more radios than the Grasshoppers, along with a two-man crew and two rocket pods, it was still slow and extremely vulnerable to damage from ground fire. The Birddog saw combat in Korea as a liaison and Forward Air Controller aircraft before being called into action again in Vietnam. Used in every theater of the war in Southeast Asia, the O-1 became the ubiquitous Forward Air Controller aircraft for the Air Force, Army, Marines, the CIA, and the South Vietnamese.[9]

Famously, in April 1975 as Saigon fell, South Vietnamese Air Force Major Bung-Ly loaded his family of seven into an O-1 and fled Vietnam. Heading out to sea without any way to navigate to his desired destination or communicate with nearby ships, the American-trained Forward Air Controller managed to find the aircraft carrier USS *Midway* supporting the evacuation of the city. Circling low over the deck, he dropped a note stating he only had one hour of fuel remaining. The ship's captain, Lawrence Chambers, cleared the deck—pushing several helicopters overboard—and allowed the aircraft to land. Once Bung-Ly made a safe landing on the *Midway*, he escorted his family clear of the airplane as Navy crewmen pushed it overboard too. After the harrowing adventure, the former Birddog pilot and his family resettled in the US, along with nearly two hundred thousand other Vietnamese refugees.[10]

Despite being highly classified and initially considered a temporary endeavor, the missions and personnel of Farm Gate steadily expanded between 1962 and 1964 to support the increase in South Vietnamese combat operations. To accommodate the workload, the 19th Tactical Air Support Squadron (TASS) activated at Bien Hoa in July 1963. Drawing its lineage from the 19th Liaison Squadron, which conducted the Grasshopper experiments during the Army maneuvers of 1942 and 1943, the 19th TASS assumed the publicly acknowledged Forward Air Controller training missions for the South Vietnamese Air Force. Like the Mosquitoes in Korea, the 19th was also responsible for training US Air Force Forward Air Controllers in Vietnam. Formally named the Theater Indoctrination School, the pilots who flew there gave it the irreverent nickname FAC University or "FAC-U" for short.[11]

The Forward Air Controller mission in Vietnam was extremely perilous, and Farm Gate and 19 Tactical Air Support Squadron FACs accounted for 13 percent of Air Force combat deaths before 1965. Only Farm Gate T-28 and A-1 pilots suffered higher loss rates.[12] On March 23, 1964, an O-1 from the 19th Tactical Air Control Squadron, piloted by Air Force Captain Richard Whitesides flying with US Army Special Forces Captain Floyd Thompson as the observer, was shot down about fifteen miles west of Khe Sanh. The crash killed Whitesides, and Viet Cong guerrillas took Thompson prisoner. His captors held him in various camps in the central highlands before moving him to Hanoi in 1970. Thompson returned to the US on March 16, 1973, during Operation Homecoming, after spending 3,311 days in captivity and earning the dubious distinction as the longest-held American POW of the Vietnam War.[13]

The creation of the 20th TASS at Da Nang Air Base, 21st TASS at Pleiku, and 22nd TASS at Binh Thuy Air Base followed the activation of the 19th in May 1965. The four squadrons provided air support for American combat forces, now officially engaged on the ground in Vietnam. Each squadron assumed responsibility for a different Corps area in South Vietnam. The 20th supported operations in I Corps—the base at Khe Sanh, the Laotian border, the ancient city of Hue, and the major airbase at Da Nang all lay in the narrow I Corps area, which stretched southward from the Demilitarized Zone (DMZ) about 150 miles to the Quang Ngai province. The 21st operated in II Corps, the largest sector of Vietnam, which held the strategically important Ia Drang Valley. Flowing from the Central Highlands along the Vietnamese-Cambodian border towards Saigon, the Seventh Cavalry famously launched one of the first major American operations into the Ia Drang in November 1965. For the remainder of the war, US special forces camps in the valley came under regular attack, keeping Forward Air Controllers of the 21st Tactical Air Control Squadron busy.[14]

In addition to operating FAC-U, the 19th retained the responsibility for the III Corps area, located around Saigon. In the III Corps, about twenty miles north of Saigon towards the Cambodian border, was the "Iron Triangle," a Viet Cong stronghold near the terminus of the Ho Chi Minh Trail. In 1967, American and South Vietnamese forces attempted to clear the area during Operation Cedar Falls. Further north in III Corps lay An Loc, site of a two-month-long struggle during the Easter Offensive of 1972. Finally, the 22nd flew in IV Corps, which straddled the Mekong River in the south-ernmost area of South Vietnam. Due to the Mekong River and proximity to Saigon, it was a hotbed for South Vietnamese and US counterinsurgency operations. When the North Vietnamese launched the Tet Offensive in 1968, it fell hard on IV Corps.[15]

Known as "In-Country" and "Free World" Forward Air Controllers, these four Tactical Air Support Squadrons formed the backbone of the commu-nity in Vietnam.[16] Forward Air Controllers earned the names "In-Country" and "Free World" as a function of the divided nature of the air war in Vietnam. Air operations in the Vietnam War covered three theaters and varied widely in scope and aircraft involved, depending on the types of missions within that theater. In the air war over North Vietnam, US Air Force and Navy fighters and bombers tangled with North Vietnamese MiGs, surface-to-air missiles, and antiaircraft guns while conducting strategic bombardment and air superiority campaigns. These operations stretched from 1964's Operations Pierce Arrow and Flaming Dart through the longest air campaign, Rolling Thunder, which lasted from 1965 to 1968. After a four-year bombing halt, President Richard Nixon authorized 1972's Operation Linebacker I in response to the Easter Offensive and, in December, Linebacker II, the American response to the failure of the Paris Peace Accords.[17]

Free World and In-Country Forward Air Controllers provided close air support for US, South Vietnamese, Korean, Australian, and Thai Army operations within the geographic boundaries of South Vietnam.[18] In addition to Forward Air Controller pilots and observers, the four squadrons also provided Air Liaison Officers and Tactical Air Control Parties to their assigned Corps ground units. Responsibility for staffing and equipping the squadrons fell on the 504th Tactical Air Support Group, located at Bien Hoa Air Base. It was a massive undertaking. By August 1968, the 504th had nearly three thousand officers and enlisted airmen on its roles and grew to more than four thousand by year's end. Eighty new Forward Air Controllers per month rotated into Vietnam.[19] Spread across seventy locations, they worked with two Field Force headquarters, four Corps, twenty Divisions, thirty-four Brigades, 119 Battalions, and sixty-three Special Forces Camps spread across forty-three Provinces.[20]

The final theater of the air war was known as the "Out-Country" war. Highly classified, these operations took place in Laos, Cambodia, and the southern part of North Vietnam. Four different Forward Air Controller missions made up Out-Country operations. The most clandestine operation were the Ravens. Ravens FACs operated as part of a CIA program that supported the Hmong and Montagnard guerrillas fighting communist forces in Laos. The 20th and 23rd Tactical Air Control Squadron also conducted the second secret operation in Laos. Operating from Nakhon Phanom Air Base, Thailand, the 23rd directed interdiction missions along the Ho Chi Minh Trail in northern and central Laos. Complementing the 23rd's operations in northern Laos, Forward Air Controllers from the 20th Tactical Air Control Squadron flew from their bases in I Corps to control attacks in southern Laos.[21]

The third element of Out-Country Forward Air Controllers was Operation Commando Sabre, who came to be known as the Misty FACs. Also highly classified, the program began in June 1968 to assess the feasibility of the F-100 Super Sabre as a FAC aircraft. Eventually expanded to the F-4 Phantom, these Fast FACs directed interdiction strikes in the most dangerous areas of southern North Vietnam and Laos. Finally—before President Richard Nixon publicly acknowledged the US invasion of Cambodia on June 3, 1970—American pilots from most squadrons directed airstrikes as part of clandestine operations there as well. Initially, these missions conducted interdiction along southern sections of the Ho Chi Minh Trail. As American and South Vietnamese operations expanded into Cambodia, Forward Air Controllers first supported special forces teams and then controlled close air support missions for the main invasion in May 1970.[22]

Whether In-Country or Out-Country, close air support requests in Southeast Asia funneled through a Theater Air Control System built on the Mosquito model from Korea and the improved Air Force-Army TACS-AAGS system. When an army unit planning a patrol or sweep operation wanted air power, they requested what was known as planned support.

Their request included the location of the operation, time on target, expected targets, likely enemy forces, and types of weapons requested. With the help of their Tactical Air Control Parties, brigade and battalion Fire Support Coordination Centers passed planned close air support requests up through the division and then corps Tactical Air Support Elements to MACV. Each increasing level of command sorted and prioritized the requests. Seventh Air Force planners chose the best aircraft available to support the Army's request and issued the support orders to the fighter and bomber squadrons scattered around Southeast Asia. On the day of the operation, the attack aircraft took off from their base, refueled if necessary, and checked in with the Direct Air Support Center (DASC) responsible for that area. Like the Pineapple Tactical Air Control Centers of World War II, the DASC controlled aircraft specifically engaged in close air support and interdiction missions in Vietnam. Meanwhile, the Air Force personnel attached to the army unit conducting the operation made themselves intimately familiar with the plan. Although their headquarters were at major air bases in South Vietnam, the Forward Air Controllers and Tactical Air Control Parties usually operated attached to the army unit in their sector, living on and flying from the same base as the division or battalion they supported.[23]

When troops without air support came under fire, they used an immediate air support request network. A Forward Air Controller on station or the Tactical Air Control Party on patrol could radio the DASC, who diverted fighters to provide support. If neither was available, the ground forces called back to their headquarters, who passed the request up the chain of command. The second method was slower, as the request transmitted through many headquarters and then dispatched to the DASC before both a Forward Air Controller and fighters were diverted to help. The Air Force's goal was a thirty-minute response time, from the receipt of an immediate air support request to the delivery of bombs against enemy forces. In most cases, immediate calls for help got a response in twenty minutes for aircraft diverted from another mission and forty minutes for aircraft scrambled from their home base.[24]

Once the Forward Air Controller was overhead, usually thirty minutes before the fighters' arrival for planned missions, the FAC coordinated with the ground forces and the DASC in preparation for the arrival of the attack aircraft. Controllers updated the fighters of any changes to the plan and then passed them off to the Forward Air Controller. When the attackers checked in, the FAC provided them with a briefing and then prepared to pass targets. Just as in World War II and Korea, the Forward Air Controller passed the fighter coordinates, target description, and other information relevant for the attack. By 1968, the Air Force created a standard form used by all Forward Air Controllers and fighters for these radio exchanges that contained thirteen lines of target information. Over time, that form evolved into nine lines of information. Aptly named "9-Line," the format is now standardized across all US and NATO (North Atlantic Treaty Organization) forces.[25] With the target information passed, the Forward Air Controller usually marked the target with smoke rockets and prepared for the fighters' attacks. Building on procedures developed in the Mosquito SOPs, Forward Air Controllers watched the fighters as they prepared to attack, and positioned their aircraft to see the target and the attacking fighters. Using this method, they could determine if the fighters were attacking the correct target. If not, the Forward Air Controller could call off the attack.

When the fighter aircraft began their attack, they called the Forward Air Controller and said "One's in—target in sight, FAC in sight"—meaning the number one aircraft of the formation was attacking and had both the target and the Forward Air Controller's aircraft in sight. At this point, the FAC replied "Cleared hot"—meaning the fighter could drop their weapons. The most iconic radio call a Forward Air Controller typically made when clearing an aircraft for an attack was "Hit my smoke!" This meant the Forward Air Controller had hit the target with a Willie Pete rocket and wanted the fighters to drop their bombs on that position. This radio call communicated, in just three words, a myriad of information—it was both descriptive and directive. In the heat of battle, it might be the only information a Forward Air Controller communicated to the attacking aircraft. Still, the quick radio call told the close air support aircraft everything they needed to know. Further, the radio call communicated to both the attacking aircraft and the ground parties that the Forward Air Controller was on their game and had everything confidently in hand even when situations were desperate, confused, and chaotic.

If the Forward Air Controller wanted to adjust the fighters' aim, they might say "from my smoke, hit the tree line just to the north," or "from your last bomb, move across the road to the east." The Forward Air Controller might restrict the fighters' bombs by saying something like "no bombs west of the river, friendlies are west of the river." Finally, if the Forward Air Controller wanted the fighter to stop the attack, they called "Abort." After the attacks were complete, the Forward Air Controller directed the fighters out of the area, checked with the ground parties, and visually checked the results of the attack. If the fighters were out of bombs or fuel, or their support was no longer needed, the Forward Air Controller radioed a battle damage assessment before they departed. The battle damage assessment included the numbers of weapons dropped, the types of targets attacked, and the results of the attacks.[26]

To orchestrate the strikes, Forward Air Controllers used three different types of radios. The first, called "Fox Mike," the Forward Air Controllers' FM radio, was used to contact ground parties, which used FM radios for communication within their unit. Despite their smaller size and lighter weight, Air Force and Navy airplanes did not use FM radios, due to their relatively short range, so most aircraft could not talk directly to the ground forces. As a result, a Forward Air Controller served as the radio bridge during a close air support mission between the ground and air forces. The second radio type most military aircraft used was Ultra High Frequency, or UHF, to communicate with other aircraft. Colloquially called "Uniform," Forward Air Controllers used their UHF radio to contact and control the fighters and bombers assigned to conduct close air support and interdiction attacks. Finally, the Direct Air Support Center, air traffic control, and other tactical control agencies usually used VHF radios, or Very High Frequency, to communicate. Known as "Victor," VHF radios had advantages and disadvantages in size, weight, power requirements, range, security, and reliability compared to UHF. Each radio stored frequencies in dozens of preset channels, and pilots loaded these channels with the frequencies they used most. Forward Air Controllers could also manually tune hundreds of separate frequencies on each radio.[27]

Maintaining awareness of who was on what frequency, where everyone was on the battlefield, and keeping records of what occurred during a mission in order to provide results to fighters and debrief with intelligence after the mission, was practically a full-time job. Forward Air Controllers had to do all of this while flying an aircraft, usually at low altitude, while dodging terrain, bad weather, and enemy fire—all in abundance in Vietnam. Observers helped with the bookkeeping, but Forward Air Controllers did not always fly with observers on board. Therefore, pre-mission planning was the first critical step in keeping tabs on everyone during a mission and providing an accurate account after the mission was complete. If Forward Air Controllers knew about the mission ahead of time, they worked with their ground counterparts to create frequency cards, maps, and photos of the area. To speed the support process once airborne, they prefilled cards with target information and data about the other aircraft assigned to the mission. Before each mission, just like the Mosquitoes in Korea, Forward Air Controllers received a brief from their intelligence section and updates about the mission from the unit they were supporting.[28]

Upon arriving at their mission location, Forward Air Controllers began the cataloging process. Most carried a small clipboard attached to their leg while flying, giving them a place to hold checklists, maps, and mission cards on a solid writing surface. However, for the most critical information, and during the heat of battle, every Forward Air Controller worth their salt scrawled notes on the windows of the cockpit in grease pencil. Used since World War II, the grease pencil on glass technique provided an easily accessible, stable writing surface for the cramped cockpits of fighter and liaison aircraft. Each pilot or observer developed a technique for organizing and recording information on the cockpit glass. They also had their own set of shorthand symbols and odd hieroglyphics to speed the recording process and preserve precious space on the window. After the mission was complete, and the aircraft parked back at base, they sat in the cockpit and transferred the grease pencil markings to paper for post-mission debrief. Although frowned upon by Air Force maintenance personnel and not an officially sanctioned technique, the grease pencil on glass method of record-keeping persists, handed down by one generation of Forward Air Controllers to the next.[29]

As the war in Vietnam dragged on, Air Force personnel policies and rules of engagement created a staffing crunch within the flying squadrons and restricted the Air Force's supply of combat aircrew to the Forward Air Controller programs. Agreements between the Army and the Air Force along with the rules of engagement restricted the types of pilots able to control close air support attacks during specific In-Country and Out-Country missions. Only FACs with previous fighter experience were able to control In-Country operations with the US Army and other Free World forces. This same restriction applied to Out-Country operations as well. Pilots labeled "B Type" Forward Air Controllers came to Vietnam without previous fighter experience, and the rules of engagement only allowed B Type FACs to control In-Country airstrikes for South Vietnamese units. Air Force combat rotation policies complicated the staffing and rules of engagement restrictions, dictating no pilot could be involuntarily sent to Vietnam until all Air Force pilots rotated into a Vietnam combat tour. Once in Southeast Asia, aircrew combat assignments lasted either one year or one hundred missions in North Vietnam, after which the aircrew could rotate home. Known as a "counter," it usually took about nine months for a pilot to reach the hundred-mission milestone. The short duration of combat rotations during the Operation Rolling Thunder era between 1965 and 1968, rapidly burned through the pool of Air Force fighter pilots qualified to fly the bulk of Forward Air Controller missions.[30]

To circumvent the staffing and qualification issue, the Air Force created a new squadron, the 3329th Combat Crew Training Squadron, at Cannon Air Force Base, New Mexico. Nicknamed "Instant Fighter School," the Air Force charged the 3329th with providing fighter training to pilots with no fighter experience. Utilizing the Air Force's first jet trainer, the T-33 Shooting Star, a derivative of the Air Force's first jet fighter the F-80, the Instant Fighter School instructed pilots in the basics of ground-attack missions, such as dive-bombing and strafing. Now with "fighter experience," the pilots went to Vietnam for Forward Air Controller training at FAC-U, which offered seven courses tailored for a pilot's experience level and aircraft assignment.[31]

To further increase available pilots for FAC duty, the Air Force created a similar training pipeline in the US specially designed for Air Force pilots recently graduated from pilot training. Newly minted Air Force pilots went

for training at a stateside course at Hurlburt Field, where they trained along-
side South Vietnamese Forward Air Controllers. The Forward Air Controllers-
in-training attended the revamped Air Ground Operations School before
completing their flying training at Hurlburt. They then attended the Instant
Fighter School at Cannon before heading off to Vietnam. Additionally, the Air
Force began stateside Forward Air Controller squadrons, such as the 704th
TASS at Shaw Air Force Base, South Carolina, to build a pool of qualified
Forward Air Controllers.[32]

Of course, the Air Force also harnessed its pool of already qualified
fighter pilots. Fighter pilots assigned in Europe or Japan could not go directly
to Vietnam from their overseas bases unless they volunteered. As a work-
around, the Air Force rotated pilots through a stateside Forward Air Controller
squadron as part of the regular Air Force assignment cycle, before deploying
them to Vietnam. Interceptor pilots assigned to Air Defense Command, the
Air Force Command charged with the air defense of North America, were
another pool of potential Forward Air Controllers. Interceptor pilots did not
fly the types of fighter aircraft involved in combat in Vietnam but were still
technically fighter pilots. The Air Force used them to augment the ranks in
Vietnam, although interceptor pilots lacked experience attacking targets on
the ground. Finally, the Air Force used backseat pilots from the two-seat
fighters of the era, such as the F-4 Phantom.[33]

Prior to acquisition of the F-4, the Air Force never had a two-seat multirole fighter. Therefore, they did not have a cadre of specially qualified
Air Force navigators known as Weapons System Operators trained for the
backseat of the F-4. Several two-seat variants of fighters existed for special
missions—such as reconnaissance, electronic warfare, or suppression of air
defenses—but these aircraft used a different subset of aviators called Electronic Warfare Officers. The first class of Weapons System Operators did not
join the Air Force F-4 community until 1968.[34] To fill the role in the interim,
the Air Force assigned newly graduated pilots to the back seat of the F-4.
Dubbed "GIBs," for Guy in Back, the new pilots joined Phantom squadrons
in Vietnam. In keeping with its policy, the Air Force could not send GIBs who
previously served a combat tour in Vietnam back involuntarily. However, the
Air Force offered the GIB pilots a fast track to the front seat of an F-4 *if* they

volunteered for Forward Air Controller duties in Vietnam. Many jumped at the opportunity.

The GIBs typically had more combat assignments in Vietnam than any other fighter pilot of the era. Most saw combat in the back seat of an F-4 as a young Lieutenant during the Rolling Thunder years, 1965-1968. They often quickly volunteered for a Forward Air Controller tour, and their previous fighter experience qualified them for all In-Country and Out-Country missions. After completion of their Forward Air Controller tour in Vietnam, they moved to the front seat of a fighter—generally the F-4. Many volunteered as a front seater for a third tour in Vietnam, usually seeing combat again in the later stages of the war, 1970 to 1972.[35]

Eager for combat, Lieutenant Hale Burr volunteered for an F-4 GIB combat assignment straight out of pilot training in 1966. After survival training and a check out in the F-4, he joined the 557th Tactical Fighter Squadron at Cam Ranh Bay, South Vietnam. Arriving in September 1967, Burr only logged twenty "counters" by the halfway point in his tour. Since Cam Ranh Bay was five hundred miles from the DMZ and nearly eight hundred miles from Hanoi, the fighter squadrons based there did not often make the long flight up to North Vietnam. Further, due to heavy American losses over North Vietnam and mounting political pressure, President Johnson slowed the pace of the bombing operations in North Vietnam in early 1968 before ending the campaign in October. Instead, Burr flew mostly close air support and interdiction missions in South Vietnam and Laos, always working with FACs.[36]

As American involvement in Vietnam deepened, demand for FACs in Vietnam grew steadily. In March 1968, a request for Forward Air Controller volunteers arrived at the 557th from Seventh Air Force, and Burr jumped on it. Instead of returning to the US, he moved mid-tour. After five flights at FAC-U at Phan Rang, he joined the 20th Tactical Air Control Squadron, supporting operations in I Corps. However, Burr did not stay with the 20th at Da Nang; instead, he moved to the forward operating base at Phu Bai, outside of the city of Hue. For the next six months, he lived with and provided air power for the 3rd Brigade of the 82nd Airborne Division. After the plush life of Cam Ranh Bay, conditions at Phu Bai were a shock. Burr lived in a two-man tent and slept on a cot. "My shower was a community cold water 55-gallon water drum with a faucet. I never did get the ever-pervasive red dust out of my skin until I rotated back to the USA! I did get hot meals for breakfast and supper but usually ate K rations at lunch. And there was limited ice or refrigeration, so I learned to like warm beer and soft drinks."[37] For a chance to sleep in air conditioning, Burr and other pilots took every opportunity to fly back to Da Nang and have a hot meal and a cold drink at the Officer's Club.[38]

Burr's previous fighter experience earned him the ability to support US forces. After two weeks of quiet orientation missions in I Corps, he faced the real test of a Forward Air Controller on May 1, 1968—emergency support for troops in contact with the enemy. Using his usual Forward Air Controller callsign, "Gimpy 35," Burr flew out to support Alpha and Charlie Companies of the 82nd's 1st/505th about fifteen miles west of Hue, near the Au Shau Valley. Slotted as the first Forward Air Controller of the day, usually an easy assignment, he did not anticipate what came next. As the patrolling companies approached a series of villages, they came under heavy fire. "We actually were facing a pretty desperate situation with two American companies hunkered down in the open flat rice paddies to the west and south, plus the recon platoon on the road to the northeast. The 'good guys' were outnumbered and under intense fire from numerous locations!"[39]

With the two companies under attack by a battalion of North Vietnamese Army regulars, Burr quickly radioed for air support. For the next four hours, he controlled thirteen flights of Air Force and Navy F-4 Phantom, F-100 Super Sabre, A-4 Skyhawk, and F-8 Crusader fighters against twenty-one targets, many "danger close"—within a few hundred yards of friendly forces. With the enemy fire silenced, he made the short flight back to base. "When I landed, I was completely drained, and my flight suit soaked with sweat! But I felt great—knowing that I had helped our American troops in the battle." Alpha and Charlie Companies suffered five killed and eighteen wounded, despite being outnumbered two-to-one, a testament to American air power.[40]

Within six months, flying nearly every day, Burr logged about four hundred hours in the O-2 Skymaster. A military conversion of the Cessna

337 Super Skymaster, the Air Force purchased the O-2 as a stopgap measure
in 1966. The O-2 featured a high wing, twin tail booms, and two engines in
a pusher-tractor set up—meaning one engine was in the front of the aircraft
and the other in the rear. Resembling a Volkswagen microbus with wings,
the boxy aircraft was faster and capable of longer flight durations than the
O-1. The Skymaster also carried fourteen rockets compared to the O-1's
eight. It could carry illumination flares for night operations and occasionally
sported a 7.62 mm minigun in a pod on the wing.[41]

Despite its increase in capabilities, the Skymaster still had many limitations as a FAC airplane. When fully loaded with fuel, fourteen rockets,
a pilot, an observer, and all their gear, the aircraft operated over its designed
weight limit. The aviator's gear alone added close to fifty pounds to the total
weight of the crew. Combat pilots flew with a survival vest, which contained
dozens of pockets stuffed with emergency equipment should they eject or
crash. Pilots also carried dozens of maps, flight manuals, checklists, and
water to drink on the extended missions. Finally, O-2 and O-1 FACs carried a
short-barreled automatic rifle and a pistol. Intended for self-protection should
their aircraft be shot down, some Forward Air Controllers instead fired their
weapons out the window at enemy positions during missions. The airplane's
four bulky radios and armored seats further contributed to the O-2's weight
problem. Besides their weight, the seats were uncomfortable and partially
obstructed the crew's view. The O-2 also had a side-by-side seating configuration versus the O-1's tandem-seat layout. When coupled with the placement
of the aircraft's high wing, the pilot and observer's view was significantly
restricted when attempting to look out the other side of the cockpit.[42]

The austere operating conditions of Vietnam were not kind to the O-2
either. Cessna designed the airplane for business travelers and upscale
personal transportation, on flights in and out of well-appointed airfields with
long paved runways and smooth taxiways. The dirt and gravel airfields of
Vietnam, with their short runways, brutalized the overloaded O-2's spindly
retractable landing gear. Further, the aircraft's low stance and odd engine
layout caused the front propeller to throw damaging rocks and other debris
into the rear engine. Unfortunately, the rear engine was the most critical of
the two, and failure of the rear engine due to damage usually resulted in a

crash, particularly if the failure happened during takeoff. O-2 pilots loathed any loss of thrust or lift to the overweight airplane in the hot and humid flying conditions of Vietnam, particularly when making a rocket pass. While the O-2 could climb to higher altitudes than the O-1 could, it still had to dive low to mark a target with a rocket successfully. After the Forward Air Controller fired their rocket, they were extremely vulnerable to ground fire as they coaxed their overweight craft back to a safer altitude. Many O-2 pilots met their demise after coming under fire during a rocket pass.[43]

Prompted by the increase of FACs squadrons, losses due to combat and hard flying in Vietnam, slow airspeeds, light payload, and vulnerability to ground fire, the Air Force sought a replacement for the O-1 in 1965. Partnering with the US Marine Corps, the Air Force purchased the OV-10A Bronco to serve as its new Forward Air Controller aircraft in 1966. However, the OV-10 did not deploy to Vietnam until 1968, forcing the interim acquisition of the O-2. The OV-10 represented a significant upgrade in capabilities compared to the O-1 and O-2. Purpose-built for forward air control and close air support, it had two powerful turbo-prop engines with massive propellers. When fully loaded, the airplane weighed more than six tons, but it was faster, nimbler, and capable of carrying a larger load than any previous FAC aircraft. For increased survivability the rugged OV-10 had armor plating around the cockpit, the engines, and critical flight control areas. The Bronco also came equipped with self-sealing fuel tanks. Self-sealing fuel tanks have a foam core that expands to fill holes caused by antiaircraft guns and missiles, stopping fuel from leaking out and preventing fires. Resembling a mechanical dragonfly, the OV-10 could loiter for several hours and had excellent visibility from the cockpit, allowing the pilot to survey the landscape effectively.[44]

Under the code name Combat Bronco and Misty Bronco, combat testing of the OV-10 began in July 1968 with six aircraft. The test team recruited ten pilots, each with a different amount of Forward Air Controller experience ranging from original members of the Farm Gate team to brand-new Forward Air Controllers straight from the stateside FAC-U. The Combat Bronco team initially attached to the 19th Tactical Air Support Squadron at Bien Hoa. To evaluate the OV-10's capabilities in the austere environments and short dirt runways of forward operating bases, they also moved to several different locations during the three months of testing. The aircraft had five weapons stations capable of carrying a wide assortment of ordnance. During Combat Bronco, the team experimented with various combinations of rocket pods, fuel tanks, and flares during night testing. The most commonly tested weapons loads, and the ones most commonly used once the OV-10 was operational, featured four rocket pods, each with fourteen rockets for day operations. For night operations the Forward Air Controllers most often used two rocket pods plus two flare pods with six flares each. A fuel tank was added to the center weapons station to extend the OV-10's range, a configuration used most often during Out-Country missions.

The Combat Bronco Team offered plenty of praise for the new airplane. In addition to its speed and durability, the OV-10 had zero-zero ejection seats, meaning the pilot and observer could safely eject from a damaged and burning aircraft even if it were sitting completely still on the ground. The zero-zero ejection seat was a significant improvement that saved the lives of many pilots. Also, the airplane had an area behind the cockpit that could hold small amounts of cargo, a wounded soldier on a stretcher, or three passengers. Army and Air Force special forces used the Bronco regularly to support isolated teams flung across austere operating bases in Southeast Asia. They even parachuted from the back of the aircraft during clandestine missions. There were a few downsides, however, to the OV-10. The cockpit was close to the two huge propellers, making it extremely loud. OV-10 pilots prized Army helicopter helmets over standard Air Force helmets because they suppressed the noise better. Also, while the expansive canopy offered excellent visibility, it created an extreme greenhouse effect in the intense Vietnamese heat and humidity. Since the OV-10 did not have any air-conditioning, and instead relied on outside air piped into the cockpit via a few small air vents, pilots were often forced to shorten their missions during the summer and always required several canteens of water. Many OV-10 pilots carried frozen water bottles in the pockets of their flight suits, which cooled them down as the ice melted and provided much-needed hydration.[45]

Night operations revealed another flaw in the OV-10 design. Night close air support and interdiction missions steadily increased as the North

Vietnamese Army and the Viet Cong used the cover of darkness to avoid attack by American air power. The OV-10's cockpit lighting, external lighting, and narrow cockpit layout made it difficult for pilots or observers to use a Starlight Scope during night operations. An early generation night vision device, Starlight Scopes were about two feet long and weighed about twenty pounds. Difficult to use during ideal conditions, the cramped cockpit rendered the device almost useless, and the airplane's lights distorted the device's image. Instead, OV-10 Forward Air Controllers relied on illumination flares for night operations. Unfortunately, the use of the flares alerted an enemy to the Forward Air Controller's presence, causing them to dive for cover.[46]

While the Combat Bronco tests fielded the OV-10 strictly in a Forward Air Controller role, Misty Bronco tested the aircraft's ability to act as both a Forward Air Controller and attack aircraft—something unexamined since the Grasshopper tests of August 1945. The OV-10 was able to carry nearly four thousand pounds of ordinance and had four internally mounted 7.62 mm machine guns. During the tests, the Misty Bronco team swapped the Willie Pete rockets for high explosive rockets and found the OV-10 could provide immediate firepower for Army forces in danger, or be able to attack fleeing North Vietnamese or Viet Cong forces—in most cases just as well as fighter aircraft.[47]

During the tests, the team also found the aircraft was able to operate at altitudes and airspeeds high enough to simplify rendezvous with fighter aircraft. The ability to meet at higher altitudes and airspeeds, coupled with a unique smoke generating system, allowed the fighters or bombers to quickly locate the Forward Air Controller and begin providing close air support to troops in need. Further, the OV-10's performance allowed Forward Air Controllers to position their aircraft to better observe fighter attacks and confirm that the correct target was in their sights. Toward the end of the Vietnam War, the OV-10 added a laser designator pod as part of the Pave Nail program, allowing Forward Air Controllers to guide laser-guided bombs dropped from other aircraft. The revolution in accuracy afforded by laser-guided bombs proved lifesaving in danger-close situations.[48]

As the principal operator of an aging O-1 fleet, the 19th TASS received sixteen of the twenty-one initial OV-10s shipped to Vietnam in

November 1968. By May 1969, the 19th operated thirty-eight aircraft, the 20th had twenty-four, the 23rd eleven, and FAC-U operated four aircraft for OV-10 qualification training—for a total of seventy-seven deployed Broncos. The Air Force eventually acquired 157 OV-10s, and they remained in service through the end of Operation Desert Storm. In 2015 the US Navy resurrected the OV-10 for operations in Iraq and Syria as part of the Combat Dragon II program. As part of a broader DOD examination of light attack aircraft, the program tested, and proved, the OV-10's capability as a Forward Air Controller, intelligence collection, and attack aircraft just as the Combat and Misty Bronco team did five decades before.[49]

The 20th and the 23rd primarily operated their OV-10s for Out-Country missions, due to its higher speed and extended range. Within a month of acquiring the OV-10, the two squadrons were averaging more than two hundred Out-Country sorties a month.[50] Forward Air Controllers were omnipresent during Out-Country interdiction missions due to US rules of engagement. The rules required a Forward Air Controller for all airstrikes in Southeast Asia that were not in self-defense or in a designated free-fire zone. The bulk of Out-Country Missions were interdiction missions in Laos, intended to stem the flow of North Vietnamese men and equipment down the Ho Chi Minh Trail into South Vietnam. Forward Air Controllers of the 20th and 23rd represented the confluence of the OV-10 and the inter-diction mission.

Breaking North Vietnamese support to the Viet Cong and stopping communist infiltration were the principal concerns of every political admin-istration and military command from the beginning to the end of US involve-ment in Vietnam. In April 1965, the Air Force and Navy began an interdiction program in northern Laos under the code name Steel Tiger, and a second program in December 1965 under the code name Tiger Hound. A more exten-sive joint interdiction program code-named Operation Commando Hunt subsumed these programs, along with others, in November 1968. This was a massive undertaking involving the insertion of Special Forces teams, defoli-ation programs, placing sensors along the Ho Chi Minh Trail to detect trucks and people, and persistent air interdiction; the seven phases of Commando Hunt operations dropped millions of tons of ordinance in Laos.[51]

Out-Country operations used a parallel Tactical Air Control System to the planned and immediate system used In-Country. Due to the secrecy, lack of infrastructure, and distances from US bases in Thailand and South Vietnam, airborne command-and-control aircraft directed Out-Country missions. Military Assistance Command, Vietnam (MACV) divided the airspace over Laos into two main operating areas: Steel Tiger, covering southern Laos along the Cambodian and South Vietnamese border; and Barrell Roll, located in northern Laos along the border with North Vietnam. They subdivided the areas into North, East, and West areas in Barrell Roll, and Steel Tiger was divided into Steel Tiger East and Steel Tiger West. Numbered subdivisions further parsed operations. Out-Country Forward Air Controller missions initially began with reconnaissance of a numbered or named operating area. Just as in Korea, Forward Air Controllers served as a critical source of intelligence for Out-Country Missions. Forward Air Controllers often patrolled the same area day after day and noticed small changes to the scene below, telltale signs that revealed where a North Vietnamese convoy might have stopped for the night. Once they located their prey, Forward Air Controllers radioed the airborne command-and-control, called "Hillsboro" or "Cricket," who tapped into the Tactical Air Control System, just as In-Country missions did. Planned Out-Country interdiction missions, like In-Country close air support, assigned Forward Air Controllers and dedicated strike aircraft to heavily trafficked sections of the Trail, known operating bases, and targets located during previous Forward Air Controller reconnaissance efforts.[52]

Not every In-Country Forward Air Controller conducted Out-Country missions. The most experienced Forward Air Controllers typically controlled these secretive missions. To disguise the nature of their mission, Out-Country Forward Air Controllers used a different callsign than their squadron normally used. Forward Air Controller callsigns evolved as the Vietnam War changed. In a variation of the Mosquitoes in Korea, Forward Air Controllers in Vietnam used callsigns specific to the unit they supported. However, Forward Air Controllers also used a different callsign when on a special mission. For example, the 19th TASS Forward Air Controllers in III Corps flew with snake names such as Kingsnake or Python. A special section of Forward Air Controllers attached to the 19th used Red Marker when supporting the South Vietnamese Airborne Brigade. OV-10 pilots of the 23rd TASS operating in Laos used the callsign Nail to distinguish themselves. Still, other FAC callsigns distinguished special missions. When the US began operating in Cambodia, Out-Country FACs supporting those missions used the callsign Rustic.[53]

A FAC's callsign was significant to the pilot, the unit they supported, and the aircraft conducting strikes. Every Forward Air Controller callsign in Vietnam signified the type of aircraft flown, the home unit of the Forward Air Controller, and for the pilots, their standing within the community. A FAC who carried the numerical designation 01 was the first commander of that unit. For example, Nail 01 was the first Nail Forward Air Controller and the 23rd commander at its inception; Nail 02 was the second pilot assigned to that unit, and so on. An Air Force tradition developed around this practice, particularly around clandestine programs such as the U-2 Dragon Lady, F-117 Nighthawk, and dozens of others. Even after the secretive missions or aircraft became public, or after an Air Force squadron transitioned to a new airplane, the practice of numbering members carried on. Two programs that used special callsigns to distinguish their unique missions and adopted the numbering tradition were the Raven and Misty FACs. Ravens operated as part of a secret CIA mission in Laos, while the Misty FACs formed the cadre of the F-100 Fast FAC program.

The Central Intelligence Agency established the Raven Forward Air Controllers in 1965 to assist the Laotians in combating North Vietnamese incursions in Laos and interdicting the Ho Chi Minh Trail. Born from the post-World War II chaos in Indochina, CIA operatives and their Laotian partners struggled to stunt the growth of the Viet Minh-supported Pathet Lao. To fill the Raven program, selected Air Force volunteers with Forward Air Controller experience in Vietnam transferred to the Steve Canyon program, named after the comic book series of the same name, where they were "sheep-dipped"—a slang term for removing Air Force personnel from the service and transferring them to the CIA. In reality, the Air Force pilots never officially left the Air Force but did so only on paper to create the illusion the Air Force was not involved in the Raven program. The Air

Force and the CIA also practiced sheep-dipping for pilots transferred to the secretive U-2 spy plane program in the 1950s. Ravens flew in civilian clothes in unmarked O-1s, T-28s, and U-18s—a military conversion of the Cessna 185 aircraft—from remote and austere fields scattered around Laos. The Ravens courted reputations as cowboys and mavericks, flying in jeans and cowboy hats, sporting handlebar mustaches, Bowie knives, Colt .45s, or whatever weapons they might get their hands on.[54]

The Laotians, the CIA, and the Ravens gradually lost ground to the steady advance of the communist forces. Enemy forces targeted the web of secret US operating bases in Laos. Called Lima Sites, these remote airfields, firebases, and radio and navigation stations supported clandestine operations in Laos and US bombing operations in North Vietnam. The Raven's main operating base in central Laos, Long Tieng, was a hub for US interdiction missions in the Plain of Jars, the main thoroughfare from North Vietnam to South Vietnam, and the unofficial heart of the Raven mission. In an effort to stop enemy incursion from North Vietnam into South Vietnam, Laos, Thailand and Cambodia, US bombing missions in the Plain of Jars decimated the area, turning whole areas into scenes of bombed-out desolation resembling a World War I no-man's-land. Falling back in the face of steadily growing communist attacks, the US gradually lost control of the Lima sites. A combined North Vietnamese and Laotian enemy force overran Long Tieng in February 1975. After nearly two decades of struggle, the Royal Laotian government fell, and the Pathet Lao established a communist government in December 1975.[55]

Ravens, O-1, O-2, and OV-10 operators during Vietnam—both In and Out-Country—were collectively called "Slow Forward Air Controllers." The slow propeller-driven aircraft were ideal for patrolling over the dense jungles, scattered villages, and flooded rice paddies of Southeast Asia. Their low speeds enabled them to survey the terrain below carefully, while their long flight durations allowed them to provide hours of support to ground forces. However, Slow Forward Air Controllers were exceptionally vulnerable to ground fire of all types: AK-47 rifles, large caliber antiaircraft artillery, shoulder-fired portable surface-to-air missiles, and even advanced radar-guided surface-to-air missiles. Areas in southern North Vietnam, along the DMZ and Laotian border, contained a particularly dense network of antiaircraft artillery and surface-to-air missiles, making operations there too dangerous even for the rugged OV-10. The ranges of North Vietnamese surface-to-air weapons allowed them to target aircraft in South Vietnam and Laos. Still, the rules of engagement dictated Forward Air Controllers control all airstrikes. To solve the issue of North Vietnamese threats and fulfill the rules of engagement requirements, the Air Force initiated the Commando Sabre program.

Officially Detachment 1 of the 416th Fighter Squadron, Commando Sabre began June 15, 1967, at Phu Cat Airbase, South Vietnam. Flying two-seat F-100 Super Sabres and using the callsign Misty, after their first commander Major Bud Day's favorite song, Commando Sabre pilots hunted North Vietnamese and Viet Cong trucks and troops moving in the border areas of North Vietnam. Operating in an area named Tally Ho, which covered the DMZ and sections of North Vietnam along the DMZ and Laotian border, the Misty Fast FACs used a combination of tactics borrowed from both the fighter and Slow FAC communities. The combined tactics were necessary because of the dangerous nature of the mission— Mistys flew well into areas where surface-to-air missiles and antiaircraft guns operated. The Misty pilots also flew at much lower altitudes than most fighter missions, which exposed them to every type of ground fire. To combat the fire, they never flew in a straight line for more than four or five seconds; instead, they made aggressive high G turns at more than five hundred miles per hour to prevent gunners from getting a clear shot.[56] Mentally and physically exhausting, Misty missions often lasted for four hours, twice the average Vietnam combat mission. Pilots returned to their base, drenched in sweat and drained after each mission.[57]

Misty FACs, like their slower counterparts, regularly flew over the same areas, learning the contours of the terrain and developing a sense of any changes to the scene. Commando Sabre pilots alternated missions between the front and back seat of their two seat F-100Fs. The observer in the back visually searched the area while the pilot avoided terrain, clouds, and ground fire. Although their first few missions during the summer of 1967 faced no North Vietnamese fire, after July nearly every Misty mission came under attack. Due to the demands of the mission and the heavy defenses of their operating areas, Misty FAC tours were shorter than other Vietnam combat tours. Misty combat tours lasted 120 days or 75 missions, whichever came first. The missions proved to be as effective as they were dangerous as Misty-controlled aircraft destroyed 79 trucks in a single day in March 1968.[58]

Operation Commando Sabre officially closed in May 1970, due to high loss rates and the Air Force's phase out of the aging F-100.[59] The newer and more capable F-4 Phantom gradually took over Fast FAC operations. Beginning in August 1968, the Air Force tested the F-4 as a Fast FAC aircraft, flying with the callsign Stormy and operating from Da Nang Air Base. The Stormy F-4 FACs checked out in eight missions. During the first five missions, F-4 instructor pilots flew in the rear cockpit of an F-100 with experienced Misty FACs. The Misty pilots then flew in the back of three F-4 missions with the Stormy pilots up front.[60]

Like other Forward Air Controller aircraft, the F-100s and F-4s flew with two rocket pods for marking targets and a fuel tank to extend their flight time. However, because the F-4D did not have an internal cannon like the F-100, the Stormy FACs also added a 20 mm gun pod to their load. The F-4 had some limitations and advantages in the mission compared to the F-100. The Phantom's design restricted visibility from the rear cockpit, making the GIB's job more difficult. However, the aircraft had better electronic protection against radar guided surface-to-air missiles and antiaircraft guns. The F-4 was also slightly less maneuverable and used more fuel than the F-100, but it compensated by being faster and capable of carrying a larger payload. After the Stormy FAC tests, the Air Force approved the F-4 as the new Fast FAC aircraft.

The mission spread to three other F-4 fighter wings. Tiger, Wolf, and Laredo F-4 FACs operated from Ubon, Korat, and Udorn Air Bases in Thailand. The additional bases expanded Fast FAC operations into the northern sections of Steel Tiger and Barrel Roll. Phantom FACs also adopted the Misty's night missions, flying with the callsign Night Owl, to support the increased night interdiction missions. During night missions, both the F-100 and F-4s added flare dispensers to their weapons load. The new Fast FACs further innovated to maximize the F-4's multirole capability. Tiger FACs developed a hunter-killer tactic, integrating into a four-ship of strike F-4s.[61] "On these missions, the jet Forward Air Controller, combining his knowledge of the terrain with his superior navigational equipment, led the strike flights directly to the targets."[62] Once over the target, the FAC located and marked targets for the fighters to attack. A second variation, the Atlanta/Falcon team, paired a Fast FAC with a reconnaissance aircraft. Flying from Udorn, Thailand, the two aircraft took off together. While the Falcon pilot refueled, the Atlanta RF-4 reconnaissance aircraft surveyed the target area, taking prestrike photos. The FAC then led a strike team into the area for attacks, before the Atlanta aircraft took postattack photos.[63]

After spending less than a year in the US, Hale Burr volunteered for a second combat tour in Vietnam. It was a dangerous choice; fifteen of the forty-four members of Burr's pilot training class were shot down in Vietnam, and eight of those fifteen died in combat. Burr wrote philosophically about his choice: "It's really easy to become addicted to the adrenaline rush of flying jet fighters and flying in combat! But that feeling is not worth much if you don't survive."[64] Still, the pull of combat drew him back. The very same traits that made him and other fighter pilots keen for battle are the same that make them successful. Echoing the ethos of pilots back to World War I, Burr said, "It is hard to explain the psychology of fighter pilots—they are self-assured over achievers, very aggressive, and with a strong competitive will to win. These traits are what it takes to be successful in air combat."[65]

In May 1969, Burr joined the 13th Tactical Fighter Squadron at Udorn Air Base, Thailand, for his second tour in Vietnam. After four months of flying interdiction missions in Laos and North Vietnam, he was selected for the F-4 Fast Forward Air Controller program. After a five-flight checkout, Burr returned to doing "exactly the same mission that the O-2's Forward Air Controller's had that I flew in Vietnam a year before. The F-4 Forward Air Controller's went out and marked targets with white phosphorus (WP) rockets for other fighters to attack."[66] When the Lieutenant Colonel in charge of the FAC program at Udorn transferred to Europe, the Wing Commander selected Captain Burr to run the program, after only two months on the job, in no small part because of his previous experience flying the O-2 in the mission.[67]

The rules of engagement requirement remained until the end of the Vietnam War. Increasingly, however, the weight of the Air Force mission shifted to the Fast FAC community and the interdiction mission. The F-4 Forward Air Controllers harnessed new technology, the laser-guided bomb, as a tool of the trade. Before the introduction of laser-guided bombs in 1972, pilots had to acquire the target marked by the FAC visually, dive towards the target while aiming at it through a heads-up display sight, and then release their unguided bombs at a specific altitude, airspeed and dive angle—all while being fired at by SAMs (Surface-to-Air Missiles) and enemy gunners, avoiding the ground, and traveling at five hundred miles per hour.

With laser-guided bombs, the F-4 GIBs located the target in a view finder and then fired a laser at the target. Attacking aircraft dropped special bombs with laser seekers mounted on the front, and the bomb followed the reflection of the laser to the target. Since the bombs followed the Forward Air Controller's laser, attacking aircraft did not have to aim the bomb as accurately with their airplane, and this alleviated the need to dive down at the target. As a result, pilots released the laser-guided bombs from higher altitudes, above the effective range of the antiaircraft guns. Since the bombs were more accurate, fewer weapons were needed to destroy targets, and fewer aircraft went into harm's way. Although other aircraft also added laser pods during Vietnam, and the use of laser-guided bombs steadily expanded during and after the war, the full weight of the smart bomb revolution did not come of age until the 1990s.

Fast FAC missions in Steel Tiger, Barrell Roll, and Tally Ho were different from missions in South Vietnam. Instead of supporting Army units, Fast FACs hunted North Vietnamese and Viet Cong forces as they moved south along the Ho Chi Minh Trail. One of the most dangerous missions was hunting and attacking North Vietnamese surface-to-air missiles. With just four days remaining in his tour, Burr was selected to lead an armada of twenty-four F-4s against a SA-2 surface-to-air missile site along the North Vietnamese-Laotian border. Using the Atlanta/Falcon and Hunter-Killer tactics developed by Phantom FACs, Burr was charged with the dangerous job of locating, marking, and directing the attack on the SAM site.[68]

On the morning of March 27, 1970, the strike aircraft took off, loaded with a mixture of Mark-82 500-pound general-purpose bombs and CBU-24 cluster bombs. The Mark-82 is an unguided high explosive bomb, and as the name suggests, it is useful for attacking almost any target from small buildings to vehicles, tanks, and troop concentrations. CBU-24 cluster bombs weigh about seven hundred pounds and resemble a general-purpose bomb from the outside. However, when the weapon is released, a clamshell mechanism opens the bomb's outer case like a blooming flower, releasing 670 smaller baseball size bomblets. The bomblets, which are similar to hand grenades, spread over the target area, destroying targets over a wider area than a general-purpose bomb. The CBU-24 was very effective against troop concentrations, antiaircraft gun batteries, and surface-to-air missiles.[69]

North Vietnamese missile sites were notoriously dangerous to attack. Using contemporary Soviet air defense doctrine, the North Vietnamese built their SAM sites in an overlapping and layered defense pattern. However, the North Vietnamese were also very adaptable to the terrain and circumstances of the Vietnam War. Vietnam combat pilots developed mixed feelings of fear, hate, and respect for North Vietnamese SAM operators. A North Vietnamese SA-2, like the one Burr and his force were charged to destroy, usually consisted of a radar guidance system and multiple missile launchers. The launcher fired a massive missile, resembling a telephone pole in flight, with a 400-pound warhead and a range over twenty miles and eighty thousand feet. The Soviets designed the SA-2 as a defense against high flying, larger, slower aircraft, like the B-52 nuclear-capable bomber. A SA-2 famously shot down Gary Powers while flying a U-2 over the Soviet Union in 1960, but during the Vietnam War the North Vietnamese proved it to be capable against fighter aircraft as well. One SAM site usually overlapped its defenses with other SAMs, and dozens of antiaircraft batteries further protected the individual sites. The North Vietnamese were also notoriously good at camouflaging sites, further complicating the job of finding and destroying the missiles.[70]

The weather for the attack was clear, and after refueling, Burr pressed into North Vietnam to find and mark the SAM site for the attackers.

I dropped down to about 200' low altitude and moved out about 10 miles ahead of the bomber formation as we approached North Vietnam. This area was a mountainous region covered by thick jungle. I pushed up my airspeed to over 500 knots and flew directly over the SAM site which was camouflaged and hidden in the trees. I could clearly see the SA-2 missile launchers, radar control vans, assorted trucks and people on the ground. While lighting both afterburners, I pulled up immediately and climbed to about 8,000 feet while confirming the target to the incoming fighters. I rolled over in a looping 135-degree slicing turn back to mark the target with a white phosphorus rocket.[71]

As Burr and his GIB bore down on the SAM, three SA-2s launched up at them. Burr transmitted "Falcon, we've got SAMs" on the radio to warn the other F-4s. When he reviewed the audio from the mission during debrief back at Udorn, the stress of the moment became evident. "The first three words were in a normal tone, but I knew the SAMs were tracking toward me and 'SAMs' came out in a much higher pitch—perhaps an octave higher!!" To avoid being hit by the missiles, Burr began aggressively maneuvering his aircraft, while still trying to mark the target with his Willie Pete rocket.[72]

As I completed the roll in, one SAM went below me and two flew over my F-4 so close I could feel the buffet or turbulence as they barely missed me. I fired 2 WP rockets into the middle of the target area … As my white smoke rose from the ground, I called out to the first flight of F-4s "Hit my smoke" and they rolled in on the target. I pulled up to get out of their way and noticed 3 white clouds where the SAMs exploded high above us. Then I watched the 24 F-4s drop their ordnance on the SAM site. It was always fascinating to see the bomb impacts and explosions. This resulted in numerous secondary explosions and fires on the ground that billowed up where the SAM site had been located. As the smoke cleared, a RF-4 recce bird from Udorn came zooming over to take post-strike photos of our attack results.[73]

Three days later, Burr flew his 525th and final combat mission in Vietnam. Hale Burr went on to serve thirty-two years in the US Air Force, retiring as a Major General. During his career, he flew more than 3,500 hours in the F-4, O-2, F-15, and F-16. He commanded a fighter squadron, two fighter wings, and earned five Distinguished Flying Crosses during his two combat tours in Vietnam.[74] Burr was fortunate to survive his two FAC tours. According to a study published by the Air Force in 1977, being a Forward Air Controller in Vietnam, of any variety, was a dangerous business. Every aircraft to perform the mission ranks in the top ten of highest loss rates in Vietnam. Further, FACs were among the least likely to reach the one hundred mission or one-year tour completion mark and among the most likely to be killed, rather than taken prisoner, as the result of a shootdown. The only aircraft with a higher loss rate that never performed the Forward Air Controller role was the F-105, which suffered the highest loss rate of any aircraft in Vietnam.[75]

Forward Air Controller losses earned other grim honors during the Vietnam War. On May 11, 1972, First Lieutenant Michael Blassie, a FAC in the 8th Special Operations Squadron, went down in his A-37 Dragonfly near the town of An Loc, South Vietnam. Blassie's aircraft crashed in a North Vietnamese-held area, and it took several months before a South Vietnamese special forces team made it to the crash site. The team recovered only a few items from Blassie's survival kit, and some human remains. Unfortunately, the airman's remains disappeared in the chaos in the waning days of an unpopular war, his family unaware of their son's fate. In 1984, President Reagan presided over the dedication of the Vietnam Unknown Soldier at Arlington National Cemetery. Later, an investigation by the Department of Defense and First Lieutenant Blassie's family discovered the government buried his remains in Arlington National Cemetery as the Vietnam War Unknown Soldier. In 1998, after spending fourteen years as the Vietnam Unknown, Blassie's remains were returned to his family and laid to rest at the National Cemetery near St. Louis, Missouri.[76]

Not only were Forward Air Controllers among the first airmen killed in Vietnam, but they were also the last. On June 16, 1973, four months after the signing of the Paris Peace accords, Captain Samuel Cornelius and Captain John Smallwood launched on a mission into Cambodia from Ubon Air Base, Thailand. While attempting to mark a target with their Willie Pete rockets, their F-4 was hit by ground fire and crashed, making Cornelius and Smallwood

the last airmen to officially die in combat in Vietnam. The tragedy of loss for
Forward Air Controllers continued beyond the war, however. On December
23, 1975, Raven Paul Jackson became the last Raven to die in defense of
Laos, nearly three years after the signing of the Paris Peace Treaty, in a place
where Americans were never officially at war, doing a job unacknowledged
by the US.[77]

In keeping with tradition, Vietnam War FACs composed a tune that
captured the realities of being Forward Air Controllers in Vietnam. The song
is still sung in Air Force fighter squadrons in honor of the sacrifices of
Vietnam Forward Air Controllers. Entitled "Dear Mom," it is sung in an
ironic bright barbershop quartet style:

> Dear Mom, your son is dead.
> He bought the farm today,
> He crashed his OV-10 on Ho Chi Minh's highway.
> It was a rocket pass, and then he busted his ass.
>
> Mmm, mmm, mmm.
>
> He went across the fence to see what he could see,
> There it was a plain as it could be,
> It was a truck on the road,
> With a big heavy load.
>
> Mmm, mmm, mmm.
>
> He got right on the horn, and gave The DASC a call,
> "Send me air I've got a truck that's stalled,"
> The DASC said "That's all right,
> I'll send you Pirate flight,
> FOR I AM THE POWER."
> The fighters checked right in, gunfighters two by two,
> Low on gas, and tankers overdue.
> They asked that FAC to mark,
> Just where that truck was parked.
>
> Mmm, Mmm, mmm.
>
> The FAC he rolled right in, with his smoke to mark,
> Exactly where that fucking truck was parked.

The rest is still in doubt,
because he never pulled out.

Mmm, mmm, mmm.

Dear Mom your son is dead,
He bought the farm today.
He crashed his OV-10 on Ho Chi Minh's highway.
It was a rocket pass, and then he busted his ass.

Mmm, mmm, mmm.

HIM, HIM, FUCK HIM!
How did he go? Straight in!
What was he doin'? 169.
Indicated? Yeah.[78]

When an aircraft went down in Vietnam, an elite Air Force Combat Search
and Rescue team called Pararescuemen helicoptered into the crash area to
rescue the downed aircrew. The mission of Combat Search and Rescue was
a dangerous business. Rescue teams helicoptered into the most dangerous
areas of Vietnam, sometimes deep into North Vietnam, to bring US airmen
home. Dedicated Combat Search and Rescue efforts began during the
Korean War, but like the Forward Air Controller mission, they fully
matured in Vietnam. Responsibility for leading these missions was the
Rescue Mission Commander, nicknamed Sandy. The name Sandy derived
from the callsign used by the pilots who filled this role in Vietnam. In Vietnam,
Sandy pilots flew the A-1 Skyraider, a rugged relic of World War II, uniquely
adapted to the Sandy mission.

A rescue mission began soon after an aircraft went down, as nearby
aircraft attempted to contact the downed aircrew. Once in contact, an aircraft
nearby was designated On Scene Commander. The On Scene Commander was
responsible for attempting to locate the survivor and coordinating the initial
steps of the rescue. Forward Air Controllers often served as On Scene
Commanders due to the number of radios on their aircraft. Further, FACs
were able to find, mark, and call in fighter attacks against any enemy
attempting to capture the downed airmen. Once the rescue mission was

on the way, the Sandy aircraft assumed the responsibilities for finding the survivor and keeping them safe. The Sandy also guided the rescue helicopter to the survivor and protected it as it hovered or landed to pick up the downed aircrew. With the survivor safe on the rescue helicopter, the whole team flew back to a friendly airbase. The similarities between the Forward Air Controller and Sandy mission blurred the roles between the two. Before becoming a Sandy, a pilot was required to be previously qualified as a Forward Air Controller. The Sandy mission combined the dangers of the Forward Air Controller's job plus the danger of flying into areas where the most advanced US fighter aircraft were lost every day. The A-1 flown by the Sandy pilots suffered the second highest aircraft and aircrew loss rate in Vietnam.[79]

Vietnamization became the buzzword of the American strategy as the Vietnam War drew to a close, including Vietnamization of the Forward Air Controller mission. The South Vietnamese Air Force gradually took responsibility for the Tactical Air Control System, increased the number of Forward Air Controllers, and added new aircraft, like the A-37, to their inventory. The American squadrons, particularly the Tactical Air Support Squadrons, were reorganized and consolidated as US troop numbers dwindled. In January 1972, the first American FAC squadron in Vietnam, the 19th TASS, closed its doors and moved to Osan Air Base, South Korea. In March, the 22nd moved its Forward Air Controller mission to Wheeler Air Force Base, Hawaii. The 20th followed a year later, moving to George Air Force Base, California, and in August 1973, the 21st left Vietnam for MacDill Air Force Base, Florida. The 23rd remained in Thailand until 1975, participating in the evacuation of Phnom Penh as Cambodia fell to the Khmer Rouge, before being deactivated in September 1975.[80]

Despite the closing of the squadrons in Vietnam, for the first time since World War I the lessons of the Forward Air Controller programs integrated into a postwar peacetime Air Force framework. Most of the TASS units reactivated at their new locations, integrating into nearby Army divisions, just as they had during the war. The Air Force created new squadrons, particularly in Germany, to deepen interservice cooperation. Further, the command-and-control structures, joint force integration, and training programs created in Vietnam to provide close air support to the Army and effectively distribute air power in a ground war became permanent fixtures of the post-Vietnam Air Force. The service recognized the need for faster, more survivable Forward Air Controller and close air support aircraft, and several joint initiatives attempted to expand the positive lessons of Vietnam. While the United States, as a whole, and the armed forces individually, took many negative experiences from Vietnam, the Air Force used the lessons of the war to remake itself. The service reorganized and modernized. At all levels of the Air Force—training and tactics, doctrine and theory, personnel and leadership—the Air Force changed, adapting the lessons of Vietnam to face the threat of the Soviet Union. Air Force Forward Air Controllers also adapted.

Airco DH-4 over France 1918. The DH-4 was a British observation
airplane built under contract in the United States and one of the primary
observation airplanes used by the United States in World War I.
Photo by United States Army Air Service.

1st Aero Squadron Salmson 2A2 over France 1918. 1st Aero Squadron
was America's first combat flying unit. The 2A2 was a French observation
airplane and one of the primary observation airplanes used by the United
States in World War I. *Photo by United States Army Air Service.*

SIGNALS FOR INFANTRY CONTACT AND ARTILLERY REGLAGE

PLANE TO P.C.		P.C. TO PLANE BY PANEL		SERVICE SIGNALS		NUMBERS
ART	Artillery	P.C. of Battalion	◀	Ready to rec ── BR		1
AVI	Enemy aviator			End of mess. ── AR		
BAV	Bat.against avia.	P.C. of Regiment	◖	Understood ── SN		
BCA	Anti tank battery			Wait ── AS		2
BTA	Battery in action	P.C. of Brig.or Div.	●	Repeat ── UD		
BTO	Battery occupied			Separation ── DA		
CAV	Cavalry			BY T.C.F.	ALPHABET	3
COV	Convoy				A	
DIR	Direction(Followed by name of local'y	Objective reached	◀━ QT		B	
DRO	Right at				C / Ch	4
EST	East of	Request barrage	◀ O		D	
FDF	Barbed wire				E	
FRO	Front(Followed by number of meters)	Request fire for attack preparation	◀‖ CE		F G	5
GAU	Left at				H	
IPC	Infantry in column	Friendly field art. fires on us	◀─ S		I J	6
IPD	Infantry deployed				K L	
IFR	Infantry massed	Friendly heavy art. fires on us	◀ V		M N	
LDX	Here available aeroplane				O	7
NOR	North of	We are ready to attack	◀◀ I		P Q	
OUS	West of				R	8
PRF	Depth(Followed by a number)	Will not be ready to attack at fixed time	◀◀ GW		S T	
QUE	Rear at				U	
RAS	Nothing to report	We wish to advance lengthen fire	◀ H		V W	9
REG	Request for orders				X Y	
RLV	Relieve me	Replacement of small arm ammunition	◀+ Y		Z	0
SUD	South of					
TAK	Friendly troops	Replacement of hand grenades	◇ Q			
TCF	Railroad train					
TET	Head at	Message understood	◀ SN			
TRA	Trenches					
VRV	Coming to rel.you					

PLANE TO P.C.		P.C. TO PLANE BY PANEL				SERVICE
01	First piece	Observe for group		Fire by piece		Fire 3 dashes
02	Second piece					
03	Third piece	Request adjustment		Fire by salvo		Right 2(I)
04	Fourth piece					
05	By piece	Observe fire on Tar (Foll'd by coordin)		Amelioration		Left 2(m)
06	Is battery ready					
07	Has battery fired	Adjust on Target you just indicated		Fire in series of 24 rounds		Short 2(h)
08	Can't see proj'tr					
09	Aim on me	First battery ready		Continue fire for effect		Over (ch)
11	Can't see panels					
12	~~Auxiliary target~~	Second bat'ry ready		Control fire		Def.correct (z)
13	~~Control fire~~					
14	From right by salvo	Third battery ready		Enemy attacks. By prev.agreement		Range cor. 2(n)close
15	Inaccurate fire					
16	~~Amelioration~~	Wait a few minutes		No further need of you		Target (b)
17	Fire 2 pcs by salvo					
18	From left by salvo	Bat.not ready,delay at least 10 minutes		Hostile aeroplane near you		Change targ 2(k)
19	How many guns fire					
21	Impossible observe	Battery has fired		Fire for precision		Error 10 dots
22	Range irregular					
23	Could not observe	Wireless O.K.but si.confused.REPEAT		Zone fire		Will land (bv)
24	Deflec'n irregular					
25	Will observe as req	Can't hear you, fire not adjusted		heard *Attention*		
26	H.E.Shell					
27	Lost	Understood. Message received		Understood. Unable but repeat To reply.		
28	Cease firing					
29	Result accompl'd	No				
31	Fr.shells fall'g o					
32	Continue the fire	Continue to adjust				
33	To much distrib'n					
34	Shrapnel,time fire					
35	Z.F.I cease observ					
36	To much consent'n					
37	By battery volleys					

8 Continuous FIRE.
I cease Regular observation.

Air burst (F)

US observation ground to air signal 1918 manual.
Henry Dwight Collection USAFA.

Henry Dwight in uniform, France 1918.
Henry Dwight Collection USAFA.

Henry Dwight with DH-4, France 1918.
Henry Dwight Collection USAFA.

A Grasshopper preparing to takeoff from the USS *Ranger* during Operation Torch, November 1942. *Navy Heritage and History Museum.*

An L-4 Grasshopper takes off from USS *LST-906* flight deck during the invasion of southern France, St. Tropez, August and September 1944. *Navy Heritage and History Museum.*

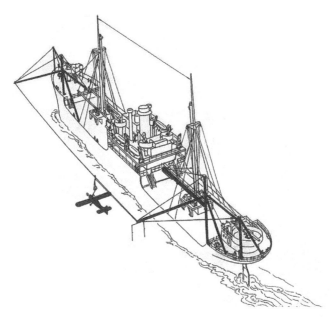

Sketch of Brodie System on the side of an LST. *Smithsonian National Air and Space Museum (NASM-9A001183).*

Grasshopper on Brodie System cable and trolley used on LSTs and remote areas. *Smithsonian National Air and Space Museum (NASM-WT-194076).*

Grasshopper on Brodie System for landings or takeoffs on very small and remote fields. *Smithsonian National Air and Space Museum (83-16835).*

Two P-47 Thunderbolt Pilots working as Rover Joes near Bologna, Italy, December 1944. *Air Force Historical Research Agency IRIS Number 0249262.*

A P-47N Thunderbolt loaded with bombs and rockets for a ground attack mission during World War II. *US Air Force Photo.*

Quentin Aanenson calling in airstrikes as a Rover Joe.
Courtesy of Aanenson private collection.

Quentin Aanenson working as a Rover Joe near Cologne, Germany.
Courtesy of Aanenson private collection.

162 Liaison Squadron Horsefly Test August and September 1945, testing
the Horsefly capability with bombs loaded under the aircraft wings.
Air Force Historical Research Agency IRIS Number sqa-lia-162-hi.

162 Liaison Squadron Horsefly Test August and September 1945, testing
the Horsefly capability with rockets loaded under the aircraft wings.
Air Force Historical Research Agency IRIS Number sqa-lia-162-hi.

A Mosquito pilot in front of a T-6 Texan with an observer in back, Korea, 1950. *US Air Force photo.*

Mosquito T-6 with a Tactical Air Control party jeep, Korea, 1950. *US Air Force photo.*

Mosquito T-6s with custom-made target-marking rockets made from white phosphorus bazooka warheads, attached to the front of rocket. *US Air Force photo.*

Original Mosquito patch artwork. Pilots, observers, and support personnel wore this patch on their uniform to identify them with the Mosquitoes. *US Air Force photo.*

An O-1, O-2, and OV-10 from the 19 Tactical Air Support Squadron's training detachment, better known as FAC-U to the pilots.
Photo courtesy of Julio Fuentes.

An O-1 Birddog over Vietnam. *US Air Force photo.*

An O-2A airborne near Pleiku, Vietnam, 1968. *US Air Force photo.*

An OV-10A Bronco firing a smoke rocket to mark a target for the F-100 Super Sabre in the background, north of Saigon, February 1969. *US Air Force photo.*

A Nail FAC OV-10 with a Wolf Fast FAC F-4 over Laos, 1970. *Photo courtesy of William Wall.*

An A1-E Skyraider pulls up after marking a target with a smoke rocket. Note the small white "Willie Pete" rocket's smoke plume (near the bottom center of photo) showing the attacking aircraft where the target is. *US Air Force photo.*

A Sandy A-1E Skyraider escorts an HH-3C Jolly Green Giant rescue helicopter on a combat search-and-rescue mission in 1966.
US Air Force photo.

A Misty Fast FAC F-100F Super Sabre prepares for a mission at the Misty FAC base Phu Cat, Vietnam. *US Air Force photo.*

Group photograph of Wolf Fast FACs of 8th Tactical Fighter Wing in front of an F-4 Phantom. *US Air Force photo*.

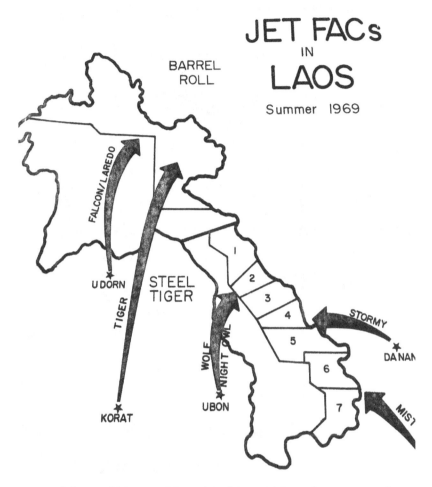

Map of the Steel Tiger and Barrel Roll Fast FAC working areas in Laos and Cambodia. *Project CHECO Report #104.*

An A-10 Thunderbolt II takes off to support a close air support mission in Iraq. The Hog carries, from its left wing to its right, a Hellfire missile, four 500-pound laser guided bombs, and a LANTIRN targeting pod. *US Air Force photo.*

An F-16 Fighting Falcon takes off loaded with two 2,000-pound general purpose bombs for a mission during Operation Desert Storm. *US Air Force photo.*

An F-16 being loaded with 2,000-pound general purpose bombs for a mission during Operation Desert Storm. *US Air Force photo.*

An MQ-1 Predator remotely piloted aircraft flies armed with a single Hellfire missile. Note the sensor ball under the aircraft's nose and the datalink antenna just behind the front landing gear. *US Air Force photo.*

An MQ-9 Reaper remotely piloted aircraft, armed with Hellfire missiles and a 500-pound laser guided bomb, flies a mission over Afghanistan. *US Air Force photo.*

Two Joint Terminal Attack Controllers observe a bomb impact during a routine training exercise. *US Air Force photo.*

Chapter 5

Vietnam to Desert Storm: FACs Rise, FACs Fade

A squabble over roles, missions, and the use of aircraft in those missions broke out between the Army and the Air Force practically the moment the services split. At the heart of the feud were the two seemingly unrelated missions of airlift and close air support. Between the National Security Act in 1947 and the Goldwater-Nichols DOD Reorganization Act in 1986, eighteen agreements, memorandums, directives, acts of Congress, and executive orders attempted to settle the roles, mission, and support issues between the two services. The Air Force argued, and the agreements repeatedly confirmed, that they held the responsibility as the nation's source of air power, to provide airlift and close air support to the Army. Integral to this responsibility, from the Air Force's perspective, was control over planning, allocation, training, employment, and perhaps most critically, acquisition of all things air power and aircraft. The Army repeatedly expressed the need, also confirmed by the agreements, for a limited battlefield airlift capability. To that end, since World War II, the Army operated a fleet of liaison aircraft, like the L-5 and the O-1. Additionally, the Army always retained the ability to control organic fire support, like artillery, within its frontline combat units. Ironically, however, it was not airplanes that deepened the conflict over airlift and fire support—it was the introduction of the helicopter onto the battlefield.[1]

Although helicopters first saw combat in World War II, they truly made their presence known in the Korean War. During and after the Korean War, the role of the helicopter on the battlefield steadily expanded, opening new tactics and operational concepts and creating doctrinal tension between the Army and the Air Force. During the Korea-Vietnam interwar period, the two services began to diverge doctrinally. The Air Force focused on nuclear-capable bombers under President Dwight Eisenhower's New Look strategy, holding fast to the ideology through the early years of Vietnam. The Army initially attempted to harness the nuclear genie with its Pentomic Divisions. However, with the introduction of President John F. Kennedy's Flexible Response, the Army shifted gears, landing on a new operating concept harnessing the helicopter's capabilities. These air-mobile tactics relied on the helicopter for the rapid movement of troops on the battlefield. The use of helicopter airlift required a light, agile force, unencumbered by armor and artillery. To fill the firepower gap created by leaving armor and artillery behind, the Army needed an increase in rapidly responding and flexible close air support. However, during the Korea-Vietnam interwar decade, the Air Force moved away from the successful close air support framework created by the Mosquitoes. As a result, the diverging doctrinal, operational, and tactical frameworks created a series of disagreements over roles, missions, and aircraft during Vietnam and its immediate aftermath.[2]

For both services, control of the resources and systems required for the close air support mission was critical to their vision of the battlefield. The Air Force wanted processes and systems adaptable to any situation, to enable a broad range of aircraft to meet the Army's need for close air support. The Army, in contrast, wanted specifically dedicated close air support resources controlled and distributed by ground force commanders. The Air Force countered that all air power must be controlled and distributed by an air commander, who understands how to use air power effectively and efficiently. The Army's argument neglected the premise that air power is a finite resource, and when put into practice as envisioned by land commanders, wasted air power's effects, as happened during Operation Torch. The Air

Force, in contrast, potentially risked the lives of soldiers if timely and effective firepower could not be delivered to places most needed in combat, as was the case in the first weeks of Korea.

The conflict between the Army and the Air Force over the role of each service and their level of control of close air support resources in combat steadily boiled between the Korean and Vietnam Wars, despite the Mosquitoes' performance in Korea. First, two Pace-Finletter Agreements, signed by the Army and Air Force Chiefs of Staff in October 1951 and November 1952, attempted to quantify size and weight limits for Army aviation while also confirming the Air Force's role as the provider of close air support and airlift for the Army. Then a series of memorandums issued between 1954 and 1958 by Secretaries of Defense Maurice Roche, Charles Wilson, and Neil McElroy attempted to quantify the limits further and clarify the roles of the military services and their aircraft. Additionally, the memorandums attempted to establish who held control over airlift and close air support missions, what constituted these missions, and what each service was responsible for providing to the joint effort. Still, the divergent operational concepts of the Army and the Air Force, coupled with the expansion of the battlefield brought about by helicopters and jets, defied readily agreed-upon solutions.[3]

In April 1962, Secretary of Defense Robert McNamara directed the Army to review current doctrine on tactical mobility and firepower, acutely exacerbating the conflict between the Army and the Air Force. In response to McNamara's directive, the Army convened in August the Howze Board, named after the study's director Lieutenant General Hamilton Howze. From the Howze Board emerged the Army's new air-mobile doctrine and tactics with their reliance on firepower and mobility. In response to the atrophy of Air Force close air support capabilities during the 1950s, as demonstrated by the Bristle Cone exercise, and to fill the gaps in fire support created by a light and mobile force, the Army turned to the helicopter as a solution.[4]

The Air Force objected to the concept and responded with a board specifically to rebut the Howze board's conclusions. The Disoway Board, named after Air Force Lieutenant General Gabriel Disoway, made its report in September 1962. The report dismissed the Army's concepts as faulty and unrealistic on the modern battlefield. In the closing months of 1962 and into

1963, both services conducted unilateral exercises to prove out their concepts of joint close air support and air mobility operations, and disprove the other service's concepts.[5] In response to the competing boards, McNamara convened a joint board named The Army and Air Force Close Air Support (CAS) Boards. It was a joint board in name only. Each service conducted separate boards, only coordinating on the final report.

The three central questions the boards addressed were: How much close air support does the Army need, who exercises control over aircraft designated to provide it, and is a specialized aircraft needed for the mission? The board published its joint report in August 1963 and offered a glimmer of cooperation and agreement. In answering the first two questions, the two services set about improving the TACS-AAGS system in Vietnam. As part of the improvements, the Air Force held responsibility for providing more Tactical Air Control Parties to the battalion level and maintenance of communication links between the Tactical Air Control Parties and the DASC. The Air Force's response led to the dramatic increase in Tactical Air Control Squadrons in Vietnam during 1965 and improvements to the immediate support request system. The Air Force also agreed to designate a set percentage of all combat missions in Vietnam to planned and on-call support, leading to an increase in the overall number of available close air support sorties and a reduction in response times in Vietnam.[6]

Answering the third question proved to be pricklier. The Air Force did not see a need for a dedicated close air support aircraft, but, ironically, neither did the Army. However, the expanding air-mobile concepts challenged the Air Force's ability to meet Army requirements with existing systems, so the Army filled the gap with helicopters. The helicopter proved both critical and revolutionary to air-mobile tactics, and the UH-1 Iroquois became its iconic embodiment during the Vietnam era. Universally called the Huey, and initially used as transport, the Army eventually armed the Huey with rocket and gun pods similar to those used on forward air control aircraft. In 1965 the Army developed the Huey's attack cousin, the AH-1 Cobra. The Air Force viewed the development of attack helicopters as a violation of the Secretary of Defense memorandums written in the mid-1950s. Adding insult to injury, the Army began operating fixed-wing aircraft larger than allowed by the Pace-Finnletter agreements. Seeking to steer the issue in favor of the Army, the Chairmen of the Joint Chiefs of Staff, General Maxwell Taylor, complained loudly about the Air Force's lack of close air support in Vietnam. The expanding disagreement drew the attention of Congress in 1965.[7]

In September and October, a House Armed Services Subcommittee, led by Representative Otis Pike of New York, held hearings examining the issue. The Pike Committee Report, issued in February 1966, took the Air Force to task, focusing on the lack of a dedicated close air support aircraft.[8] The Air Force countered with a series of articles in the March issue of Air Force and Space Digest extolling the virtues of tactical air power in Vietnam.[9] Pike, a Marine Corps fighter pilot during World War II who went on to gain notoriety for his anti-Vietnam stance and investigations of the Central Intelligence Agency and National Security Agency, responded in kind on the House floor:[10]

> We seem to have struck a nerve with the Air Force brass, however, for they have been screaming more like sick seagulls than wounded eagles ever since. If it were not for the lives at stake it might be mildly amusing. As it is, the attempt to make a great success out of what has been a slighted, downgraded, under-financed close air support role in the Air Force is not only pathetic it is dangerous.[11]

Ironically, the Air Force's close air support capabilities matured in Vietnam, with most of the necessary improvements made before the Pike Committee met. Further, as the Vietnam War progressed, the Army and the Air Force repeatedly studied and improved the joint TACS-AAGS network. The system was well-honed, adaptable, and diverse, able to provide timely and decisive air power in support of ground operations twenty-four hours a day anywhere in South Vietnam. Thanks to the development of the Tactical Air Support Squadron structure, the Tactical Air Control Parties, Air Liaison Officers, and Forward Air Controllers deeply integrated air power into Army maneuver units, and in turn, provided an invaluable understanding of the Army back to the Air Force.[12] Still, the Air Force never developed an aircraft focused on the close air support and Forward Air Controller mission. The OV-10 and the A-1 nearly fit the bill, but as Congressman Pike noted while he further skewered the Air Force's rebuttals in his floor remarks, "The article listed just four aircraft: the A-1E, the A-7, [the OV-10], and the F-4. Not one of those aircraft is a plane developed by the Air Force. Three were developed by the Navy, the fourth by the Marines."[13]

Attempting to put the issue to rest, and in light of ongoing combat in Vietnam, Air Force Chief of Staff Joseph McConnell and Army Chief of Staff Harold Johnson began meeting in private to iron out the issue. An agreement followed quickly on the heels of the Pike Report. The service chiefs signed the aptly named McConnell-Johnson agreement on April 6, 1966. In the agreement, the Army relinquished control of all its existing and future fixed-wing tactical airlift aircraft, handing the mission to the Air Force. In return, the Air Force agreed to the Army's use of helicopters for organic airlift and fire support. Further, the Air Force gave up helicopters for most tactical airlift missions, retaining their use for Special Operations and Combat Search and Rescue missions.[14] Although the agreement addressed the control and allocation of airlift capabilities, aircraft, and helicopters, it had far-reaching consequences in the close air support debate.

The agreement expanded the Army's opportunity for the use of helicopters on the battlefield in a fire support role. Throughout the process, the Army pursued a more ambitious attack aircraft, the AH-56 Cheyenne, and began utilizing a new term, Direct Aerial Fire Support. Both the Cheyenne and direct aerial fire support concept blurred the limits of the new agreement in favor of the Army. The AH-56 was a hybrid rotary and fixed-wing aircraft; meaning it used rotors for flight but also had short wings and a pusher propeller to increase its performance. Because it was, in name, a helicopter, the Army was not technically in violation of McConnell-Johnson. Direct aerial fire support, the Army attempted to argue, was different from close air support. Instead, it was more closely related to other forms of organic Army fire support, such as artillery. The Army argued that the McConnell-Johnson agreement implied the use of helicopters to accomplish the direct aerial fire support mission. The Air Force countered that direct aerial fire support was just close air support by another name, and the Air Force held responsibility for the mission.[15]

In the wake of the McConnell-Johnson agreement, critiques in Congress, and the Army's expansive view of the helicopter's role on the battlefield, General McConnell directed an Air Force study of the close air support issue. The RAND Corporation completed the study, publishing a report in August 1966. First, the Air Force examined if close air support provided to the Army fell short of needs and expectations. Second, if support was inadequate, then General McConnell ordered new systems or aircraft to be examined, tested, and fielded to address the shortfalls. The RAND study concluded that the Air Force was very good at providing planned or on-call close air support directed by Tactical Air Control Parties and Forward Air Controllers. Further, it was adequately meeting close air support needs in Vietnam, despite the haranguing the service endured from Congress suggesting the opposite. The study also found the Army was intentionally withholding some close air support requests from the Air Force, choosing instead to use artillery or helicopter fire support to bolster its arguments for armed helicopters and the direct aerial fire support concept. Most important, the Air Force lacked a platform with the ability to provide helicopter escort and suppressive fire support during air-mobile operations.[16]

Helicopters are particularly vulnerable to attack during landing or hovering to drop off or pick up soldiers. During air-mobile operations, suppressive firepower was used broadly in areas near a landing zone, rather than against specific targets, to protect the helicopters. The Air Force did not generally use suppressive fire in close air support tactics but instead attempted to attack specific targets. There were exceptions in Vietnam, such as the use of napalm, cluster munitions, or during B-52 Arc Light strikes. Arc Light attacks used B-52 strategic bombers from high altitude to saturate swaths of the jungle with hundreds of high explosive general-purpose bombs. The Air Force used Arc Light strikes for close air support extensively during the siege of Khe Sanh in 1968.[17]

Helicopter escort was a different matter. While the Air Force used the A-1 Sandy to provide helicopter escort and suppressive fire during combat search and rescue missions, it provided no such service to the Army during air-mobile operations. The Air Force's argument that the direct aerial fire support concept was close air support by another name further opened the

door for critics who claimed the Air Force was failing its statutory requirements to provide close air support. Further, the Air Force never specifically developed an aircraft that flew slow enough to follow helicopters to
and from the battlefield and still carried enough firepower to defend the
helicopters. The Army, therefore, bridged this firepower gap with Huey and
Cobra attack helicopters.[18]

The RAND report made two suggestions to help the Air Force overcome
the bad press and fill the gaps in mission capabilities. First, the service needed
to highlight close air support doctrine and tactics, focusing on the Air Force's
ability to accomplish the missions performed by Army helicopters. Second,
the Air Force needed to field a dedicated close air support aircraft also capable of conducting helicopter escort.[19] The study envisioned a simple, rugged,
and low-cost aircraft to be fielded after 1970 to replace the A-7D Corsair II as
the Air Force's primary attack aircraft. The decision to field the A-7 came in
1965 after a force structure review by the Air Force recognized a gap in attack
capabilities. However, the project quickly became embroiled in the close air
support mission criticism.[20]

As Congressman Pike noted, the Corsair was not originally an Air Force
project but was adopted from the Navy. Further, the aircraft was ill-suited
for the helicopter escort mission, and the project fell behind schedule and
ran over budget.[21] McConnell quickly implemented the RAND recommendations, and the Air Force began the development of a new aircraft project,
the A-X. The design requirements envisioned the A-X as a single-seat,
lightweight aircraft specializing in ground-attack, close air support, combat
search and rescue, and helicopter escort missions. The project directed the
use of available technology to speed development, procurement, testing, and
fielding of the aircraft. Using the A-1 Skyraider as its muse, the A-X was
designed to be rugged, to survive dangerous missions, and to carry a wide
array of weapons.[22]

In January 1973, after years of contract negotiation, haggling with the
Army and Congress, and a competitive prototype fly-off, the Air Force
selected the Fairchild A-10 Thunderbolt II as the winner of the A-X contract.
The A-10 is exceptionally rugged with redundant flight control systems and
titanium armor around the cockpit, making it capable of withstanding hits

from 23 mm antiaircraft guns. Its fuel-efficient engines allow it to loiter for hours over the battlefield, and it can also operate from forward bases with dirt runways. Its design allows it to fly more than four hundred miles per hour, but it is still slow and maneuverable enough for close air support missions. Finally, the A-10 brings an impressive amount of firepower to the battlefield.

The airplane can carry up to sixteen thousand pounds of weapons on eleven weapons stations and uses a massive 30 mm GAU-8 Gatling gun that occupies most of the internal space of the fuselage. The A-10 exquisitely merged the lessons the Air Force learned from the Slow FAC, Fast FAC, OV-10, and A-1 during Vietnam into one of the most capable ground-attack aircraft of all time. The Air Force even drew its name, Thunderbolt II, from another superb ground-attack aircraft, the World War II P-47 Thunderbolt, but the A-10 has earned another, more common, nickname—the Warthog or Hog because of its ugly, rugged, purpose-built appearance, and its ability to fly and maneuver at low altitude, or "root around in the dirt," to accomplish its mission.[23]

Ironically, the Air Force's decision on the A-10 did not quell the debate between the Army, the Air Force, and Congress, but in fact, it further complicated the issue. The A-10 program faced criticism from its inception, including from within the Air Force. In the aftermath of Vietnam, the service internally wrangled over the direction of doctrine, training, budgets, personnel, and aircraft. Internal service wars erupted over the bitter lessons of Vietnam, exacerbated by a service facing staffing and budget cuts but in desperate need of modernization after a decade of combat. Adding to the program's woe, Fairchild Aircraft, the company producing the A-10, began to run into production and financial difficulties, delaying the aircraft's fielding. Finally, the tightened purse strings of the post-Vietnam era brought scrutiny from Congress to the A-10 program.[24]

Throughout the development of the A-X requirements and initial stages of A-10 production, the Army continued to pursue the AH-56 Cheyenne. In January 1970, General William Westmoreland, Army Chief of Staff, wrote a letter to the Chief of Staff of the Air Force, which contained the results of an Army review of close air support requirements. In a reversal of course, the Army now desired multirole aircraft over the single mission A-X. In response, the House Appropriations Committee again reviewed the close air support requirements and capabilities of the services and called for a Department of Defense review in 1971. As part of the review, the Committee required Secretary of Defense Melvin Laird to make a recommendation on which aircraft, the AH-56 or the A-10, best fit the close air support role. Laird directed Deputy Secretary of Defense David Packard to head a board charged with answering the House's inquiry. In the process of the review, Packard's board rescinded all the memorandums published between the Pace-Finletter and McConnell-Johnson agreements, further muddying the water and removing constraints on the Army's attack helicopter and direct aerial fire support concepts. Performance issues and budget cuts led to the cancellation of the AH-56 program in August 1972, but just two weeks later, the Army began a second program called the Advanced Attack Helicopter. The second program produced the successful and capable AH-64 Apache. First fielded in 1982, the Apache remains the Army's frontline attack helicopter.

Obstacles to fielding the A-10 continued to appear in the Air Force's way. During the review of the 1974 budget, Congress called for a fly-off between the A-10 and the A-7, the same aircraft Congressman Pike criticized the Air Force for purchasing in the first place. The test, conducted in April and May 1974, proved the A-10 superior in every respect—more lethal, more survivable, and more cost-effective. Still, the Air Force faced sharp criticism from all quarters. Despite the A-10's victory, a crowd of politicians, news organizations, and the Piper Aircraft Company pressured the Air Force to purchase the cheaper Piper Enforcer instead of the A-10—even though the Enforcer was also inferior to the A-10 in combat capabilities.[25]

All the while, the Army continued to press forward with the Advanced Attack Helicopter program, and the similarities between the A-10 and the Army's proposed attack helicopter once again attracted the attention of the House Armed Services Committee. The Committee sought clarification on the intended roles and missions of the two aircraft, and in effect, clarification on the broader roles and missions debate. In response, General David Jones, Chief of Staff of the Air Force, and General Fred Weyand, Chief of Staff of the Army, signed a joint memorandum delivered to the Committee chairman

Melvin Price on April 7, 1976. Based on an agreement reached by the two generals in September 1975, and worth quoting at length, the letter finally settled the helicopter's place in the roles and missions debate. Most important, it established the modern understanding of close air support allocation and control:

> It is our view that the attack helicopter is organic to the Army ground maneuver unit and is integral to organic firepower. It is to be used with, or to the rear of, ground forces along the Forward Edge of the Battle Area (FEBA) to counter enemy armor lines, and to provide helicopter escort and suppressive fire. The Army attack helicopter is a mobile weapons system capable of providing organic fire support to local Army units. Because of the limited range, speed and firepower of the attack helicopter as compared to Air Force fixed wing close air support, we do not consider the attack helicopter as duplicating Air Force close air support.
>
> Air Force close air support capabilities, including the A-10 resources, are centrally controlled in order to be able to respond to theater-wide close air support requirements. Conversely, the attack helicopter elements are integral to each of the Army's combined arms teams, and therefore, are under the direct command-and-control of the ground command. The Army philosophy of decentralized control of attack helicopter resources is appropriate to Army operational functions as is centralized control of the Air Force close air support resources. In this regard, the principle of centralized control of Air Force resources at the highest control system level, with decentralized execution of air tasks to the lower echelon Tactical Air Control Systems elements provides the Air Force with the optimum control capability to conduct air operations theater-wide.
>
> The attack helicopter, as an integral part of the Army's organic firepower, and the Air Force close air support offer the ground commander a complementary capability in terms of a wider spectrum of fire support, enhanced responsiveness, flexibility and capability. It is our view that both the Army Advanced Attack Helicopter and the Air Force A-10 close air support aircraft are essential.[26]

After parrying multiple attempts to derail the A-10 program, the Air Force finally fielded the aircraft in 1977. Despite the two decades of struggle, each service came through the doctrine, roles, and missions debate with a favorable outcome. The Air Force retained centralized control over planning, allocation, training, employment, and acquisition of close air support and airlift resources. In the close air support mission, this allowed for a flexible system that harnessed any available aircraft to the mission. The service could pursue advanced multirole aircraft into the next decade and simultaneously use those aircraft to fulfill its statutory close air support requirements.

Further, the Air Force also got its dedicated close air support platform, the A-10, which harnessed the lessons of all its predecessors since World War II. Although considered a single mission aircraft designed only for ground-attack, the A-10's capabilities enabled it to perform the missions that required four separate aircraft in Vietnam: FAC, combat search and rescue, close air support, interdiction, and even strategic attack. The Army retained control over fire support and airlift nearest the front by exploiting the capabilities of the helicopter. Thus, the service could treat helicopters as it did tanks or artillery, organizing them and using them in the manner best suited for the Army's mission. The framework and agreements established in 1975 enabled development of operating concepts and doctrine, for both the Army and the Air Force, which harnessed the lessons of Vietnam, capitalized on the new understanding of service roles and missions, modernized the US military, and shifted how the Air Force used Forward Air Controllers.

The disagreement between the Army and the Air Force around close air support, ultimately, is a philosophical one. Their philosophies stem from their perceptions of the world, more specifically the depth of their battlefield horizon. There are physical restrictions created by the limits of human vision, geography, and environment placed on a soldier's view of a battlefield on the ground. A soldier standing on an elevated position on a clear day might be able to see a few dozen miles across flat terrain below. Similarly, on a good day with hard marching, clear weather, and easy terrain, an infantry soldier can cover maybe twenty miles in combat. While mechanized soldiers are capable of two or three times that distance, they are reliant on good roads. However, even the German Army's famed Blitzkrieg, revolutionary for its mechanized speed, covered only about two hundred miles attacking across the Low Countries between May 10 and May 21, 1940. The assault averaged less than seventeen miles per day during the twelve days. Technology improves a soldier's ability to see and travel greater

distances. For example, during Operation Desert Storm, the Allied invasion force advanced 150 miles in just five days of combat, nearly double
the pace of the Blitzkrieg. Still, these distances pale in comparison to how
airmen experience a battlefield.[27]

In contrast, a World War II fighter or bomber covered two hundred
miles in forty-five minutes or less, and a jet fighter in the Gulf War traveled the invasion route in less than twenty minutes. A soldier in a firefight
in Vietnam might engage an enemy soldier at five hundred meters, and a
sharp-shooting sniper two or three times that distance. In contrast, Forward
Air Controllers in Vietnam routinely hit targets at ranges greater than three
thousand meters with Willie Pete Rockets. In air-to-air combat, the horizons expand even further. One thousand meters is roughly the *minimum*
range of the Aim-9 Sidewinder air-to-air missiles used in Vietnam, and
the Aim-7 air-to-air missile used in Vietnam were capable of ranges out to
twenty miles. A combat pilot could fly in one mission an area larger than
an infantryman saw in a year in Vietnam. The pace and distance offered to
airmen in combat allows an Air Force Captain leading a flight of four fighters to destroy in minutes or hours an enemy unit an Army division might
not encounter and defeat for days or even weeks.

One is not better than the other; they are just different—so different
that the speed, distance, and time horizons of the two services are nearly
incompatible. The Air Force's ways and means of providing support to the
Army is where these two different philosophies come into conflict. With
a broader view of the battlefield, the Air Force sees interdiction of enemy
forces well beyond the front as equally useful to the Army as close air
support. However, enemy forces actively engaged against friendly forces
attracts the attention of Army commanders, even if those enemy forces pose
little threat. When faced with the choice of engaging a closer, if less serious, enemy force with air power or utilizing air power to destroy or delay a
more dangerous enemy force further away, the Army historically chose the
former. The critique of Air Force support in Vietnam serves as an excellent
example of the conflict.

However, in the aftermath of Vietnam, the AirLand Battle concept
brought the two philosophies into near harmony for the first time.

First appearing in Army doctrine in 1976, the AirLand Battle concept recognized the interdependence of the Army and the Air Force, embracing the ability of the Air Force to expand the battlefield and still support Army tactical and operational objectives. The 1976 version of the Army's *FM 100-5 Operations* doctrine document articulated it as follows: "Modern battles are fought and won by air and land forces working together. The interaction and cooperation between air and land forces extends into almost every function of combat."[28] The manual additionally states, "But neither the Army nor the Air Force can fulfill any one of these functions by itself."[29] Neither of these statements are particularly earth-shattering and, at first blush, reflect a boilerplate nod to interservice cooperation found in most US military doctrine. However, the next sentence is truly revolutionary, almost heretical, in the context of a service doctrine document: "Thus, *the Army cannot win the land battle without the Air Force.*"[30] With that statement, the creators of AirLand Battle set their sights at nothing short of upending the US military's way of warfighting.

AirLand Battle, like most military doctrine of the era, was partially a byproduct of Vietnam. The war's legacy left the Army facing aging equipment, budget cuts, and the end of the peacetime draft. In the aftermath of Vietnam, the service scrutinized everything about its existing and future force structure, training, personnel, tactics, and doctrine, projecting ten to twenty years into the future in an attempt to discern a way ahead. More influential on the Army's creation of AirLand Battle, however, was the October 1973 Yom Kippur War. The war featured modern state actors wielding cutting-edge American and Soviet technology and using contemporary tactics in a manner unseen since World War II.[31]

As Major Robert Doughty wrote in 1979 for the US Army Combat Studies Institute, "The unexpected level of violence in that war convinced many observers that future wars would be remarkably more violent and lethal than those of the past and that the successful outcome of the war would depend on the results of the first crucial and violent battles."[32] In light of the Yom Kippur revelation, the newly created US Army Training and Doctrine Command (TRADOC) critically evaluated the service's existing operating concepts and began anew. Led by General

William DePuy, TRADOC staff reimagined the Army's method of warfare, but it was no easy task.[33]

FM 100-5 spelled out the problem. A major conflict with Warsaw Pact forces in Central Europe, though unlikely, represented the most dangerous scenario for the United States. NATO Allies faced the threat of invasion, and such a conflict would undoubtedly escalate to a nuclear war, posing an existential threat to the US. Warsaw Pact armies outnumbered NATO armies and operated both massive infantry and armor forces shielded by a dense, mobile, and modern air-defense network. NATO land and air forces needed to cooperate to suppress or destroy the air-defense network before close air support missions began assisting the outnumbered NATO armies. The US Army could not expect to have airpower constantly overhead due to the air-defense threat. Instead, Forward Air Controllers and attack aircraft needed to hold back from the battle area, rushing in only to attack before moving back to safe airspace. Still, *FM 100-5* predicted close air support and forward air control missions would certainly face sustained attack by Warsaw Pact antiaircraft artillery, surface-to-air missiles, and fighter aircraft.[34]

For Air Force Forward Air Controllers, the task was daunting and dangerous, unlike any situation they faced since World War I. The problem required new tactics, techniques, and procedures. As the 1976 manual wrote, "The density and lethality of enemy air defenses force us to adopt new procedures to direct close air support. The airborne Forward Air Controller is still necessary but ... *It is the ground FAC* [TACP] *or Army forward observer who identifies the target.*"[35]

This statement represented the first step in the Air Force's overhaul of forward air control and close air support operating concepts. The next steps came as the Army, under new TRADOC commander General Donn Starry, revamped AirLand Battle. The Army refined the concept, shifting from a more defensive mindset in 1976 to a more offensive mindset harnessing the air mobile cornerstones of maneuver and firepower.[36] Revised versions of AirLand Battle produced new *FM 100-5*s in 1983 and again in 1986. Each successive version pressed the Army's operating horizon further, aligning it closer to that of the Air Force.[37]

The staff at the Air Force's Tactical Air Command, working in conjunction with TRADOC, reimagined close air support and interdiction missions, and the Forward Air Controller's role in them in light of AirLand Battle and a potential massive NATO-Warsaw Pact showdown in Europe. An article in the July/August 1977 issue of Air University Review by Air Force Lieutenant Colonel Robert Dotson first articulated this reimagined concept, calling it Battlefield Interdiction. Soon after Dotson's article, the term morphed to Battlefield Air Interdiction in the 1979 Air Force doctrine document *AFMAN 1-1*. Dotson imagined battlefield air interdiction as a hybrid between traditional close air support and interdiction missions.[38]

Historically—such as during World War II's Operation Strangle, the Air Pressure campaign in Korea, or Commando Hunt in Vietnam—the Air Force's interdiction missions were conducted without regard to specific Army objectives and well beyond the Army's area of immediate concern.[39] At times the operations reached hundreds of miles beyond the front and more closely resembled strategic bombardment. On the plains of Central Europe, however, the weight of Warsaw Pact forces was concentrated much closer to NATO forces, requiring a different philosophy for interdiction. Critical in this dilemma was the Bomb Line, renamed the Fire Support Coordination Line, or FSCL, in 1965. The difference between the Bomb Line and the FSCL was more profound than the name. The Bomb Line traditionally represented a hard boundary between air and land forces. The Air Force owned the space beyond the Bomb Line, and the Army owned everything behind it. However, Army commanders needed a means of influencing the battlefield beyond the Bomb Line, and Air Force commanders needed faster coordination measures between the Bomb Line and areas where close air support missions flew.[40] Dotson articulated the battlefield air interdiction concept as follows: "A shift in emphasis from the familiar close-air support (CAS) mission—where aerial ordnance is delivered near friendly ground forces—to ground attack in support of friendly ground forces beyond the effective range of weapons organic to those ground forces (the so-called 'battlefield' interdiction mission)." Dotson did not intend to abandon the close air support mission, but instead wanted to push the weight of air attacks to the enemy's follow-on forces.[41]

NATO Air Forces seized Dotson's concept immediately, but the US Air Force was more reluctant to follow suit. Fearing too many levels of bureaucracy to coordinate, the US Air Force pressed for a less restrictive definition of battlefield air interdiction during NATO and US Air Force staff working groups. A compromise emerged, the Joint Targeting Process, which offered the most freedom to the Air Force and, simultaneously, the most influence for the Army. Under the new targeting construct, battlefield air interdiction missions remained under the control of the air component, but the new process enabled land commanders to nominate targets on both sides of the Fire Support Coordination Line. Similarly, the process enabled battlefield air interdiction missions to cross the Fire Support Coordination Line with minimal coordination with land forces.[42]

Thanks to the efforts of TAC and TRADOC, deeper Army-Air Force cooperation became a positive outgrowth of the AirLand Battle doctrine, battlefield air interdiction concept, and the threat of war in Europe. Beginning in 1976, the Army and the Air Force agreed to share new acquisition requirements. From this agreement, and critical to the development and improvement of AirLand Battle, battlefield air interdiction, and other projects, the service chiefs signed a joint memorandum titled "Joint USA/USAF Efforts for Enhancement of Joint Employment of the AirLand Battle Doctrine" on April 21, 1983. Thirteen months later, a second agreement spelled out 31 Initiatives the services needed to improve interoperability and combat effectiveness. The 31 Initiatives served as the critical foundation for bringing a host of joint programs into being. However, the central driving force of the Initiatives was always AirLand Battle.[43]

Putting the AirLand Battle and battlefield air interdiction concepts into practice hinged on processes and capabilities already in use in the Army and the Air Force, as well as harnessing new concepts and new technologies developed through the 31 Initiatives. First, TRADOC's creators and developers of AirLand Battle embraced the TACS-AAGS system created and improved during Vietnam. Both the planning and execution functions of the system readily adapted to the battlefield in Europe and the AirLand Battle and battlefield air interdiction concepts. Command-and-control, intelligence, communication, and targeting capabilities developed under the programs of the 31 Initiatives further enhanced these systems.[44] Second, the Air Force's Tactical Air Support Squadron structure created in Vietnam, as the 1976 version of *FM 100-5* noted, provided the Tactical Air Control Party with the critical interface between frontline Army and Air Force combat units.[45] Finally, the battlefield air interdiction concept, as envisioned by Dotson, relied upon the injection of the A-10 and precision-guided weapons onto the battlefield. Dotson envisioned utilizing the procedures and weapons in a carefully conducted symphony to maximize air power's effects at the most critical points of the battlefield. In his conclusion he wrote that "we must conduct battlefield interdiction and close air support skillfully, so that we conserve our forces while exploiting the great firepower potential only tactical air power can bring to the battlefield."[46]

Implementing the AirLand Battle and battlefield air interdiction concepts had far-reaching consequences for the Air Force's use of Forward Air Controllers. Critical to this shift was the threat umbrella faced by Allied air forces in a NATO-Warsaw Pact war in Europe. Just as the threats in North Vietnam led to the creation of the Fast FACs, the more advanced surface-to-air threats in Europe and the additional threat of Warsaw Pact fighters prevented the use of Slow Forward Air Controllers during the critical early phase of the battle. Instead, Slow FACs filled Tactical Air Control Party and Air Liaison positions in US divisions stationed in Germany.

Tactical Air Control Parties served as the Air Force's ground arm since the time of the Rover Joes of World War II, but the injection of Forward Air Controllers increasingly shifted control of close air support and interdiction airstrikes to the air. Vietnam was the pinnacle of Forward Air Controller responsibilities in combat, as all airstrikes during the war required their control. The advanced threats found on the battlefield of Europe did not offer US forces the same luxury. The combined surface-to-air and air-to-air threat was beyond anything Air Force Forward Air Controllers encountered stretching back to World War I, where Cavalry Reconnaissance and Observation patrols faced German antiaircraft guns and Fokker fighters simultaneously. Recognizing the potential danger, both the Army and the Air Force shifted back towards the Tactical Air Control Party as the primary controller of close air support attacks.

Although the Air Force shuttered Tactical Air Support Squadrons at the end of the Vietnam War, they resurrected with the development of AirLand Battle. To meet the demands of a potential war in Europe, the Air Force reestablished the 20th TASS with OV-10s at Sembach Air Base, Germany, in June 1974. Two years later, the 704th TASS, also equipped with OV-10s, joined the 20th at Sembach. In becoming Forward Air Controllers, Air Liaison Officers, and OV-10 pilots during the height of the Cold War, David Keyes and Mike Peters followed similar paths to Sembach. After completing a year at pilot training, both joined A-7 squadrons, where they gained about three to four hundred hours of flight experience—a process that took a bit less than two years. Following their A-7 tour, they attended the Air Ground Operations School at Hurlburt and OV-10 training at MacDill Air Force Base, before joining the 20th at Sembach. Keys was assigned to the squadron from 1976 to 1979 and just overlapped with Peters, who flew there from 1979 to 1981.[47]

Upon arriving, Forward Air Controllers in Germany, like those in Vietnam, joined a Tactical Air Control Party team assigned to a US Army division stationed in Germany. Unlike Vietnam, however, FACs in German Tactical Air Support Squadrons did not expect to fly during combat because the OV-10s would not survive an all-out war in Europe. If they saw aerial combat at all, it would only be in a desperate losing situation as Allied armies were driven back in Western Europe or, in a rear-guard role, securing a victory as NATO forces pressed deep into Eastern Europe. Instead, their primary responsibility was learning, in detail, their assigned division's area of operations and scheme of maneuver should war erupt in Europe. OV-10 Forward Air Controllers expected to serve as part of Tactical Air Control Parties directing airstrikes along the East-West German border as enemy armor and infantry thrusts poured in. Assigned to the 3rd Squadron of the Eleventh Armored Cavalry Regiment, Peters's Air Liaison responsibilities placed him in the heart of the famed Fulda Gap, the anticipated route of a Warsaw Pact invasion through Central Germany. Both Keyes and Peters realized the Tactical Air Control Party mission in Western Europe was both critical and dangerous, and without air power NATO land forces would be quickly overwhelmed.[48]

In the AirLand Battle concept, Tactical Air Control Parties became the primary controllers up to the Fire Support Coordination Line. Air power used inside the Fire Support Coordination Line was close air support. Though the Fire Support Coordination Line had no prescribed distance in placement from friendly forces, the Army preferred the line to be as far out as practical, to give ground commanders maximum influence. Air commanders, in contrast, preferred it to be as close as possible to friendly forces, giving the Air Force maximum flexibility. Ultimately, placement and movement of the Fire Support Coordination Line, as it moved with the shifting of the front, was based on an agreement between the most senior air and land commanders.

In order to contain, coordinate, and synchronize air efforts beyond the Fire Support Coordination Line with ground efforts up to it, *FM 100-5* introduced the concepts of close and deep battle. Essentially, the Army separated the battlefield into "who we are fighting now" and "who we will fight in the future." In theory, the boundary between close and deep operations coincided with the Fire Support Coordination Line. In reality, the battlefield was a much too complex and fluid environment for such a neat division. Forward Air Controllers seemed to be perfect for optimizing interdiction in the deep battle and providing close air support for Army units currently engaged in combat by bridging the potential gap at the Fire Support Coordination Line. However, Dotson eschewed the use of Forward Air Controllers. He feared the necessary coordination between the aircraft slowed the process and was vulnerable to Soviet electronic warfare capabilities.[49] Further, newer versions of *FM 100-5,* produced in 1983 and 1986, never mention the role of Forward Air Controllers. In short, the anticipated pace and depth of the battlefield left FACs out of the picture altogether as AirLand Battle developed between 1976 and 1986.

All the theoretical and doctrinal developments associated with AirLand Battle, along with the expanding dangers on the battlefield, blurred the close air support and interdiction missions together in both the Air Force and the Army. This merger produced a desire in both services for faster and more advanced multirole fighters for the ground support mission. The vision of a major war in Europe reopened the debate over the Army's air power support requirements, which the creation of the A-10, AirLand Battle, and the 31 Initiatives closed just a few years before. The role of Air Force Forward Air Controllers again became caught in the fray.[50]

The dangers of the European battlefield negated a Forward Air Controller's ability to remain over friendly forces, creating a reliance on Tactical Air Control Parties to direct close air support. As a result, the European Tactical Air Support Squadrons all closed again by 1985, and their aircraft were shipped back to the US mainland. Forward Air Controllers disappeared from Europe. The structure of Air Force close air support capabilities in Europe reverted to a pre-Korean War construct. The Air Force consolidated radio operators and Air Liaison Officers into Tactical Air Control Squadrons, Groups, and Wings. Despite their demise in Europe, however, the TASS structure remained in the US, Latin America, and Pacific commands until the end of the 1980s.[51]

The resulting force informally developed two battlefield operating concepts for the Air Force's conventional capabilities. The first centered in Europe, defined by advanced threats and focused on battlefield air interdiction and deep battle. This war featured America's most advanced frontline fighter aircraft, such as the F-15 and F-16, engaging equally advanced Soviet MiGs and surface-to-air missiles to create an umbrella for close air support and battlefield air interdiction against the massive, combined Warsaw Pact armies. The second focused on less capable adversaries, such as North Korea, and low-intensity conflicts supporting counterinsurgencies. Both the Army and the Air Force envisioned the continued use of Forward Air Controllers and the A-10 in these lower threat conflicts.[52] However, as the 1980s progressed, the number of Forward Air Controller aircraft in the Air Force inventory steadily dwindled.

By 1985 the Air Force all but eliminated the O-2 fleet, reduced the A-37 fleet to one squadron operating in Panama, and cut the OV-10 fleet in half.[53] Complicating the problem for Air Force Forward Air Controllers, the proliferation of advanced surface-to-air missiles, portable shoulder-fired missiles known as Man Portable Air Defense Systems (MANPADS), and advanced mobile antiaircraft-artillery, all made slower aircraft obsolete in combat except in the most permissive counterinsurgency scenarios.

Even though the threat systems in North Korea were far less sophisticated than in Europe, they were still more capable and dense than those used by the North Vietnamese.[54]

More advanced threat capabilities and the scope of battlefields envisioned in AirLand Battle, particularly in Europe, prompted both the Army and the Air Force to question the survivability and practicality of the A-10. Both services argued the aircraft's slow speed made it less responsive to the large and fast-paced ideas of AirLand Battle. Further, lack of advanced long-range air-to-air capabilities made it vulnerable to the air threats in Europe. The argument limited the A-10 solely to close air support against Warsaw Pact armor divisions, a mission for which it was uniquely suited. The A-10's 30 mm gun and Maverick air-to-ground missile were superior antitank weapons. European based A-10 squadrons trained relentlessly on rapidly deploying to forward operating bases to conduct close air support against enemy armor. The aircraft's radio suite, ability to maneuver at slower speeds, and ability to operate at low altitudes also allowed it to pair easily with Army helicopters to form Joint Air Attack Teams.[55] However, as the AirLand battle concept evolved, both the Army and the Air Force placed priority on battlefield air interdiction over close air support.[56]

Outside of Europe, however, the story was different. By the 1980s the A-10 was the only aircraft in the Air Force inventory performing the combat search and rescue mission. It also began slowly assuming the forward air control role. Since the Air Force did not keep fighters in the Fast FAC role after Vietnam, and steadily reduced the increasingly obsolete Slow FAC aircraft fleet, the A-10 became the only feasible Forward Air Controller platform on a contested battlefield. Ironically, the A-10 was not originally designed for the mission. Only after 1987 did the Air Force modify the aircraft, creating a variant known as the OA-10 specifically for the Forward Air Controller mission. As the A-10 program report says, "The alterations necessary were relatively minor and inexpensive to implement. These were mostly internal although there was modification in the pylon loading to allow phosphorus marker rockets to replace the Mavericks and bombs."[57] These modifications were additive to the aircraft's ongoing avionics, weapons delivery systems, night capabilities, and display improvements made during the 1980s.[58]

The introduction of the FAC mission into the A-10 community also unofficially split it into two camps: the European A-10 pilots and everyone else. With the advanced threats and focus on battlefield air interdiction, European A-10 pilots trained almost exclusively in close air support, and the A-10 squadrons in Europe did not qualify their pilots as Forward Air Controllers. In contrast, A-10 squadrons everywhere else maintained a robust number of FAC qualified pilots on their rosters, particularly in Korea, and the Air Force even designated some A-10 squadrons in Korea specifically for the mission. US military planners in Korea expected a modern war on the peninsula to mirror the 1950s conflict. They anticipated the North Koreans to launch an infantry-centric campaign supported by massive quantities of artillery and less technologically advanced armor and air defenses. The relatively small number of American fighter aircraft stationed at Korean bases were vulnerable to North Korean artillery. Therefore, much of the American air power used in a new Korean War would fly from Japan or further, just as in 1950. The terrain on the peninsula still loomed large, though, and for the same reasons the Mosquitoes thrived, it solidified the need for Air Force FACs in any potential Korean conflict.[59]

In 1984 the Air Force began contemplating the A-10's replacement, and by 1985 it started work with defense aircraft manufacturers to assess different options. Based on the threat of war in Europe, the Air Force proposed in 1986 to replace the A-10 with an attack version of the F-16. The study and proposal met resistance in both the House and Senate Armed Services Committees, who forced yet another round of studies and contemplated another close air support competition. The Secretary of Defense directed the Air Force to study the issue, and then created two working groups to comply with the order.[60] Additionally, the General Accounting Office initiated a separate study that "reviewed the Army's CAS requirements and the extent to which these mission requirements have been defined." The General Accounting Office "also reviewed the Air Force's and OSD's plans and processes to identify alternatives to the A-10."[61]

The report found the Air Force desire to replace the A-10 two years ahead of schedule fell in line with the Army's operating concepts spelled out in AirLand Battle. However, neither Congress nor the Office of the

Secretary of Defense agreed that a need to replace the A-10 existed. Further, no one involved in the process agreed on the project's projected budget.[62] In early 1989 the Secretary of Defense scheduled a Defense Acquisition Board intended to resolve the conflict. Initially slated for April 1990, it did not meet until the fall. The board settled on a plan that retained two wings of A-10s, a dramatic cut to the A-10 force leaving approximately one-quarter of the original 713 aircraft. To fill the gap, the board proposed modifications to four wings of F-16s for the close air support and battle-field air interdiction missions. This compromise plan upset the Air Force's vision for future battlefield air interdiction aircraft and its Forward Air Controller fleet as well. The dramatic cut in the overall number of A-10s affected the service's ability to replace the OV-10 and A-37.[63] However, the whole process was quickly overtaken by world events and altered the path of Army-Air Force cooperation, the A-10 replacement plan, and the future of Air Force Forward Air Controllers.

Chapter 6

Desert Storm
to 9-10-2001:
FACs Fractured

n November 1989, the Berlin Wall fell. Eight months later, the Warsaw Pact formally ended, and in December 1991, the Soviet Union ceased to exist. These events justified a complete review of the entire Department of Defense structures and operating concepts. The breakup of the Warsaw Pact alone seemed to remove the *raison d'etre* for AirLand Battle and battlefield air interdiction. However, sandwiched amid the collapse of European communism, on August 2, 1990, Saddam Hussein invaded Kuwait. The Iraqi invasion of Kuwait was not a complete surprise. In the months leading to the invasion, Saddam made clear his desires to gain control of the islands Warba and Bubiyan in the mouth of the Euphrates River, favorably adjust the Iraq-Kuwait border, and receive relief from a $40 billion debt to Kuwait which financed his war with Iran.[1] The US response to Iraq's invasion displayed America's emerging role as the lone superpower and the results of a decade of US military reform.

A year before the invasion, attempting to head off a potential conflict in the strategically important region, President George H. W. Bush's national security advisors contemplated a variety of military, economic, and political efforts to engage Iraq to "moderate its behavior and increase our influence."[2]

Similarly, General Norman Schwarzkopf's Central Command eyed the Iraqi dictator with suspicion. Central Command began internal exercises and revamped regional contingency plans in the spring of 1990, addressing the possibility of an Iraqi invasion of its neighbors. For the next few months, Central Command constructed the framework for the defense of the Arabian Peninsula, OPLAN 1002-90, completing the work in July 1990. In the days before the invasion, General Schwarzkopf and his staff presented President Bush, Secretary of Defense Dick Cheney, and the Joint Chiefs of Staff with possible American responses to an invasion.[3]

Across the Potomac in the Pentagon, Colonel John Warden's Directorate of Warfighting Concepts, known as Checkmate, was creating an air power response to the crisis as well. On a previous assignment in the mid-1970s, Warden served in the Pentagon as part of the Middle East Plans Division and was intimately familiar with the capabilities and vulnerabilities of Iraq.[4] When Iraq invaded Kuwait, Warden directed his staff to formulate a plan based on the ideas he articulated two years earlier in his book, *The Air Campaign*. His intent radically diverged from the defensive-minded framework of OPLAN 1002-90 and its AirLand Battle roots. Although AirLand Battle relied on the initiative and aggressive attacks by rapidly moving NATO forces, it was a strategically defensive ideology designed against a Soviet invasion of Western Europe. Battlefield air interdiction, as AirLand Battle's air power extension, was similarly designed to support the defense of NATO by halting the flow of Warsaw Pact armor and infantry to the battlefield. Neither idea inherently expanded into an offensive inside enemy's territory. The air component of OPLAN 1002-90 similarly harnessed air power as a defensive mechanism, primarily supporting the ground forces' defense of Saudi Arabia with close air support and battlefield air interdiction.[5]

As events unfolded during the week following the Iraqi invasion, General Schwarzkopf asked for assistance from the Air Force staff at the Pentagon. The President wanted the ability to conduct punitive and retaliatory strikes against Iraqi political and military targets. He needed options to coerce Saddam Hussein to withdraw from Kuwait, not just defend against further incursions. Warden and Checkmate anticipated Bush's request, drafting a plan that harkened back to the Allied Combined Bomber Offensive of World War II: an offensive air campaign designed to cripple Iraq's ability to fight. Harnessing the ideas Warden articulated in his book, *The Air Campaign*, they built an offensive-minded campaign designed to achieve air superiority while attacking enemy centers of gravity. On August 10, 1990, Warden briefed the plan, called Instant Thunder, to General Schwarzkopf.[6]

Instant Thunder went through several revisions during August, and the Air Staff at the Pentagon generally embraced it as the best course of action. Air Force General Chuck Horner, however, was among the critics of Instant Thunder. Horner was the Central Command Air Component Commander, and he ultimately held responsibility for executing the air war against Iraq, so undoubtedly his vote mattered. Chief among his complaints against Instant Thunder was its neglect of the Iraqi Army. As the Gulf War Air Power Survey recalled, "The Instant Thunder plan that General Horner received on 20 August used new technologies to refurbish ideas about strategic bombing that could be traced at least to the Army Air Forces in World War II."[7] Further, Warden's personality did not lend itself to compromises on his ideas, and he felt justified in ignoring the Iraqi Army.[8] In his mind, "There would be no need to divert other aircraft from the strategic air campaign. Whatever might happen, Warden insisted that the strategic air campaign should go forward, for it was the strategic air campaign that would win the war."[9] Instead of adopting Warden's plan and ideology as written, Horner sent Warden back to the Pentagon and tapped Brigadier General Buster Glosson, Lt. Col. David Deptula, and others from the Checkmate staff. The new team formed a planning entity enigmatically dubbed The Black Hole and created a hybridized plan that phased elements of Instant Thunder's offensive campaign into an AirLand Battle-based interdiction campaign.[10]

On paper, the Iraqi Army and Air Force were a formidable threat. The Iraqi Army boasted one million men on its rosters and employed modern Soviet tanks, armored personnel carriers, and artillery. They enjoyed a three-to-two advantage in personnel and equipment over the assembled Coalition. There were nearly as many Iraqis in Kuwait alone as US Army and Marine Corps combat troops in the entire theater. Finally, the Iraqi Army had recent combat experience garnered from its eight-year war against Iran. In contrast, the US Army had not fought in a major engagement since Vietnam.[11] Similarly, the Iraqi Air Force theoretically matched up well against the US Air Force. It operated a slightly larger and equally modern Soviet fighter, attack, and multirole force, though the addition of US Navy, Marine, and Coalition aircraft created a two-to-one numerical Coalition advantage.[12] However, Iraq operated the French-designed KARI air defense system, one of the densest and most sophisticated surface-to-air defense networks on the globe. The defenses included 3,600 surface-to-air missiles and 8,500 antiaircraft guns arrayed in mutually supporting batteries. A layered network of state-of-the-art radar and command centers controlled the entire system. The Iraqi air defense network was one of the air planners' primary concerns.[13]

While the paper statistics were worrisome, and occasionally the Iraqis proved to be both capable and dangerous, the reality of Desert Storm turned out to be quite different than Saddam Hussein warned about and American leaders feared.[14] After the six-month build-up and anticipation of Operation Desert Shield, Schwarzkopf, Horner, and the Central Command planners designed Operation Desert Storm as a four-phase operation. The first two phases solely harnessed air power, resembling the original construct of Instant Thunder. Coalition Air Forces would first gain air superiority of the Iraqi and Kuwaiti airspace. The plan then called for a systematic destruction of the Iraqi Air Force, the dismantling of the KARI air defense network, and cutting the Iraqi leadership in Baghdad off from their fielded forces. In reality, while the Iraqi Air Force and air defenses were decimated, Warden's dream of cutting off Iraqi leadership and winning the war with air power alone never materialized.

In phase three, Coalition air power would turn on the Iraqi Army with an interdiction campaign against everything in the Kuwait Theater of Operations, the area in southern Iraq along the Kuwaiti and Saudi border and all of Kuwait. Saddam's Republican Guard units, in particular, were to receive relentless Coalition bombardment. The objective was the reduction of Saddam's combat power by 50 percent. Following phase three, Coalition ground forces would launch into Iraq and Kuwait, in a wide, rapidly moving arc towards the Euphrates River pivoting around a US Marine assault directly into Kuwait.

While the first three phases showcased Air Force technologies and tactics developed since Vietnam, phase four showcased the same for the Army. In total, Desert Storm lasted little more than a month. Phases one through three were executed from January 16, 1991 onward, until phase four began February 23. The ground campaign lasted a mere one hundred hours, halting on the morning of February 28.[15]

The true success of the Desert Storm air campaign was not the strategic bombing idea articulated by Warden, but in the sustained air assault against Iraqi ground forces. The portion of an air campaign deemed to be least important by Warden ironically became the central focus of Desert Storm. Both ground and air staffs, from the highest levels, put a remarkable amount of faith in air power to affect the ground war. Although the goal of 50 percent attrition of the Iraqi Army was an idea Schwarzkopf seemingly conjured from thin air, Horner and his planners seriously worked to achieve it. Amazed that both the Army and the Air Force bought into the idea, the Gulf War Air Power Survey observed that "senior army commanders, including Schwarzkopf, assigned air power the mission of taking the Iraqi military forces down by half; what is perhaps even more surprising was the willingness of air commanders to accept this charge."[16]

The shift in mindset and level of cooperation between the Army and the Air Force expressed in the planning for Desert Storm deeply reflected the dynamic developed in the AirLand Battle doctrine during the 1980s. In fact, the shift was so significant that Iraqi Republican Guard formations, imagined by Saddam as the ultimate backstop to the main Iraqi Army arrayed along the Kuwaiti-Iraqi-Saudi border, were classified as strategic targets instead of tactical ground force targets. Both air and land planners placed the Republican Guard squarely in the crosshairs of the deep battle assault. Unfortunately, planners' prewar estimates on air power's ability to degrade or destroy the Iraqi forces in the Kuwait Theater of Operations proved wildly optimistic. Coalition air power was never able to achieve the reduction of the Iraqi Army by 50 percent.[17]

Nevertheless, air power provided several critical functions for the ground forces. First, during Phase I and II of the war, Coalition air forces so dominated the battlefield that the Iraqi Air Force was effectively grounded after the first week of the war. To prevent further destruction of his air force, on January 26 Saddam ordered the remaining aircraft buried in the Iraqi desert or flown to Iran. The few Iraqi Air Force attempts to attack Coalition ground forces rapidly met their demise. Similarly, the Coalition's systematic targeting of key infrastructure and command networks created delays and uncertainty within the Iraqi high command—though the leadership attacks never delivered on Warden's promises and they certainly never completely disabled the Iraqi's command-and-control architecture. However, the air superiority and disruption of Iraqi communications networks did give Coalition ground forces unprecedented freedom of movement and security from attack.[18]

This freedom and security critically allowed the movement of the French Sixth Cavalry, US Eighty-Second Airborne, and US Twenty Fourth Infantry Divisions to the far western end of the Coalition line. The Coalition ground forces needed two weeks to shift adequate forces to the western desert for the wide-sweeping advance on the far left of the Coalition front. The western assault needed personnel and hundreds of thousands of tons of supplies and equipment critical to the attack to be repositioned by Coalition airlift. The dominance provided by air power completely screened the move west, and when the assault began in late February, the attack into the right flank of Iraqi forces so stunned Iraqi commanders at every level that their resistance collapsed.[19]

The second critical function air power provided to the Coalition ground force was the attrition of the Iraqi Army. Although planners initially overestimated the ability of Coalition air forces to destroy Iraqi ground forces, the destruction visited on the Iraqi Army was significant. The specific level of destruction, and how to measure air power's effects, sparked a debate which raged from Washington, D.C., to the frontline commanders in the Saudi desert. Politicians and military leaders worried that intact Iraqi forces would inflict heavy casualties on the Coalition assault. Interestingly, General Schwarzkopf seemed less concerned over the exact level of attrition, recognizing instead the cumulative demoralizing effect on the enemy under the sustained fury of air attacks.[20]

As the war progressed, the total number of missions allocated to the battlefield air interdiction campaign increased exponentially, from just over nine hundred sorties in the first week to more than four thousand by the fifth week. As the Gulf War Air Power Survey recounted, "During the last two weeks [of the war], Coalition air forces sent more than 90 percent of their strike sorties against the Iraqi army."[21] This included diverting F-111F Aardvark and F-15E Strike Eagle missions from their normal strategic attack and scud hunting missions to "plink tanks" with their Low Altitude Navigation and Targeting Infrared for Night (LANTIRN) targeting pods and laser-guided bombs. While the capability of the Iraqi Army prior to the war is debatable, the destruction of men, equipment, and morale by air power rendered it utterly incapable of resisting the ground assault.[22]

Finally, when Coalition ground forces began their attack, air power continued the sustained battlefield air interdiction effort but also added close air support missions. The air assault suppressed Iraqi firepower and thwarted the Iraqi Army's ability to effectively maneuver—either defensively or offensively—to counter the rapidly advancing Coalition line. Unsurprisingly, the means of close air support provided by the US Air Force to the US Army reflected the influence of AirLand Battle doctrine and the shifting force structure of the Air Force. As the Gulf War Air Power Survey reported, "Seventh Corps used its air power resources in accordance with the army's 'air-land battle' doctrine—as a tool to fight the deep battle." Despite the success of the plan, the absence of determined Iraqi resistance never allowed the concept to be fully tested during Desert Storm.[23]

To meet the Army's demand for aerial firepower, air planners, perhaps unknowingly, borrowed tactics from the World War II Horsefly experiments in Italy. Echoing the 324th Fighter Group's use of P-47s at one-hour intervals to support the 1st Armored division's breakout south of the Anzio beachhead, the Central Command's Air Operation's Center fed fighters to the front at seven-minute intervals. They called the tactic "Push CAS." Flights of Coalition fighters moved over the front and checked in with Tactical Air Control Parties or Forward Air Controllers. If the ground forces did not need support, the aircraft moved to planned or on-call interdiction targets. In this way, Coalition armies received a constant supply of close air support to keep the assault moving and minimize casualties, while also applying pressure to echelons of the Iraqi Army in the deep battle.[24]

From Saddam's invasion of Kuwait until well into the war, Warden insisted the Instant Thunder campaign could win the war alone.[25] This claim is, of course, unsurprising and in line with a long line of air power theorists and their beliefs about air power. Critically, despite Warden's insistence to the contrary before, during, and after the war, the strategic air campaign envisioned in Instant Thunder did not win the war.[26] Air superiority and strategic attack missions constituted a fraction of the missions in Desert Storm. Air-to-ground missions in support of Coalition land forces, either as Forward Air Controller, battlefield air interdiction, or close air support, were the unequivocal master of the air power effort during Desert Storm. Of the 112,235 sorties flown, 44,789, or nearly 40 percent, were directed against Iraqi combat power, the most of any single mission category in the war. Air refueling missions were a distant second at 18,455.[27] Assessing the close air support and battlefield air interdiction effort against other air-to-ground missions, 67 percent of all air-to-ground missions engaged nearly twenty-seven thousand Iraqi Army targets during the war. For comparison, Iraqi airfields, the second-most-attacked target category, endured only 3,400 attacks by Coalition air forces. Iraqi leadership targets were subjected to just over three hundred attacks, the least among all air-to-ground target categories.[28]

When compared to previous wars, this is an increase in both number and percentage weight of close air support and interdiction. It is particularly staggering when considering Coalition air forces used a much higher number of air refueling, airlift, and special operations missions than ever before. Evaluating the effort in the broader historical context, during World War II and Korea about 30 percent of all sorties were close air support and interdiction. Only Vietnam has a higher percentage of close air support and interdiction strikes, totaling 79 percent of all missions. However, this came in a war where all air-to-ground missions in South Vietnam, Laos, and Cambodia theoretically supported ground forces, and the land campaign lasted for more than a decade.[29]

Despite the resounding success of air support to Coalition armies during Desert Storm, there were struggles directly related to a decade of decisions made by the Air Force as part of its modernization, battlefield air interdiction concept implementation, and reduction of the Forward Air Controller force. Chief among the problems were rapidly shifting fronts, difficulty locating and identifying targets, lack of adequate communication between ground forces and aircraft providing air support, and conflict in the placement of the Fire Support Coordination Line. Many of the issues echoed those of World War II and Korea that originally necessitated the development of Forward Air Controller capabilities.[30]

Forward Air Controllers should have been able to solve these problems deftly, but this did not happen for several reasons. First, the KARI air defense system and proliferation of advanced shoulder-fired surface-to-air missiles prohibited the use of more vulnerable aircraft like the OV-10. The Air Force never attempted to use its OV-10s, although the US Marines did use the venerable Bronco, losing two aircraft to shoulder-fired missiles during the war. Two of the Marine Bronco aircrew were killed, and the other two captured, proving yet again the dangers of the mission.[31]

The KARI system also exposed the vulnerabilities of the A-10 on a battle-field defended by advanced surface-to-air missiles. To mitigate the threat, A-10s operated above ten thousand feet, but this significantly diminished their bombing accuracy and prevented the use of their 30 mm cannon and Maverick missiles against Iraqi armor. Further, they lacked laser guidance pods to use precision weapons, and A-10 pilots did not yet use night vision goggles during Desert Storm. When A-10s did begin operating at lower altitudes, they suffered withering defensive fire from the Iraqis. Rightly, the Iraqi Army judged a persistent A-10 orbit over their position as a harbinger of an impending attack. After suffering several A-10s damaged and two shot down on one day, General Horner limited the Hogs to operations within twenty miles of the Coalition front. During the final two weeks of the war, the A-10 flew almost exclusively in its close air support role, just as NATO planners envisioned for a major European war.[32]

Limited A-10 operations, coupled with a neglected Fast FAC program, hampered the ability of Coalition air forces to find targets and coordinate on portions of the battlefield straddling the Fire Support Coordination Line. The dearth of easily distinguishable terrain features in the trackless deserts of southern Iraq and northwestern Kuwait compounded the struggle. To solve this second issue facing the close air support and battlefield air interdiction campaign in Desert Storm, air planners used a previously created grid system over the Kuwait Theater of Operations called Kill Boxes. Consisting of thirty by thirty-mile squares with an alphanumeric name for reference, the boxes were further divided into four quadrants. The system served simultaneously as a planning, airspace deconfliction, and attack allocation tool. To that end, attack flights or Forward Air Controllers might patrol a whole Kill Box, a subdivision of one box, or a series of boxes together. Still, the Kill Boxes comprised a large search area; even a single quadrant was 225 square miles. With their ability to loiter over the battlefield, A-10s tasked into Kill Boxes served as Forward Air Controllers for interdiction missions similar to the Misty Fast FACs of Vietnam.[33]

Once commanders withdrew the A-10s from deep operations, however, the Coalition needed a substitute to help attacking aircraft quickly locate targets. To solve the problem, the planners at the Air Operations Center asked if the F-16 equipped 388th Tactical Fighter Wing had pilots with previous FAC experience. The 388th's 4th Fighter Squadron possessed an abundance of pilots with previous Forward Air Controller or A-10 experience, and so these pilots transitioned to the new squadrons as the F-16 expanded throughout the 1980s. As General Mark Welsh—who flew F-16s in the 4th—wrote in Air Force Magazine, "through a process of elimination ... the question was referred to the 4th [Fighter Squadron]. The squadron, it turned out, had sixteen pilots with FAC experience, A-10 close air support experience, or both."[34]

To direct the flow of air power into Kill Boxes and search out the Iraqi Army, F-16s developed "Killer Scout" tactics. Though not technically Forward Air Controllers, they essentially performed the Fast FAC mission. The killer scouts reconnoitered an area and then directed fighters against any Iraqi forces. Since the F-16s did not carry rocket pods, they often marked targets with 500-pound general-purpose bombs. Unfortunately, it required three flights of F-16s to accomplish the mission of a single flight of A-10s. While A-10s could remain on station for more than an hour, the F-16s rotated out of a kill-box every twenty to thirty minutes. One flight conducted reconnaissance and directed attacks, while the second conducted air-to-air refueling, and the third flight shuttled between refueling and the Kill Box.[35]

When the ground war began, air planners added close air support missions to the Kill Box and Killer Scout melee. To ensure control, ground and air commanders agreed to place the coordination line about thirty miles north from the Saudi border. Kill Boxes within the immediate zone of advance became exclusively used for close air support. A-10s occasionally controlled operations in these kill boxes, but Forward Air Controller missions account for a pittance of all missions flown during the war. In a reversal from Vietnam, the rules of engagement dictated ground-based Tactical Air Control Parties control all close air support. The next row of Kill Boxes north, which were just beyond or split by the Fire Support Coordination Line, fell under the direction of the Killer Scout F-16s. Ground commanders nominated targets in these Kill Boxes, enabling them to mold the deep battle to match the invasion plan. Finally, air planners controlled the third layer of Kill Boxes north. There, flights of fighters and bombers conducted uncontrolled interdiction against Iraqi reinforcements, road systems, supply chains, and command networks.[36]

Still, the rapid advance of Coalition ground forces, and the corresponding collapse of Iraqi resistance, surprised commanders and planners of all branches, creating tension within the Coalition. Tracking the progress of individual units became difficult, particularly on the far left of the Coalition line. To minimize the chances of friendly fire, ground force commanders moved the Fire Support Coordination Line several times. Eventually, they pushed it well beyond the ability of friendly ground forces to reach the line, without coordinating with the air staff. It finally settled along the Euphrates River near the Iraqi city of Nasiriyah, about 150 miles north of the Saudi border. In a repeat of the Italian campaign during World War II, the wide gap between the forward line of Coalition ground forces and the Fire Support Coordination Line created a sanctuary for Iraqi Republican Guard units.[37]

As the Coalition forces advanced, the retreating Republican Guard Units halted south of the river, arrayed in a defensive posture but still trapped by the river. Despite their exposure, air power could not be brought to bear against them. Initially, the five Kill Boxes that overlaid most of the Republican Guard units on February 25 and 26 were beyond the Fire Support Coordination Line, and 272 strike sorties attacked targets in those five Kill Boxes on those days. However, as the Fire Support Coordination Line moved east on February 27, just fifty-four strikes went into the Kill Boxes containing the Republican Guard.[38]

Due to the distance between the US VII Corps, their embedded Tactical Air Control Parties, and the trapped Iraqis, they could not provide positive confirmation of targets needed to control air strikes. Exacerbating the situation was the lack of Air Force Forward Air Controllers. Unfortunately, due to the informal division of the A-10 Forward Air Controller forces, not every A-10 squadron had Forward Air Controllers on their roster. Further, there were only twelve A-10s modified for FAC duty in the entire theater.[39] Finally, since the F-16 Killer Scouts were a hastily employed stop gap, they were not qualified to control close air support. As a result, there were few Forward Air Controllers able to assist the VII Corps Tactical Air Control Parties. The war ended before VII Corps and the Republican Guard units engaged in any serious fighting, which minimized potential Coalition casualties, but also prevented VII Corps from cutting off the retreating Iraqis. Unfortunately, the gap between the two forces, coupled with an hours-long Coalition pause before the official cease-fire, allowed the Republican Guard to escape north to Baghdad, preserving one of Saddam's key mechanisms of power in the years after Desert Storm.[40]

The swift victory of Desert Storm, its mere one hundred hours of ground war and low number of Coalition casualties, "to a great extent represented the achievement of air power," as part of the Gulf War Air Power Survey concludes.[41] As decisive as the Instant Thunder inspired Phase I proved to be in enabling the three other phases, it was not the primary air power effort. Indeed, the final two phases highlighted the core idea of the AirLand Battle concept articulated in *FM 100-5* in 1976; without air power the Army cannot succeed in its mission.[42] Truly, Coalition air power—and the US Air Force in particular—proved overwhelming from the beginning. Soon after the war, however, air power advocates began touting the victory as an Air Force-only achievement. Chief of Staff of the Air Force, General Merrill McPeak, said during a press conference in March 1991: "This is the first time in history that a field army has been defeated by air power."[43]

Such statements seem to reflect a vanity of overselling air power, which McPeak himself accused previous air power advocates of doing just a few minutes before his own statement on the subject.[44] However, McPeak acknowledged that despite all of the Air Force's success, there are some things which air power cannot do. Chief among those, he said, was, "move in on the terrain and dictate terms to the enemy."[45] Soon, however, the second half of McPeak's comment was forgotten. The lessons learned and relearned, from World War II, Korea, Vietnam, and Desert Storm, about the friction points between the ground and air interface during combat so often solved by Forward Air Controllers, were forgotten for the new air superiority and strategic attack narrative.

In the years between Vietnam and Desert Storm, the Army drove the warfighting vision of the United States while the Air Force reconstructed its capabilities and institutions, building the force that proved critical to victory in Iraq. During that time, however, it lacked an overarching doctrinal ideology of its own. Instead, the Air Force followed the lead of the Army and the doctrinal ideology of AirLand Battle. Following Desert Storm, with the focus on what the Air Force had wrought, the service seized the initiative on the strategic and doctrinal conversation. Air power was now ascendant. There was no Soviet armored hoard to halt at the Fulda Gap, no Soviet Navy to challenge the sea. Instead, regional powers led by strongmen, in the vein of Saddam Hussein and Kim Jong Il, emerged as the primary concern of the post-Cold War world. Their modern, but less sophisticated militaries, the Air Force argued, could be readily dismantled by air power alone, just as in Desert Storm.[46]

Significant to the ideas driving the Air Force's campaign were the influence of John Warden and the concepts captured in *The Air Campaign* and Instant Thunder. As a RAND study of the Air Force's strategic planning during the decade stated, "For some, the Gulf War validated Warden's theories and 'the Gulf War was a harbinger of a "New Way of War"'—what some would even dub 'An American Way of War,' in which airpower proved decisive."[47] In the mid-1990s the Air Force published a series of "vision documents," written by many Air Force officers who worked for and with Warden, that were designed to sell the Air Force's ideas. Each document contained the word *Global* in the title, a marketing strategy intended to sell the Air Force's global capabilities during a decade of budget austerity.[48]

The timing of the first vision document, *The Air Force and US National Security: Global Reach—Global Power*, could not have been better for the Air Force. Published in June 1990, air power advocates seized on the connection between the ideas articulated in *Global Reach* and the success of Desert Storm. The document touted that "the strengths of the Air Force rest upon its inherent characteristics of speed, range, flexibility, precision, and lethality—characteristics which are directly relevant to the national interest in the future." The document went on to extoll the Air Force's influence on strategic partners, swift mobility, ability to control the high ground, inherent flexibility, and sustained nuclear deterrence posture.[49]

An additive driver to the Air Force's campaign was a keen desire to seize the lion's share of falling military budgets. Congressional Republicans and Democrats alike, as well as the outgoing Bush and incoming Clinton administrations, were eager to capitalize on the fall of the Soviet Union and reduce Defense budgets.[50] Fortuitously, the Air Force's campaign lined up with predictions, planning, and reviews conducted within the Department of State, Department of Defense, and the Joint Chiefs of Staff. There was consensus across the board that regional conflicts driven by expansive strongman states were going to be the new norm. The newly inaugurated President Bill Clinton and his advisors were eager to engage regional conflicts, promote democracy, and provide humanitarian relief backed by the might of the US military.[51]

The US military needed to be ready to confront several of these conflicts in different parts of the world simultaneously. As the Vice Chairman of the Joint Chiefs of Staff, Admiral David E. Jeremiah wrote in the first publication of *Joint Forces Quarterly* in 1993, "The world is going through an incredible metamorphosis. Some changes are directly related to the end of the Cold War; others have no connection with the late East-West conflict. The sum total of these changes, whatever their source, is a world teeming with nascent crises."[52] Of course, the military's ability to conduct such operations hinged on budget allocations, and the Air Force was winning the internal budget promotion battle.[53] Events in the world seemed to prove these ideas. In the 1990s, a series of conflicts erupted, driven by the crumbling social constructs and economic support lost in the collapse of the Cold War order.

Before leaving office after his defeat in the 1992 presidential election, President George H.W. Bush authorized US military force to assist United Nations security forces distributing aid to the disintegrating nation of Somalia, leaving the unfinished mission to the new president. Within the year, the mission expanded to include taking down the Somali warlord Mohamed Aidid. On October 3, 1993, a US special operations task force, known as Task Force Ranger, launched a raid under the code name Gothic Serpent to capture two key members of Aidid's organization. Instead of a surgical operation lasting only forty-five minutes, the situation devolved quickly into a running gun battle that raged in the streets of Somalia's capital city Mogadishu for fifteen hours. The results were eighteen American soldiers killed and seventy-three wounded. Rocket-propelled grenades downed two US Army Blackhawk helicopters, and Army pilot Mike Durant was captured and held for eleven days. An unknown number of Somalis, perhaps thousands, were killed in the process, and outraged Somalis dragged the bodies of dead Americans through the streets. For the new Clinton team, it was a public relations disaster, for "the same news networks that broadcast the start of the peaceful humanitarian mission less than a year earlier now ran horrific footage of Aidid supporters desecrating the corpses of US soldiers."[54] Clinton withdrew US forces from Somalia, and the hangover of Operation Gothic Serpent directly influenced his decision not to intervene in the Rwandan genocide a year later.[55]

The disaster that befell Task Force Ranger in Somalia stood in stark contrast to the use of air power in the Balkans over the next few years. In the aftermath of the Soviet collapse, countries of Eastern Europe began fitfully transitioning away from communist governments. By April 1992, Yugoslavia split into several independent republics divided along its many ethnic and religious lines, sparking a civil war. Bosnian Serbs, supported by the former Yugoslav government holding power in Serbia, attacked the newly independent Bosnia from within. The UN intervened to halt the violence, sending a peacekeeping force to Croatia in the spring of 1992. The NATO alliance, in turn, intervened with the blessing of the UN. For the first time in its history, the alliance conducted combat patrols, observing flights moving in and out of Bosnia. This mission became known as Operation Sky Monitor and began in October 1992. In April 1993, NATO transitioned to the more aggressive Operation Deny Flight, which permitted alliance aircraft to shoot down unauthorized flights under UN Security Council Resolution 816. NATO began allowing close air support airstrikes to protect UN peacekeepers beginning in July. By August, NATO authorized aircraft to conduct offensive interdiction airstrikes against Bosnian Serb forces.[56]

Over the next two years, NATO offensive air operations steadily increased, as did US involvement. In July 1995, Bosnian Serb forces attacked and occupied the city of Srebrenica, driving out or executing its Muslim population. Further attacks ensued in Bosnia, and on August 20, NATO launched Operation Deliberate Force. Compared to Desert Storm, Deliberate Force was a minuscule undertaking. NATO Allies flew only 3,535 total sorties during the month-long operation. Still, the air campaign reflected many of the lessons of the war in Iraq. Just as in Iraq, the 1,365 close air support and battlefield air interdiction missions made up the bulk of Allied sorties, just over 38 percent of the total.[57] Similarly, Allied air planners overlaid Bosnia with a Kill Box grid, though the whole theater consisted of just eight boxes. The A-10 pilots were the sole US Air Force FACs in the conflict, but F-16 Killer Scouts again served as a stop-gap, although their combined missions made up a small fraction of the NATO total.

However, Forward Air Controllers still served as a critical air power enabler. All A-10 and some F-16 sorties during Deliberate Force, and its predecessor Deny Flight, were designated as Forward Air Controllers for troops in contact missions. Additionally, for the first time, Air Force AC-130 Spectre Gunships were designated as Forward Air Controllers. Actual missions were few, and NATO only launched one hundred missions to support UN requests. Further, only four of those close air support missions employed weapons: two during Deny Flight and the other two on August 30 and September 10, 1995. The August mission was the most significant, as an A-10 directed six other A-10s, two F-16s, and a Mirage 2000 against sixteen targets. The A-10s made several attacks with their 30 mm cannons. The aircraft also hit twelve targets with a combination of 500-pound general-purpose bombs, cluster bombs, Willie Pete and high explosive rockets, and one 500-pound laser-guided-bomb dropped by an F-16.[58]

F-16 Killer Scouts proved particularly helpful in fulfilling the rules of engagement during missions. Deliberate Force rules of engagement required pilots to visually identify targets before using weapons, a restriction that reflected the desire of NATO political and military leaders to minimize civilian casualties. Often, to fulfill the rules of engagement, aircraft made a high-speed low-altitude pass over the target to visually confirm it—a dangerous proposition due to the threat of antiaircraft fire and shoulder-launched surface-to-air missiles. To solve this problem, F-16 Killer Scouts searched out and identified targets before marking them and clearing other aircraft to attack them.[59] Even with the assistance of Forward Air Controllers and Killer Scouts, weather and the restrictive rules of engagement prevented many missions from using weapons at all. The Allies dropped just over 1,000 weapons during the campaign, but for the first time in aerial warfare, a majority, 708, were precision guided.[60]

The air campaign lasted until September 15, and it succeeded in halting further Serb attacks and aiding the Croat-Bosnian coalition forces in their offensives. Coupled with a herculean diplomatic effort by Richard Holbrooke, the Dayton Peace Accords ended the conflict on November 21, 1995. A sixty-thousand member NATO force went to Bosnia to maintain the peace agreement.[61] Unfortunately, NATO's involvement in Balkan conflicts did not end with the Dayton Peace Accords. Part and parcel of the internal strife in the former Yugoslavia was a deep-seated conflict between the Serbian government, led by Slobodan Milosevic claiming title to the former Yugoslavia, and Kosovar separatists. Under Marshal Tito's regime, largely Albanian Kosovo maintained a semiautonomous status, which Milosevic revoked in 1989. For the next decade, Kosovo descended steadily into violence and reprisals between the Kosovo Liberation Army, fighting for Kosovo independence, Milosevic's Interior Ministry Police, and the Serbian Army. In 1998, the Interior Ministry Police increased attacks against Kosovar civilians. In response, the UN Security Council passed Resolution 1199 in

September condemning the violence. As part of Resolution 1199, the two sides agreed to unarmed monitors from the Organization for Security and Cooperation in Europe to monitor compliance with the Resolution.[62]

The agreement only prevailed for a short time. By the beginning of 1999, after a particularly egregious killing of forty-five Albanian civilians, calls for more aggressive action to stop the violence rose in NATO capitals. Encouraged by the success of Deliberate Force and burdened by the failure to intervene in Rwanda, yet wary of committing ground forces, President Clinton's Secretary of State Madeleine Albright led the push to intervene. For Clinton and his administration, the situation was tailor-made for US action, an opportunity to rewrite "the rules for how the international community would respond to a government that refused to stop repressing its own people."[63] After diplomatic efforts fell apart, Operation Allied Force began on March 24, 1999. In initiating Allied Force, NATO leaders rejected the use of ground forces, relying solely on coercive air power to force Milosevic's forces and the Serbs out of Kosovo.[64]

NATO planners laid out a three-phase air campaign that mirrored Desert Storm. The first phase focused on Serbian air defense systems, which, while less numerous than those in Iraq, were densely packed in the confined Balkan nation's airspace. Further, the network used more advanced missile and radar systems than the KARI system. Finally, the Serbian forces were far more adaptable compared to the Iraqis—they changed tactics based on the lessons of both Desert Storm and Deliberate Force. The second phase of the air campaign focused on Serbian-fielded forces and infrastructure in southern Serbia and Kosovo, south of the 44th parallel. The third phase shifted back north, focusing on strategic leadership and infrastructure targets in Serbia—including targets in the capital, Belgrade.[65]

Hoping for a rapid capitulation by Milosevic, Clinton and other Allied leaders were quickly disappointed. On March 27, just three days into the first phase of the war, Serbian air defenses shot down an F-117 stealth aircraft, an unimaginable feat, given that these aircraft flew unimpeded in 1991. During Desert Storm, the Iraqi radars were unable to even track the aircraft and fired blindly into the air after impact of the F-117's laser-guided bombs.[66] Defiantly, Milosevic stepped up his campaign against the Kosovars, forcing

NATO to shift into phase two, then quickly to phase three after only one week of air attacks.[67]

NATO and President Clinton received criticism from all quarters. Despite the bravado of air power advocates in the years leading up to Allied Force, doubts quickly surfaced about the ability to succeed using air power alone. As Benjamin Lambeth wrote, "Deep doubts that the air attacks alone would suffice in forcing Milosevic to knuckle under, however, soon prompted a steady rise in military pressure—notably from some US Air Force leaders directly involved with the air war—for developing at least a fallback option for a ground invasion."[68] Further, after the initial wave of attacks, the Serb Army dispersed their forces, hiding them in caves, tunnels, forested areas, and in villages. Serbian military units moved into populated areas, forcing NATO leaders and planners to reconcile the military need of attacking the marauding Serbs against the political concern to prevent civilian casualties. Using these human shield tactics, Milosevic sought to drive a wedge into the Coalition.[69]

Without friendly ground forces, NATO again turned to the skills of Air Force Forward Air Controllers to hunt down and destroy Serbian forces while minimizing collateral damage. American fighter squadrons in Europe maintained a near-continuous presence in the Balkans beginning in 1993, conducting combat air patrols and maintaining aircraft ready to conduct Combat Search and Rescue missions should an Allied aircraft go down. US F-16 and A-10 squadrons also rotated Forward Air Controllers on call to support the UN and NATO peacekeeping forces. Since a ready supply of Air Force FACs were available when Allied Force began, Major Phil Haun developed a concept to maximize their skills.[70] Similar to the hunter-killer tactics used by Hale Burr to destroy a North Vietnamese SAM site, A-10s and F-16s led large strike packages into Kosovo to hunt down and destroy Serbian forces. The NATO Air Operations Center put the plan into action, with the first successful mission flown on April 6, 1999, just two weeks into the war.[71]

A-10s from Haun's 81st Fighter Squadron, F-16s from the 31st Fighter Wing, and F-14s from the aircraft carrier USS *Theodore Roosevelt* operated as the mission commanders for sometimes as many as forty other aircraft. Lasting for typically around four hours, these missions consisted of several flights of attack aircraft, a flight or more of air-to-air fighters, two or more flights of aircraft dedicated to the suppression of Serbian air defenses, air refueling support, command-and-control aircraft, and the Forward Air Controllers. Just as in Korea, Vietnam, Desert Storm, and Deliberate Force, the Forward Air Controllers moved into Kosovo ahead of the strike force and began searching for signs of Serbian forces. To assist in their search, the Forward Air Controllers in Kosovo had new technologies at their disposal, such as Joint Surveillance Target Attack Radar System (JSTARS) and the RQ-1 Predator unmanned aircraft. The JSTARS used a huge radar array under the aircraft capable of locating moving and stationary vehicle convoys. The RQ-1 loitered for hours, employing infrared and optical sensors to feed real-time video to intelligence analysts, which allowed the operators to find and identify targets or verify the absence of civilians. Possible targets were passed via radio to the A-10 pilots for visual identification and targeting. Just as before, the Forward Air Controllers verified and marked the target for the strike aircraft to attack. Occasionally, if the target appeared ready to move or the situation required a rapid response to halt possible Serb attacks on Kosovar civilians, they blasted the targets themselves.[72]

After seventy-eight days of bombardment, the war abruptly ended on June 9, 1999. As Milosevic remained defiant against NATO attacks for nearly three months, the reasons he capitulated when he did are the subject of much debate. Although extreme air power enthusiasts cite the NATO victory as a win for air power alone, given that NATO political leaders never anticipated the use of ground forces, this seems too simplistic an answer. Several factors likely contributed. First, the tactical campaign, led by Air Force Forward Air Controllers against Serbian ground forces, interrupted but never destroyed Milosevic's key mechanisms of power and leverage over the Kosovars. Although weather and rules of engagement, along with the adaptability of Serbian forces, prevented air attacks from completely stopping the killing, their persistent aerial onslaught destroyed the Serbian-fielded forces. While the actual level of destruction is debatable, and likely only around 10 percent, what is clear is the destruction accelerated as the campaign persisted, thanks in no small part to the Forward Air Controller led strike teams. Further, the roving strike forces prevented the Serbs from assembling and moving in large formations and also hampered their ability to use tanks, artillery, and other mechanized weapons. Additionally, Allied air attacks forced the Serbian air defense network to shut down to save their forces from destruction, allowing NATO air power to roam unimpeded in most of the theater.[73]

Second, beginning in May after the NATO Fiftieth Anniversary Summit, the alliance deliberately increased attacks against Serbian infrastructure, particularly in and around Belgrade. This included airstrikes against targets NATO commanders left untargeted for political reasons early in the campaign. The attacks targeted the Serb electrical grid and oil infrastructure, as well as military and industrial manufacturing capacity. Besides hampering Serb military capabilities, the attacks around Belgrade also had a direct effect on the city's populace—most notably on the city's military and political elite. By attacking the elites' sources of power and income, NATO also undermined Milosevic's hold on power. The alliance attacks took Serbian state TV off the air and deprived much of the country of electricity and water for days at a time. NATO stepped up attacks against government buildings and political and military leadership, leaving the Serbian leader with a dwindling number of political allies and few places to hide.[74]

The final factor for the Serbian surrender was the loss of Russian support in the face of a determined NATO alliance. On April 25, the last day of the NATO Fiftieth Anniversary Summit, Clinton and Russian President Boris Yeltsin had a seventy-five-minute telephone call. During the call, Clinton made clear the alliance's determination to continue the course and Yeltsin offered to send Russian Prime Minister Chernomyrdin to meet with the Serbs personally. Following this phone called, and the visit of Chernomyrdin, it likely became clear to Milosevic that NATO was resolute in its course and the Russians were no longer on his side.[75] As Benjamin Lambeth concluded about the reasons Milosevic finally capitulated, "At some point it was clear to Milosevic that he wasn't going to be able to wait out the bombing, that NATO wasn't going to go away, and that progressively Serbia was being destroyed, [so] he chose to get the best negotiated settlement he could."[76]

As part of the peace settlement, Serbian forces withdrew from Kosovo, and NATO halted its bombing campaign. A fifty-thousand member NATO force secured the agreement, which included a five-mile buffer zone along

the Serb-Kosovo border.[77] Allied Force lasted much longer than NATO
leaders envisioned and opened both NATO leaders and air power to criticism.
The two NATO air campaigns in the Balkans also left many questioning
the ability of air power to singlehandedly win wars.[78] Even though no
NATO ground forces engaged in combat, it still required their presence
to secure the peace in Bosnia, just as General McPeak noted in his press
conference in 1991.

Still, the two air campaigns in the Balkans succeeded in their aims, if less
efficiently than advertised or hoped.[79] Similarly, air power enabled the lopsided
success of Desert Storm. While FACs did not win either war, the skills they
brought to the air wars in the 1990s caused the Air Force to expand the available pool of Forward Air Controllers. In 1997 the Air Force officially began
training F-16 pilots for the mission. The first class graduated on October 27,
1997, in time to make critical contributions to Allied Force. Now the F-16s
were Forward Air Controllers in both name and deed, giving credence to their
role as Killer Scouts in Desert Storm and Deliberate Force, and resurrecting
the Fast FAC mission begun by the Misty pilots thirty years earlier.[80]

Nevertheless, as the twentieth century closed, the capabilities of the Air
Force Forward Air Controller, and their critical function in providing air
power support to friendly ground forces, were pushed aside in favor of
air-centric missions. For air power advocates, Desert Storm hypothesized air
power alone could win wars, and the two wars in the Balkans seemed to
confirm that theory.[81] As Michael Lamb wrote in his article for the Maxwell
Papers, "Operation Allied Force provides the United States and our allies
an opportunity to glean valuable insights about military operations as this
century unfolds."[82] Echoing Lamb's sentiment and the air power-centric strategic concepts of the 1990s, Secretary of Defense William S. Cohen wrote
in his statement to the Senate Armed Service Committee in October 1999,
"NATO demonstrated both the unwavering political cohesion and unmatched
military capability that will be required to overcome the complex and unpre-
dictable security challenges of the 21st century."[83] Both of these statements
proved false. Soon after the turn of the century, the US became embroiled
in two ground-centric counterinsurgency and nation-building campaigns in
Afghanistan and Iraq.

Chapter 7

Since 9-11-2001: FACs Forgotten

I f Operation Desert Storm represented the ideal scenario to vindicate the US Air Force's post-Vietnam transformation, then the wars since 9–11 are the antithesis. The ground-centric wars did not fit the mold anticipated by American political and military leaders at the close of the twentieth century. They proved to be different from any previous American conflict both in scope and duration. US military forces were not engaged in a battle against another superpower or regional power with advanced military technology. Instead, American soldiers, sailors, and airmen fought in all corners of the globe against stateless terror organizations and proxy forces using irregular guerrilla forces armed with little more than light weapons. As one observer wrote about the nearly two-decade-long war in Afghanistan, "the war does not fit neatly into how we traditionally understand the use of force. In counter terrorist campaigns like the one against al-Qaeda, victory is not a matter of forcing enemies to surrender or destroying their ability to resist."[1]

In a flashback of Vietnam, the conflicts featured light infantry units and special forces teams, supported by air power, engaged in close combat with insurgent forces distributed among civilian populations.[2] With Vietnam as the model for air power in counterinsurgency operations, it follows Air Force

Forward Air Controllers should have played a significant role in the wars after 9-11. Yet they have not. In American conflicts since 9-11, both air and land commanders and planners only used them sporadically, by accident, or by the initiative of individuals to fill the void. In their absence, air-ground cooperation struggled. To fill the gaps, soldiers and airmen attempted technological workarounds and Forward Air Controller-like substitutes. However, when fighting became the most intense, or during the most dangerous missions, no adequate substitute existed. Examining the conflicts between the US invasion of Afghanistan in 2001 and the recapture of the Iraqi city of Mosul from the Islamic State of Iraq and Syria (ISIS) in 2017 reveals how Air Force Forward Air Controllers still serve a critical role on the battlefield.

Al-Qaeda and their Taliban sponsors did not match any mold US forces expected to face as the twenty-first century dawned because they lacked any air-to-air threats, surface-to-air threats, or traditional industrial centers of gravity. Beginning October 7, 2001, the US required just ten days of airstrikes to destroy the few fixed targets in the country. Through October and November, US air power provided support to US Special Forces units embedded with the loose coalition of Afghan warlords known as the Northern Alliance. Backed by American airstrikes, they captured the city of Mazar-e-Sharif in northern Afghanistan on November 9, and by early December, the Northern Alliance held the capital, Kabul. Driven out of the cities, Osama bin Laden, the remnants of the Taliban, and around one thousand fighters sought refuge in the remote White Mountains southeast of Kabul along the Pakistani border, in the Tora Bora cave complex.[3] The swift and overwhelming victory of the conglomerate ground forces supported by American air power delivering precision-guided weapons convinced many observers that Afghanistan represented a new model for American combat, but bin Laden managed to slip out of the trap and into Pakistan, where he remained for the next decade.[4]

Though American precision air power proved critical in the rapid defeat of the Taliban, tightly constrained resources and extreme flight distances effectively left Forward Air Controllers off the battlefield. Only eight Air Force FAC missions flew between October and December 2001, out of just over 2,300 close air support sorties. Significantly, no Air Force Forward Air

Controllers flew during the few critical days bin Laden holed up in Tora Bora.[5] The conduct of the initial campaign in Afghanistan seemed to make their role irrelevant. New technology-enabled Air Force Joint Tactical Air Controllers were attached to special forces teams to provide devastatingly accurate air power in all weather, day and night. Advanced satellite communications allowed ground forces to communicate with command centers thousands of miles away, who in turn could use the same technology to communicate with and direct close air support aircraft.

Once aircraft were over a ground team's position, laser devices allowed them to accurately generate coordinates for enemy positions, which pilots could use to cue LANTIRN targeting pods. LANTIRN targeting pods look a bit like a *Star Wars's* R2-D2 on its side and mounted on an aircraft. They contain an infrared camera and a laser designator, which directs laser-guided weapons to their targets. When the pilot inputs the Joint Tactical Air Controller's target coordinates into the aircraft's onboard navigation and targeting systems, the pilot can quickly point the pod at the desired target area. The pilot can then see the image of the target area on their cockpit display, forgoing the need for an elaborate talk on using surrounding terrain features. Once cleared to release their bombs, the pilot maintains the target on the display in the LANTIRN pod's crosshairs and fires the pod's laser designator to guide the weapon to the target.

While extremely accurate, laser-guided bombs require the pilot to continuously track a target under the LANTIRN pod crosshairs. If the pilot moves the crosshairs off the target, the bomb will hit where the crosshairs are pointing and miss the target. If clouds, dust, smoke, or terrain come between the aircraft and the target while the bomb is in flight, it will block the laser, and the bomb will miss. However, if the aircraft is carrying GPS-guided weapons, aircrew can input the coordinates received from the Joint Tactical Air Controller directly into the bomb's guidance system, allowing the pilot to release the bomb directly to those coordinates. Using this bomb-on-coordinates method, a pilot never needs to see the target—either with their eyes or in their LANTIRN pod—and does not need to keep the target under the crosshairs. Further, GPS-guided bombs can still find their target through smoke, dust, or bad weather.

The technological advances in communications, targeting, and weapons guidance were a significant step in the use of air power, even compared to Allied Force just two years before. The new methods also bypassed the FACs' role as a communication link, liaison, and translator of the ground party's view of the world to the pilot. Air power's accuracy and speed allowed for reduced casualties and less collateral damage. This system quickly and accurately passed target information to aircraft overhead, swiftly multiplying a ground force's firepower and reducing the number of ground personnel needed to take an objective. Further, fewer aircraft were needed for close air support, since more accurate bombs reduced the number of weapons required to destroy a target.

While the initial campaign in Afghanistan was not without friendly fire or collateral damage, the handful of cases proved to be the product of human error in the use of technology or misidentification of targets, rather than a result of errant weapons.[6] Although 1,500 US Marines were on the ground in southern Afghanistan as bin Laden and the Taliban leaders retreated to the White Mountains, less than five hundred Coalition special forces soldiers and CIA operatives embedded with the Northern Alliance fought in the opening campaigns of Enduring Freedom.[7] Unfortunately, bin Laden slipped into Pakistan sometime around December 14, as American political and military leaders failed to inject sufficient forces into the battle to ensure his capture.[8]

Despite bin Laden's escape, the first three months of operations in Afghanistan demonstrated just how effective Joint Tactical Air Controllers could be when harnessing fighters and bombers dropping GPS-guided weapons.[9] As a result, air planners showed little proclivity for using Forward Air Controllers—despite their recent success in the Balkans and their historic success in unconventional warfare, as demonstrated in Vietnam. Instead, since the inception of AirLand Battle, the Air Force demonstrated a growing preference for ground-based controllers.[10] However, the first major conventional US operation in Afghanistan—Operation Anaconda—tragically revealed the shortfalls brought about by the sole reliance on Joint Tactical Air Controllers and the makeshift close air support system developed during the early days of Enduring Freedom. The incident also highlighted the neglect of air-ground cooperation, the impact of personnel losses in Forward Air Controller programs in the decade since Desert Storm, and neglect of the mission by both the Air Force and the Army.

Conceived as an American-led operation and born out of the failure at Tora Bora, US leaders hoped to kill, capture, or drive out the remnants of al-Qaeda and the Taliban holding out in the Shah-i-Kot valley in rugged eastern Afghanistan. Anaconda was initially a US special forces operation, but it steadily grew and morphed into the largest conventional US Army operation since Desert Storm. Parts of the 10th Mountain Division, 101st Airborne Division, and multiple special forces task forces amalgamated into an organization known as Combined Joint Task Force (CJTF) Mountain. Simple in concept, "CJTF Mountain's plan was to surround the Shah-i-Kot valley with overlapping rings of US forces aimed at bottling up and then capturing or killing the several hundred al Qaeda fighters who were thought to be hiding in the area."[11]

Conceptualized in late January 2002, scheduled for February 28, and envisioned as a two- or three-day operation, Anaconda began two days late and lasted more than two weeks. Although ultimately successful in achieving its objectives, Operation Anaconda resulted in eight Americans killed and more than fifty wounded. American intelligence analysis anticipated light opposition from a few hundred fighters who were likely to retreat at the coordinated American assault. However, as Benjamin Lambeth wrote of the operation, "Anaconda would instead prove to be a series of intense individual firefights in which al Qaeda holdouts, rather than retreating as before at Tora Bora, would stay on and fight to the death."[12] Beyond the intelligence failure, poor planning and little interservice coordination plagued Anaconda from the outset. Combined Joint Task Force Mountain launched the operation in a complex environment further mired by the nature of the organization itself. Command relationships within it and other task forces participating in the operation were not well delineated, and outside of Anaconda's main force, an unaffiliated Special Operations Task Force simultaneously launched an overlapping raid into Shah-i-Kot.

Finally, because Task Force Mountain did not become the lead unit for Anaconda until mid-February, the planning staff did not grasp the complex operational environment of Afghanistan while formulating the ever-expanding assault. Critically, its planners and leadership did not understand Enduring Freedom's air-ground interaction and support realities. The operational concept, which was so successful during the first three months of the war, proved to be the Achilles' heel to Anaconda's air support plan. Enduring Freedom was a special forces war, with small teams distributed across the vast landscape supported by precision air power. The teams only occasionally coordinated operations, and the Air Operations Center readily supported and deconflicted their requests. However, the model produced a system of complex airspace control measures, air support request procedures, and highly restrictive rules of engagement.[13] The entire system defied decades of air-ground coordination doctrine and procedures. As an Air Force after-action study of Anaconda summarized:

> In effect, the old doctrinal concepts of control lines and area ownership did not apply. Dozens of [Joint Special Operation Areas], engagement zones, special engagement zones, restricted fire areas, no fire areas, off-limits sites of interest, and constant unknowns about friendlies created a jigsaw puzzle of battlespace control measures. It was all very different from the phase lines, corps boundaries and fire support coordination lines of a doctrinally conventional battlefield. Adding to the confusion, each set of players had their own preferences for handling the control measures for territory where they were operating. Special operations teams on the ground liked to declare whole areas off limits. Army conventional forces were used to owning a defined operating area and being able to call in airstrikes on their own authority.[14]

The situation was worse than just muddled planning, neglected doctrine, and rules of engagement ill-suited for more conventional land combat operations. The 10th Mountain Division—the lead unit for the mission—did not deploy with its assigned Air Support Operations Center personnel. Task Force Mountain's commander, Major General Franklin Hagenbeck, saw little utility in assigning personnel to the Air Support Operations Center; an unforeseen consequence of personnel reductions in the post-Cold War era, air-centric operations of the 1990s, and political influence on personnel levels in Afghanistan.[15]

After the collapse of the Soviet Union, and amid the post-Cold War force reductions, the Air Force consolidated and restructured Tactical Air Control Parties and permanently joined them with Air Liaison Officers to create Air Support Operations Squadrons. Aircrew designated as Air Liaison Officers no longer joined the Tactical Air Control Party team at the last minute as they had since World War II. Instead, the Air Force permanently assigned all the members of the Tactical Air Control Party to the Air Support Operations Squadron—radio operators, Air Liaison Officers, and a new career field known as Enlisted Tactical Air Controllers. The Air Force created the enlisted controllers in response to Air Force pilot reductions during the 1990s, which left fewer pilots able to fill Air Liaison Officer positions. By the mid-2000s, the term Joint Tactical Air Controller encompassed all Air Force ground controllers—both officers and enlisted.

Since World War II the Air Support Operations Squadron and its predecessors served, and still serves, as the central point of Air Force air-ground integration expertise for Army combat units.[16] Air Support Operations Squadron enlisted personnel and officer air liaisons also fill duties at the Air Support Operations Center during contingency operations, assisting their supported division during planning and execution of missions.[17] Put another way, the Air Support Operations Squadron is the Air Force's personnel and logistical unit that houses Air Force personnel tasked with integrating air power with ground forces—the Air Support Operations Center is both the mechanism and the place for that integration within an Army division.

Lacking an Air Support Operations Center, Combined Joint Task Force Mountain initiated planning for a major operation without consulting their air power experts. Critically, they also nearly began the operation without the support structure required to control and coordinate air support requests successfully. Only intervention by senior Air Force officers in Central Command and the 18th Air Support Operations Group enabled the 10th's Division's Air Liaison Officer and some support staff to join Task Force Mountain's Operation Center at Bagram Air Base, Afghanistan. Unfortunately, they did not arrive until February 20, after most of the initial planning was complete and just a week before the planned execution date. Even after the Air Support Operations Center personnel arrived, the Division Operations

Center at Bagram did not have any communication equipment nor computer systems capable of communicating with aircraft, the Air Operations Center, or able to display the air picture. Instead, the team created a makeshift Air Support Operations Center, developed ad hoc procedures, and use borrowed equipment to find a minimally workable solution.[18]

Even with some air power experts in place, few in Task Force Mountain or their Air Support Operations Center appreciated the difficulty in providing air power during Anaconda. First, the airspace was extraordinarily complex and compact, encompassing only a nine-mile by nine-mile area. The primary helicopter landing zones covered just three miles by five miles. Civilian airline routes, aid agency relief flights, military supply flights, attack helicopters, transport helicopters for the assault forces, command-and-control aircraft, fighters, bombers, and reconnaissance aircraft all operated in or passed through the combat zone. Within the compact battlefield, the rugged terrain of the Shah-i-Kot valley dominated, blocking ground-to-ground, air-to-ground, and air-to-air coordination. The mountains also prevented soldiers in one part of the battle from seeing or communicating with units in other areas.[19] Finally, widely scattered Coalition air power complicated the ability to coordinate and communicate the ground situation to aircraft—particularly once they were en route to Afghanistan.

The nearest source of air power were Navy carrier-based fighters operating off the coast of Pakistan, a one-way flight of nearly six hundred miles. The furthest flights were made by Air Force bombers who flew a nearly six-thousand-mile round trip from Diego Garcia, an island in the Indian Ocean, to provide close air support to Anaconda. Most missions flown in support of Anaconda lasted more than eight hours and some longer than twenty-four.[20] The extreme flight distances and a difficult battlefield taxed aircrews' endurance. The situation seemed tailor-made for Forward Air Controllers. However, no one involved in the planning process, from senior Army and Air Force commanders down to the air power experts at the Air Support Operations Center, considered utilizing them. They were never included during any of the planning discussions.[21]

Instead, when the operation began on March 2, 2002, the ground-based controller model used for the previous five months went into play. Unfortunately, things did not go as anticipated, for the initial wave of assaults met higher than expected resistance. Additionally, the assault conducted by the unaffiliated special operations task force ran into withering fire. Casualties began to mount, and multiple helicopters were either severely damaged or shot down. As the fight's intensity increased, air support requests began to overwhelm the makeshift Air Support Operations Center's ability to provide air power.

Thirty-five different Joint Tactical Air Controllers supported the various assault forces in the miniscule battlespace. As the Air Force's account of the operation noted, "In comparison, a US Army division operating on a traditional, conventional battlefield might have 1/5 as many [JTACs] in an area the same size."[22] Confusion reigned from the outset. The ground force abruptly called off preassault airstrikes because part of the assault force was unaware of these strikes, and believed they were taking fire. Friendly Afghan forces mistakenly came under fire from an Air Force AC-130, killing one Afghan, one American, and wounding five other Afghan soldiers. Due to the imposing terrain, assault plan sequence, and confusion on the ground, some Joint Tactical Air Controllers had difficulty providing timely support to elements of their assigned units. Further, several others were without adequate battlefield maps or laser target designators and could not generate accurate coordinates for close air support attacks.[23] "The first day's operations also exposed stress points in the command and organization of Coalition forces. Operation ANACONDA's true air support requirements were nothing like previous Operation ENDURING FREEDOM experience or the initial plan in CJTF Mountain's [concept of operations]."[24]

Only heroic fighting by those on the ground and a herculean effort by the airmen overhead prevented the operation from devolving into a complete failure. Aircraft dropped weapons dangerously close to friendly ground forces, often making strafe and bomb attacks within a hundred yards of friendly positions. There were multiple instances where aircraft were not able to make attacks because other aircraft, supporting other parts of the assault force, were below them and in the fall line of the bombs. There were several near mid-air collisions.[25] The battlefield desperately needed the Forward Air Controllers' skills. As a report cataloging lessons learned from Operation Anaconda said, "the Air Force lacked airborne battlefield command-and-control centers or a similar aircraft, an adequate number of Airborne FACs (Forward Air Controllers), and other resources needed for a large close air support operation."[26] Benjamin Lambeth came to a similar conclusion in his history of Operation Enduring Freedom: "The only way the large number of aircraft operating simultaneously over the battlefield could be safely deconflicted was through the use of Airborne Forward Air Controllers."[27]

After the first forty-eight hours, the Combined Air Operations Center employed some solutions. First, air component commanders diverted an experienced A-10 pilot from the Air Forces Central Command Strategy Cell to Bagram to shore up the drowning Air Support Operations Center. Second, to overcome the Joint Tactical Air Controllers' inability to generate accurate coordinates, commanders allowed the use of cluster munitions and general-purpose bombs instead of laser- and GPS-guided bombs. This allowed Joint Tactical Air Controllers, Forward Air Controllers, and close air support aircraft to use "talk-on techniques" used since World War II to quickly attack Al-Qaeda and Taliban forces in the open. Third, A-10s moved to a temporary operating base closer to the Shah-i-Kot valley, finally injecting FACs into the fight.[28] By March 5, forward-deployed A-10 and a handful F-16 FACs helped stabilize the situation over the beleaguered American forces. At last, the Combined Air Operations Center and the Air Support Operations Center began implementing new procedures to provide more sustainable air support to the operation.[29]

Both the ground and air components overcame the initial struggles in Operation Anaconda, and after seventeen days of hard fighting Task Force Mountain achieved its objectives. The battle, and the early struggles in particular, opened old wounds and accusations. Major General Hagenbeck publicly criticized the Air Force's performance as substandard. Yet it was Hagenbeck's own ignorance of airpower, and unwillingness to use the Air Support Operations Center at his disposal, that significantly contributed to the near disaster. Still, in response to Hagenbeck's remarks, the Air Force Chief of Staff, General John Jumper, initiated an Air Force after-action review. The Air Force published its findings, in classified and unclassified formats, in 2005. Similar reviews of joint planning and execution doctrine took place across the Department of Defense, resulting in current Joint Doctrine publications and, eventually, a new Forward Air Controller qualification process.[30]

However, the lessons of Anaconda arrived at the fielded forces slowly, with most of the urgency lost by mid-2002. Within months of Operation Anaconda, Coalition ground forces began focusing on building a new Afghan government and securing the country against the remnants of the Taliban. As the mission shifted, the use of air power settled into a new model. Instead of focusing on destroying the Taliban and Al Qaeda forces, Coalition air power provided on-call close air support and intelligence collection to Coalition ground forces attempting to rebuild and secure the Afghan government.[31] Further, political urgency shifted the Department of Defense's focus to Iraq again. Forward Air Controllers again became a lesson relearned in Afghanistan, just as in World War II, Korea, and Vietnam.

In 2003, US-led Coalition forces once again rolled across the desert of Iraq. However, Operation Iraqi Freedom unfolded differently than the 1991 war. Most of Saddam's ground forces did not recover after Desert Storm. A decade of airstrikes conducted in Operations Northern and Southern Watch and UN sanctions further depleted Iraqi defenses. As a result, Iraqi forces could not mount significant resistance. Further, there were no phased air and ground campaigns. Instead, the ground and air assaults began simultaneously on March 21, 2003.[32] Despite the lack of significant resistance, the invasion did not unfold smoothly.

Abbreviated planning and shifting force structure—heavily influenced by political decisions—hampered the ability of military staffs to mesh the air and land components successfully. Influenced by the rapid overthrow of the Taliban, Secretary of Defense Donald Rumsfeld pushed for a minimal invasion force, constraining resources from the outset. The Turkish government denied the use of their country to stage a second invasion front, and ongoing operations in Afghanistan further limited both the air and land commanders' resources.[33] While these political limitations manifested into negative long-term results in Iraq, during the initial invasion military planners also repeated mistakes made in Desert Storm, resulting in preventable operational struggles. Most of these decisions revolved around air-to-ground control measures and inefficient air support to ground forces. More robust use of Forward Air Controllers on the battlefield could have prevented many of the conflicts and assisted in more effective distribution of air power. Instead, although both A-10 and F-16 formations flew as Forward Air Controllers during the early phases of Iraqi Freedom, their skills were not well used.

Smooth air-to-ground integration was hampered from the start as planners placed the line for coordinating fire support 85 miles ahead of friendly ground forces. Unfortunately, they also maintained it at or near that distance for the duration of the invasion.[34] As in 1991, this placement of the Fire Support Coordination Line created a refuge for Iraqi forces and caused havoc for air-ground coordination measures. Attempting to offset the gap, air planners implemented an expanded and improved system of Kill Boxes. The new system subdivided Kill Boxes into nine smaller numbered "keypads," instead of the four quadrants used during Desert Storm. The new keypad system enabled air and ground planners to open or close smaller parts of Kill Boxes to airstrikes, theoretically allowing for more flexible air operations. Similar to Desert Storm, the Kill Boxes closest to the Coalition advance were used for close air support, while in Kill Boxes near the Fire Support Coordination Line, interdiction missions attacked Iraqi forces. Still, compared to Desert Storm, the control measures implemented were complex, and the rules of engagement more restrictive. The air control measures system more closely resembled Kosovo or Afghanistan, reflecting a political desire to limit collateral damage, but also limiting the technological ability of the Air Operations Center to control airstrikes on the battlefield directly.

Air planners opened Kill Boxes beyond or near the line to interdiction missions similar to the battlefield air interdiction used during Desert Storm. However, the doctrinal and tactical terms for battlefield air interdiction no longer existed. Instead, the planners at the Air Operations Center coined a new term—Strike Coordination and Reconnaissance—and called them SCAR missions. The term SCAR was first used in Vietnam for airstrikes away from friendly forces, like those along the Ho Chi Minh Trail. However, the SCAR missions used in Iraq in 2003 were an expanded and renamed version of the F-16 Killer Scouts of Desert Storm. In Iraqi Freedom, many different Coalition multirole fighters, such as RAF GR-4 Tornados, US Air

Force F-15Es and F-16s, and US Navy F-18 Hornets filled the role. The solutions provided only a small amount of relief to the muddled air-to-ground support system.[35]

A decade of air-centric operations coupled with the neglect of any air-ground interoperative concept, such as AirLand Battle, left Army-Air Force cooperation rusty. Close air support missions got jammed up over Coalition forces, waiting for targets. However, the rules of engagement required Joint Tactical Air Controllers on the ground to verify targets before aircraft attacked them. Due to the wide gap between the Coalition front and the Fire Support Coordination Line, thinly scattered Tactical Air Control Parties could not confirm every target for the pilots. Also, the ground commander's operational concept bypassed many Iraqi units, leaving potential threats behind the leading edge of the advancing Coalition force. The result was a complex ground picture, fraught with difficulties for Joint Tactical Air Controllers and aircrew. Poor weather and sandstorms further hampered smooth air-ground operations, preventing aircraft from seeing the ground to locate friendly and enemy positions.[36]

Unfortunately, air planners at the Air Operations Center made no concerted effort to harness the available Forward Air Controllers to alleviate the confusion and enhance air power support to the Coalition advance. Instead, they fed Forward Air Controllers to the front sporadically. Similarly, Army fire coordination centers and Air Support Operations Centers never developed a coherent plan to fill ground controller gaps or efficiently manage Kill Boxes with Forward Air Controllers.[37] Finally, the rules of engagement prevented the few available Forward Air Controller missions from independently engaging targets without Joint Tactical Air Controller or Air Operations Center permission. The lack of efficient control mechanisms, confusing and restrictive rules of engagement, and rapidly changing use of Kill Boxes created frustration among the aircrews attempting to support Coalition ground forces. The confusion and restriction caused many Coalition strike aircraft to return to base without expending munitions.[38]

Better use of Forward Air Controllers would have provided an essential interface for air-to-ground coordination and resulted in more efficient operations. Instead, only 230 Forward Air Controller missions flew during Operation Iraqi Freedom between March and April 2003—less than 1 percent of the sorties flown against Iraqi ground forces, despite a potential abundance of Forward Air Controller aircraft and aircrew.[39] Based on aircraft available and aircrew qualification calculations (described later in this chapter), roughly twelve F-16 and twenty-six A-10 FACs were likely available during the first month of Iraqi Freedom. Over the twenty-six-day war, as many as 1,700 Forward Air Controller sorties were possible. Accounting for more sustainable aircrew rest cycles and required ground duties, a more realistic eight hundred sorties were likely available. Yet, neither air nor land planners accounted for them, and the rules of engagement further prevented an efficient use of their skills. Instead, frustration abounded.[40]

Like Afghanistan, the conventional phase of the air war concluded relatively quickly. By April 5, US forces entered Baghdad, and by April 14 they occupied the city of Tikrit in northern Iraq. On May 1, 2003, President George W. Bush declared an end to combat operations. However, the situation in Iraq soon devolved, and by the end of 2003 it was clear Iraq was not secure. An insurgency and sectarian war steadily brewed within the country. The Bush administration attempted to shore up the fragile situation, committing a steadily growing number of US forces to the region and engaging the US in two nation-building and counterinsurgency operations. Just as in Afghanistan, American combat air power in Iraq shifted to "on-call close air support" for the counterinsurgency campaigns. However, the Air Force did not harness Forward Air Controllers as in Vietnam. Instead, the Air Force relied on Joint Tactical Air Controllers.[41]

Unfortunately, Air Force Joint Tactical Air Controllers and their home Air Support Operations Squadrons experienced the same personnel reductions faced by Air Force pilots during the 1990s. During the 1990s, the Air Force cut Air Support Operation Squadrons as the Army also lost frontline combat units, further reducing the points of air-ground cooperation. With decreased personnel, Air Liaison Officers moved from the brigade or battalion level up to the division level or higher. This reduction in expertise prompted the Air Force to qualify enlisted personnel as Joint Tactical Air Controllers, providing a slightly expanded personnel pool to distribute below the division level. Compounding the issue, the Air Force dramatically reduced its ability to produce Forward Air Controllers during the 1990s. The reduced personnel, coupled with shifting ideas of close air support during the 1980s and 1990s and the focus on air power after Desert Storm, resulted in fewer air-ground integration experts in both the Air Force and the Army. The early phases of the wars in Afghanistan and Iraq exposed this gap in personnel, expertise, and integration, prompting a concerted effort by both the Army and the Air Force to improve the situation.[42]

However, the two struggling counterinsurgency campaigns and the resulting troop surges of 2006 and 2009 further strained the system. As the Army increased the number of troops first in Iraq and then Afghanistan, the requirement for Joint Tactical Air Controllers dramatically increased as well. By 2009 the Air Force had one thousand fewer Joint Tactical Air Controllers than required to meet the Army's needs.[43] As a result, the Air Force centralized Joint Tactical Air Controllers to division or brigade operations centers, instead of having controllers in the field supporting infantrymen on patrol. Technology again made this possible. Joint Tactical Air Controllers harnessed video downlinks to stream video from aircraft targeting pods directly to their workstations. Advanced communications enabled them to communicate with both far-flung aircraft and ground units. Further, unmanned platforms provided persistent surveillance and a limited ability to provide fire support, saving the fewer fighter and bomber aircraft to cover the most urgent close air support taskings.[44]

But the new model brought new challenges. Joint Tactical Air Controllers now controlled larger areas, potentially supporting dozens of aircraft over multiple ground units. This model was adequate until friendly troops became heavily engaged by insurgents. Personnel not on the scene often had difficulty discerning the ground situation, slowing their ability to orchestrate aerial response. Further, one controller supporting many field units was often overwhelmed while attempting to control close air support for multiple simultaneous firefights. Joint Tactical Air Controllers and aircrew created ad-hoc means to reduce the workload when situations became heated. One such creation was known as "the Warden." Although not an officially sanctioned doctrinal practice, the role of the Warden was borrowed from special operations forces out of necessity—a substitute for situations typically using

Forward Air Controllers. Joint Tactical Air Controllers usually delegated airspace control to the Warden over a target area, dictating aircraft altitudes and positions around the target. The Warden's role required both radio equipment and aircrew capable of managing the complex process of deconflicting aircraft and weapons from each other, again a responsibility historically charged to Forward Air Controllers. As the Warden became common practice, a new aircraft, the MC-12 Liberty, became the go-to platform for Warden duties.[45]

Conceived in 2008 to fill gaps in Air Force surveillance needs and built on a modified civilian King Air 300 airframe, the MC-12 performed similar missions to unmanned surveillance aircraft like the MQ-1 Predator and MQ-9 Reaper. It had multiple radios, an advanced sensor with a laser target designator, and was capable of four-hour-long missions. The MC-12 also used satellite communication and digital datalink, enabling communications with the Combined Air Operations Center, or other aircraft, thousands of miles away via voice and text messages. The laser designator in the MC-12 sensor could guide laser-guided weapons dropped from other aircraft or generate coordinates for GPS guided weapons.[46]

Despite its many capabilities and its frequent use as the Warden, the MC-12 never became a Forward Air Controller platform. The Air Force simply did not have the time nor the capability for MC-12 aircrew, most of whom were on temporary assignment to the program, to meet the DOD and Air Force training requirements to gain the necessary qualifications. Additionally, the Air Force saw the MC-12 as a temporary stopgap in the unmanned surveillance program, not a close air support aircraft. As a result, the program remained an immature shoestring operation, which the Air Force divested in 2014.[47] Finally, the Warden was a position Air Force leaders at the Air Operations Center were never comfortable with; ultimately, in 2013, guidance from the Combined Air Operations Center banned the use of the Warden.[48] Warden duties blurred the lines between close air support and Forward Air Controller aircraft, giving pilots without proper training a Forward Air Controller's responsibilities of deconflicting aircraft, ground parties, and weapons.[49]

A desperate battle for Combat Outpost Keating in Afghanistan highlighted just how blurry the lines were. Isolated in the rugged and remote

Nuristan province in northeast Afghanistan, near the Pakistani border, the tiny collection of makeshift shelters lay at the junction of two valleys at the base of three prominent ridgelines rising more than two thousand feet above the valley floor. On the morning of October 3, 2009, the eighty-odd Americans of Bravo Troop, 3rd Squadron, 61st Cavalry Regiment at Keating came under attack by more than three hundred insurgents. Due to its small contingent of Americans with minimal organic fire support, the Army relied on close air support to defend the outpost. However, as one report on Bravo Troop's ordeal recounted, "Although close air support was a critical element to the defense plan, there was no [JTAC] team assigned to the outpost."[50] Instead, the Regiment's JTAC was located about twenty miles away at their operations center.[51]

Due to Keating's remote location deep in the mountains, the Joint Tactical Air Controllers were unable to communicate directly with B Troop's fire support officer. Two F-15E Strike Eagles, Dude 25 and Dude 26, arrived first to the scene and began providing close air support to the beleaguered outpost, acting as Warden and radio relay for the Joint Tactical Air Controllers. A third Strike Eagle, Dude 01, arrived about two hours into the attack and took over Warden duties as more close air support aircraft and thunderstorms began to arrive on the scene. Soon after Dude 01 arrived, the other two F-15s returned to base—one due to low fuel and the other because of an aircraft malfunction. The three Strike Eagles helped identify targets, and on more than one occasion intervened to prevent other aircraft from dropping bombs onto friendly positions. After eight hours overhead Dude 01 returned to base as more aircraft and ground reinforcements, including a Joint Tactical Air Controller team, arrived. Unfortunately, eight Americans died and twenty-two were wounded during the battle, despite the efforts of American air power.[52]

Significantly, nowhere in the mission summaries, after-action reports, news articles, or citations awarded to the aircrew is the term Warden mentioned. Instead, the term "FAC," or "Tactical Air Controller," is used in describing the Strike Eagles' role during the fight.[53] Strike Eagle aircrew are not qualified for either mission, nor did the aircrew involved receive any training as the Warden. Ironically, the situation at Keating, and the role of the Warden during its brief lifespan, is precisely what Joint Doctrine cites as the purpose for a Forward Air Controller. They were conceived to act as an airborne extension of the ground-based controller, and the ground commander, to bridge the air-ground gap.[54] The incident at the outpost reveals further weakness in the dispersed Joint Tactical Air Controller-only model, and how much a Forward Air Controller contributes to the battle when situations become confused and desperate.

Examining the state of the Air Force's Forward Air Controller program provides some insight into why the Air Force engaged in the primarily land-based model in Iraq and Afghanistan, instead of the air-ground team used historically. First, the Forward Air Controller force engaged in Iraq and Afghanistan was a shell of the force from 1991, despite the addition of F-16s to the mission. In the decade following Desert Storm, the Air Force saw a 30 percent decrease in nearly every statistical category, including personnel, aircraft inventory, and training hours. Forward Air Controller aircraft suffered one of the largest reductions, losing 52 percent of the fleet. The entire fleet of OV-10, A-37, and O-2s were eliminated by 1995, leaving just the A-10 and the F-16. Both of these aircraft and pilot pools were also reduced, including a loss of 65 percent of all A-10s. Despite intentions to replace the A-10 with F-16s, or a new attack aircraft, budget restrictions and political posturing prevented the acquisition of a new aircraft.[55]

The reductions in the decade between Desert Storm and Enduring Freedom, combined with training programs that levied a larger and more complex training burden, crippled the ability of the Air Force to produce Forward Air Controllers. Some of this complex requirement was a product of the revamped Air Force training in the post-Vietnam era. In the decade after Vietnam, the Air Force centralized and standardized advanced flight training programs with more robust requirements. This revamped training featured more training missions, along with more complex and realistic training, to produce more combat-capable aircrew. The Air Force also implemented a more robust tracking system for all programs, including recurring training requirements. Unfortunately, the more complex regimen increased the number of missions required to produce an Air Force Forward Air Controller. Every level of an Air Force pilot's training required more resources to become qualified and maintain qualifications. The overarching effect was a reduction in the ability of the Air Force to produce Forward Air Controllers at the volume achieved in Vietnam.[56]

Adding to the Air Force's increased training requirements, Forward Air Controller programs across the armed forces became subject to a joint service standardization program. As an outgrowth of both the DOD reorganization under the Goldwater-Nichols Act and the failures during Operation Anaconda, the Joint Chiefs implemented and revamped common training, tactics, and terminology in planning and execution—including both close air support and Forward Air Controller missions.[57] Various joint publications govern the programs, such as *Joint Publication 3.09.3 Close Air Support* and a joint memorandum for FAC training. The joint memorandum, or J-MOA as it is known, articulates the specific requirements to become a qualified Forward Air Controller, and the procedures are common across every branch of the US military. It also dictates the recurring training for aircrew to maintain currency in the mission.[58] These current and qualified terms are important distinctions and require some further explanation.

Beyond the training required to become an Air Force pilot, in order to perform specific missions the pilots must first receive training in that mission. This initial mission qualification applies to tasks such as air-to-air refueling, low-altitude flight, flying while using night-vision goggles, and many others—including the ability to perform the role of a Forward Air Controller. This initial qualification training is considered a basic level of proficiency in these tasks, and further practice is required to improve a pilot's skills for combat. Additional training and experience are also required for a pilot to instruct others in these tasks. To maintain their ability to perform a mission after becoming qualified, a pilot must perform the task at regular intervals. Within a given year, Air Force pilots must perform these mission tasks a certain number of times and within specified time intervals: this is known as currency. Experience and mission difficulty determine the number and frequency of training events required to gain qualification and maintain a pilot's currency in the various missions performed by their aircraft.[59]

The J-MOA standardizes qualification for the Forward Air Controller mission across all services, and the minimum requirements are the same for every aircraft. To become a Forward Air Controller, a pilot must complete twelve attack controls under various conditions, including a minimum of two at night, one control of a helicopter attack, and four utilizing live weapons. Once qualified, a Forward Air Controller must make six attack controls every six months to maintain currency in the mission.[60] The Air Force levies additional Forward Air Controller training requirements for both initial qualification and currency. These requirements differ slightly for A-10 and F-16 pilots.

A-10 pilots must receive two hundred flight hours of experience before being eligible to become a Forward Air Controller. Typically, it requires an A-10 pilot about one year to reach the required number of hours. When selected for the program, the pilots attend a joint academic course and complete several hours of simulator training before undergoing six training flights. F-16 pilots require slightly more experience before being eligible. They must first be qualified to lead two-ship formations, which requires at least three hundred hours of flight experience and a separate qualification process. Typically, it takes slightly less than two-years for an F-16 pilot to reach the required number of hours and complete the two-ship flight lead qualification before being eligible for the FAC qualification process. F-16 pilots must attend the same academic course, and pass simulator training prior to accomplishing six training flights. During these flights, the pilots must meet the training objectives and requirements spelled out in the joint memorandum, along with additional mission-specific objectives dictated by the Air Force. In addition to the memo's currency requirements, the Air Force requires one attack control every sixty or ninety days, depending on the pilot's experience level, to maintain currency.[61]

While this initial and recurring training does not seem like much, Air Force pilots may fly as few as eight missions per month. They must divide these sorties between a wide variety of mission tasks, all having similar or perhaps higher requirements. Further, several of the Forward Air Controller training objectives require the use of outside resources, such as helicopters or Joint Tactical Air Controllers, not usually co-located with Air Force fighter squadrons. Adding in poor weather, mechanical issues, additional ground duties, and other required training, the pilots may get less than the prescribed missions in a month. More likely, they may not have sufficient opportunities to maintain currency in every mission assigned to their squadron.[62]

In the US Air Force, a squadron's Designed Operational Capability Statement, colloquially called the DOC Statement, dictates the missions performed by that squadron. Aircrew in multirole aircraft, such as the F-16 or F-15E, maintain qualifications and currency in a wide array of air-to-air and air-to-ground missions. Further, due to the different models of F-16 in the Air Force inventory, individual F-16 squadrons maintain different missions on their statement. Therefore, some F-16 squadrons have a Forward Air Controller requirement, while others do not. Instead, these squadrons may specialize in other missions, such as suppression of enemy air defenses. As a result, an F-16 pilot may complete their career without being qualified as a Forward Air Controller. In contrast, A-10 squadrons maintain fewer missions on their DOC Statement because they are focused exclusively on air-to-ground specific missions—like FAC, combat search and rescue, and close air support.[63]

The variation in mission requirements resulted in the A-10 and F-16 communities weighting the importance of the Forward Air Controller mission differently. For an A-10 pilot, for example, qualification as a FAC is a prerequisite to becoming a Sandy combat search and rescue mission commander. For A-10 pilots, Sandy qualification is an essential milestone in their career, and the ability of an A-10 pilot to perform that mission influences future job positions within the Air Force. This, in turn, shapes a pilot's opportunity for promotion. In contrast, the Forward Air Controller mission does not serve as a prerequisite for any additional qualification in the F-16. Instead, F-16 squadrons emphasize a pilot's ability to lead large formations and perform instructor duties. F-16 pilots may become Forward Air Controllers in their path to becoming an F-16 instructor or at the USAF's elite Fighter Weapons School, but it is not fundamentally important to an F-16 pilot's career. As a result, F-16 squadrons are less likely to spend training resources qualifying pilots and maintaining their FAC currency. In short, maintaining FAC qualifications in the multirole F-16 is low on the priority list for pilots and Air Force leadership.[64]

The means with which the A-10 and F-16 squadrons manage Forward Air Controller training also reflects the cultural difference in the two communities. Since the A-10 has a more robust pool of current and qualified FAC instructors, and the mission has a higher career status within the community, A-10 training is all accomplished at the pilot's squadron. Further, due to the air-to-ground mission focus of the A-10, resources are readily available for close air support missions—including the required munitions to meet the joint memorandum and Air Force training requirements. In contrast, a multi-role F-16 squadron must cycle through recurring training programs throughout the year. Therefore, during certain times of the month or year, the training focus might be on air-to-air specific missions. During these training cycles, an F-16 squadron may not have the required munitions available to complete Forward Air Controller training, necessitating additional outside assistance. Further, depending on the variant of F-16 flown by a squadron, they may have no requirement to maintain Forward Air Controllers on their staffing roster. As a result, not every USAF F-16 squadron has current Forward Air Controller instructors available to provide the necessary training to a new Forward Air Controller.

Instead, the F-16 community centralized training for all F-16 Forward Air Controllers at the 310th Fighter Squadron in 1997. The centralized training allowed for standardized instruction across the Air Force and created a single source for F-16 FAC tactics, techniques, and procedures. However, the cost of sending pilots to the centralized training fell on individual F-16 squadrons. This limited the number of Forward Air Controllers an F-16 squadron could send to training and depended on the variations in defense budgets, squadron DOC Statement priorities, and other training scheduled for the squadron. The 310th also performed F-16 initial and requalification training for Air Force F-16 pilots. This competing mission further limited the number of Forward Air Controllers the squadron produced each year, due to the larger requirements of F-16 pilot production.[65]

The reduction in FAC aircraft, increased qualification requirements, reduced training opportunities, and lack of emphasis on the mission in multirole F-16 squadrons resulted in fewer Forward Air Controllers available to support combat operations in Iraq and Afghanistan. As the campaigns dragged, the ability to produce Forward Air Controllers competed against the increased requirement to support the combat effort, other training priorities, and Air Force operational requirements. Despite the burden of two counterinsurgencies, the Air Force did not acquire dedicated FAC aircraft or create Tactical Air Support Squadrons as in Vietnam. Further, the Air Force again reduced its total personnel in the mid-2000s. Between 2005 and 2013, the deepest years of American involvement in the wars, the Air Force fell four hundred fighter pilots short of the number needed to fill cockpits, creating a massive gap in its sole source of Forward Air Controllers.[66] Assessing the availability of Air Force FACs, it becomes clear just how limited their ability to support combat operations was, particularly during the troop surges.

In 2010, during the Iraq and Afghanistan surge, the Air Force fielded forty-one A-10 and multirole F-16 squadrons with the Forward Air Controller mission. These twenty-six F-16 and fifteen A-10 squadrons encompassed the Active Duty, National Guard, and Reserve components.[67] Although there is not a required number of qualified Forward Air Controllers within each squadron, historically F-16 squadrons maintained about three current and qualified pilots on their roster, while there were around seven per A-10 squadron.[68] During the dual troop surge years of 2009 and 2010, there were typically four F-16 squadrons and two A-10 squadrons in theater, split between Afghanistan and Iraq. As a result, a maximum of twenty-six current and qualified Forward Air Controllers were possibly deployed. Typically, six two-ship flights of F-16s and two flights of A-10s were airborne in Iraq and Afghanistan at any time. Each flight lasted about four hours, which allowed for thirty-six F-16 flights and twelve A-10 flights covering Afghanistan and Iraq each day. Further, both F-16 and A-10 squadrons maintained aircrew on alert, ready to support emergency close air support requests and combat search and rescue events, adding the requirement for an additional eight F-16 and four A-10 pilots per day.

In the best case, nine F-16 and seven A-10 FACs were potentially airborne during an entire day in Iraq and Afghanistan. However, even with optimum use, these potential FAC missions represent only about 17 percent of all Coalition sorties. Because neither ground nor air planners tried to intentionally use Forward Air Controllers during the surge, it is far more likely that only a small fraction of all missions had one available.[69]

The last deliberately planned Air Force Forward Air Controller mission in Iraq or Afghanistan was scheduled and flown on April 16, 2003. A flight of two A-10s, Kahuna 61 and 62, worked with ground forces south of Baghdad before returning to base without controlling any attacks or expending any of their weapons.[70] Since that day, neither the Air Force nor the Army considered regular use of Forward Air Controllers. This dearth left few peacetime opportunities for soldiers and airmen to train with them. From headquarters staff down to combat units, a hole developed in everyone's knowledge for using a FAC's skills. As one Joint Tactical Air Controller opined, "Nobody's trying to get outside of their comfort zone in combat in regards to close air support ... guys will always fall back on whatever is ingrained in them, at their base level ... If utilizing a [Forward Air Controller] is not part of that base level they've built you're crazy to think that they're going to pull that one out."[71] Still, they sporadically made the difference when a situation on the ground escalated beyond the ability of the Joint Tactical Air Controller to handle it alone.[72]

On the morning of January 28, 2007, nearly seven hundred Iraqi insurgents of the Shiite group known as the Army of Heaven ambushed Iraqi political officials, elements of the Iraqi police, and an Iraqi military unit as the Iraqis attempted to serve arrest warrants at a compound outside the town of Al Najafa, in south-central Iraq.[73] The group "opposed the mainstream Shiite religious establishment in Iraq and were hostile to the US government and Sunni Arab insurgents."[74] Under heavy fire and taking casualties, the Iraqi forces asked for assistance from their US Special Forces advisors. A team of Green Berets also came under heavy fire as they attempted to rescue the embattled Iraqis. The situation escalated further when the insurgents shot down an American Apache attack helicopter providing fire support. As additional special forces teams joined the fight, the scene seemed to be careening towards another disaster akin to Somalia.[75]

Two American F-16s arrived quickly to provide strikes against the determined insurgent resistance and, within a few hours, the Air Operations Center began feeding planes to the battle. However, due to the layout of the battlefield, the Joint Tactical Air Controllers attached to the special forces teams struggled to locate and identify targets for a growing contingent of aircraft. Fortunately, two A-10 FACs, Tusk 5 and Tusk 6, happened to be among the airmen overhead, and they quickly put their skills to work. For the next several hours, the two pilots worked to bring order to the chaos on the battlefield. The first order of business was deconflicting the aircraft over the battle, making sure no airplanes ran into each other. Second, due to the layout of the battlefield, the Joint Tactical Air Controller on the ground could not see and identify the friendly ground parties, dramatically increasing the likelihood of friendly fire. Tusk flight built a picture of the battlefield and ensured no bombs inadvertently dropped on friendly positions. Finally, with everyone safely deconflicted, Tusk flight helped the ground forces locate and generate coordinates for enemy positions the aircraft could use their weapons against.[76]

Night soon arrived, and with it came an Air Force AC-130U Spooky Gunship. The AC-130 is a C-130 Hercules cargo aircraft heavily modified with three side-firing guns—a 40 mm Bofors cannon adapted from WWII era naval antiaircraft guns, a 105 mm Howitzer, and a 25 mm Gatling gun. Paired with the A-10s, the three aircraft began laying down devastating fire, keeping the Soldiers of Heaven pinned down and allowing the US and Iraqi forces reprieve from the fire. The next morning, after the arrival of additional US and Iraqi reinforcements, Coalition forces took the compound. Three hundred seventy members of the Army of Heaven died in the fight, with four hundred taken prisoner, at the cost of four Iraqis killed and two dozen wounded. The Americans suffered ten casualties, including two killed and eight wounded. The two American Apache pilots were the only two US servicemen killed, found dead in their helicopter when a rescue team reached it.[77]

Still, despite the utility, and occasional necessity, of the Forward Air Controllers' skills on the battlefield, they remained an untapped tool of air-ground integration. As American air power expanded operations from Iraq and Afghanistan into Libya and Syria, technological advances allowed even greater amounts of battlefield information to flow directly to senior leaders at the Combined Air Operations Centers. Coupled with concerns over civilian casualties and collateral damage, this abundance of information prompted senior leaders to adopt tighter controls and rules of engagement, restricting the tactical decisions normally made by Forward Air Controllers and Joint Tactical Air Controllers to the highest echelons of command. Streaming video, digital networks, and satellite communication enabled the increasingly centralized control of all airstrikes.

Streaming video was the primary tool enabling senior air and ground commanders to inject control directly on the battlefield. Unmanned aircraft provided the first streaming video capability when the MQ-1 Predator began combat operations during Operation Allied Force in 1999. About the size of a small civilian aircraft and able to stay airborne for more than twenty hours, the Predator provided persistent intelligence collection of sensitive and high-value targets. In a pod resembling a disco ball mounted under the aircraft's nose, the MQ-1 carried a color video camera for daytime operations, an infrared sensor for use at night, and a laser designator. Satellite links enabled Predator operators sitting in ground stations in the US to fly an aircraft anywhere in the world. The same satellite links fed video shot by the sensor ball back to intelligence support personnel for analysis and, when requested, to air and ground operations centers and headquarters. The long endurance streaming video also allowed commanders to watch ground operations and airstrikes as they unfolded in real-time.[78]

In October 2001, the MQ-1 began carrying Hellfire Missiles supporting missions in Afghanistan. About five and a half feet long and weighing in at one hundred pounds, the small laser-guided missile gave the Predator the added ability to strike small targets. Streaming video and the ability to strike targets allowed commanders to bypass Joint Tactical Air Controllers or Forward Air Controllers and direct Predator attacks from thousands of miles away. (Eventually the Air Force phased the MQ-1 out in 2018, replacing it with a growing force of larger and more capable MQ-9 Reapers.)[79]

An ever-growing and diverse fleet of small unmanned and autonomous aircraft complemented the larger aircraft and provided near-continuous surveillance coverage of vast swaths of the battlefield. Since 1999 the number of unmanned aircraft on the battlefield, both large and small, grew exponentially. Commanders' demands for real-time battlefield video proved insatiable. As an

Air Force news article about the closing of the Predator program noted, "In 2011, the MQ-1 and MQ-9 enterprise achieved a monumental milestone: Aircrew flew 1 million combat hours." Significantly, as the article goes on to say, though the one-million-hour mark required more than a decade to reach, the second million was achieved just two years later, "highlighting the demand for [Remotely Piloted Aircraft] operations and support."[80]

Additional streaming video capabilities enabled ground parties in the field to see videos from both surveillance and fighter aircraft. No longer confined to headquarters or operations centers, the Remotely Operated Video Enhanced Receiver (ROVER) transmitted directly to Joint Tactical Air Controllers and ground commanders' laptops and other portable devices. Ground parties could now see what the close air support or reconnaissance aircraft orbiting overhead saw. ROVER's advancement coupled superbly with advanced target pods, such as the Lightning and Sniper pods, first fielded by Air Force fighter and bomber aircraft in the mid-2000s. Designed to replace the less capable LANTIRN pods, the advanced Lightning and Sniper pods have high-definition infrared and color TV cameras, laser trackers, image stabilization, multiple zoom levels, and digital recording capabilities. The advanced sensors and ROVER video links reduced the chance for collateral damage or friendly fire, as both aircrew and ground parties could positively identify enemy combatants from civilians. The video feed allowed ground parties to see the situation from above, providing a view of the battlefield even when under enemy fire.[81]

Digital networks were the second step in revolutionizing how the Army and Air Force conducted close air support and expanded operations centers' control capabilities, allowing them to bypass both Forward Air Controllers and Joint Tactical Air Controllers. Akin to a Wi-Fi network in the sky, these digital networks allow participating aircraft and headquarters a voiceless means of communication. First developed by the US in the 1970s, digital tactical datalink networks offered a means of communication resistant to Soviet jamming and interference in the event of a NATO-Warsaw Pact war in Europe. The short data burst communication directly between aircraft allowed aircraft on the network to quickly and securely see where other aircraft were, and as the capability expanded, communicate via brief fixed format text messages. By the late 1990s, the US adapted the networks for air-to-ground use. The networks allowed aircraft to see the location of friendly ground forces, enabled Joint Tactical Air Controllers to send friendly and target locations via text messages, and allowed aircraft to pass the same data to other aircraft via text. By 2010, Joint Tactical Air Controllers and operations centers could send complete close air support tasking messages, target photos, and clearance to engage targets directly to the cockpit without any voice communication.[82]

In the decade following 2010 the United States engaged in two air campaigns that featured both air-centric operations, like the NATO operations in the Balkans, and the mix of special operations forces supported by precision air power, like the first months of Operation Enduring Freedom. However, the volatile setting for the two operations was born out of the social upheaval of the Arab Spring and was unlike any operation the US military engaged in since the end of World War II. The first, Operation Odyssey Dawn/Unified Protector, attempted first to check Muammar Qadhafi's attacks against rebellious factions in Libya, before ultimately seeking to overthrow the regime or kill Qadhafi. The second, Operation Inherent Resolve, sought to stop the rise of ISIS in the deserts of Syria and Iraq.[83]

During Operations Odyssey Dawn/Unified Protector and Inherent Resolve, strict rules of engagement and a confused ground picture led to tight control of airstrikes by Air Force leaders. Significantly, these strict operational controls would not have been possible without the technology developed in the decades before, and the increasing use of that technology in close air support missions supporting counterinsurgencies in Iraq and Afghanistan. In short, technology increasingly allowed senior leaders to bypass Joint Tactical Air Controllers and Forward Air Controllers and intervene directly on the battlefield, which, in turn further incentivized Air Operations Centers to do so. Interestingly, however, several proposals and debates emerged in both conflicts around the use of Forward Air Controllers on the unconventional battlefields of North Africa and the Levant.

For example, during Operation Odyssey Dawn in Libya in 2011, planners proposed to pair AC-130s and F-15Es with A-10 Forward Air Controllers— a reprised variation of the hunter-killer tactics used in Vietnam and Kosovo.

Political events eclipsed the idea as NATO took control of the mission, and US offensive combat forces withdrew, replaced by air refueling, intelligence, and other support capabilities. Senior air officers also debated the idea of using Forward Air Controllers as Operation Inherent Resolve escalated in 2015. Some saw favorable risk-reward odds by introducing Forward Air Controllers to help stop the ISIS advance; others favored a highly centralized control mechanism. Ultimately, using Forward Air Controllers was rejected again. In both situations, concerns over collateral damage and a chaotic ground situation overruled the efficiency and flexibility offered by Forward Air Controllers. Instead, battlefield decisions became highly centralized, enabled by the technological tools that allowed an unprecedented level of control of air power.[84]

What emerged as a substitute for the traditional means of airstrike control were nondoctrinal entities called "strike cells." The strike cell was an outgrowth of a process known as Time Sensitive Targeting. The term came into use after Operation Allied Force in 1999 and is a target "requiring immediate response because it is a highly lucrative, fleeting target of opportunity or it poses (or will soon pose) a danger to friendly forces."[85] The targets might be leadership, like Osama bin Laden, or weapons, like Iraqi Scud Missiles. Due to their fleeting and elusive nature, such targets require vast intelligence resources to locate and strike rapidly. In America's wars against terrorism, the vast majority of Time Sensitive Targets were local insurgent leaders or other key members of terrorist and insurgent organizations, colloquially known as HVIs, or High Value Individuals.[86]

When ISIS rose in the Iraqi desert in 2014, the process for hunting ISIS leadership became centered in strike cells. As US Army Major General Dana Pittard, who served as a director of these operations, wrote, "The strike cell soon became the dominant method of hunting and killing one of America's most brutal and elusive enemies."[87] Strike cells brought together the various specialties required to evaluate, authorize, and control a time-sensitive strike. Typically, a general officer, either Army or Air Force, served as the target engagement authority and led the strike cell. Several deputies, usually colonels, ran the daily business of the strike cell, with the senior officers stepping in to authorize the most critical or risky attacks.[88]

Intelligence analysts, targeting experts, a legal team, Joint Tactical Air Controllers, and unit representatives rounded out the strike cell personnel. The size of strike cells varied from a handful to a few dozen and depended on the size of their area of responsibility. Using intelligence sources and video feeds from unmanned aircraft, the intelligence personnel of the strike cell tracked down and confirmed the target's location. A team of legal experts vetted potential attacks against standing rules of engagement and any potential collateral damage. When the ranking officer approved a strike, the team passed the information to waiting attack aircraft and then cleared them for the attack. From start to finish, the whole process might last anywhere from a few minutes to days or even weeks depending on the quality of the intelligence, the importance of the target, and the potential for collateral damage. Critically, strike cells worked in headquarters located hundreds or even thousands of miles from the site of the attack and would not have been possible without streaming video, satellite communications, and network enabled aircraft.[89]

The US also used these capabilities to provide close air support to a hodgepodge of Kurdish, Iraqi, and Syrian partner forces as they pushed back against ISIS beginning in 2015. The most critical battles occurred in 2017 during the fights to retake the Iraqi city of Mosul and ISIS's declared capital of Raqqa, in northeastern Syria. In a situation similar to the early months of Operation Enduring Freedom, typically the American ground force amounted to just a handful of US Army combat advisors or special forces teams, and their Air Force Joint Tactical Air Controllers, paired with the local partner forces. As the local forces pressed their attacks, US air power provided critical intelligence via unmanned aircraft overhead, as well as aerial fire power. The two missions paired well together; they were closely coordinated by American JTACs and monitored by senior air officers at the Combined Air Operations Center and strike cells. Requests for close air support were carefully weighed against likely civilian casualties and collateral damage. Uncertainty about the locations of friendly forces, the true urgency of requests for US air power, and an ISIS enemy that notoriously sacrificed the lives of innocents weighed on every decision. The result was a deliberately slow, and occasionally frustrating, air campaign with relatively few airstrikes.[90]

Beyond the strike cells and close air support for partner forces, the Combined Air Operations Centers used similar processes and control measures for planned airstrikes. For example, during NATO's Operation Unified Protector, planned airstrikes on government and military buildings in Tripoli required prestrike video surveillance fed by MQ-1s back to the Combined Air Operations Center. Before the strikes could proceed, the Joint Force Air Component Commander, the most senior air component officer in the campaign, personally watched the video feed to ensure no civilians were nearby and all collateral damage concerns were mitigated.[91] This method of control became increasingly common and was used for planned strikes during both Unified Protector and Inherent Resolve.[92]

Even with a deteriorating or confused ground situation the technological tools available usually allowed a rapid flow of critical information between aircrew, Joint Tactical Air Controllers, and command-and-control agencies. Strike cells directed attacks against high value and fleeting targets with a high degree of confidence, low risk of collateral damage, and few civilian casualties. Intelligence analysts monitored targets for hours or even days before aircrew received attack approval. Aircraft quickly flew to troubled spots and from one target to another. Immediate close air support response times dropped from the twenty to forty-minute responses during Vietnam to an average of just twelve minutes in Iraq and Afghanistan during 2010.[93]

Unfortunately, technology also created poor habits in Army-Air Force coordination. A pattern repeated, similar to the failures of the Operation Anaconda debacle and the struggles during the first month of Iraqi Freedom. As Mike Benitez wrote in a 2016 article about close air support integration, "whereas the Army would spend a week or more planning an operation, fixed-wing aircraft were spread so thin they would often bounce around the country supporting three to four of these operations for an hour or two at a time in a single flight." Both the Army and the Air Force lacked the resources or failed to make the effort for consistent detailed integration. As a result, he continues, operations were "reactive to contact on the ground, rather than proactively synchronizing resources for a ground-based objective." He estimates that "less than 20 percent of close air support sorties were briefed to the aircrew supporting them." Benitez concludes: "Almost every airstrike civilian casualty and fratricide report from the war shares common themes: Air support was un-briefed prior to the mission, reactive in nature, and was not integrated in detail sufficient to prevent tragedy."[94]

While Forward Air Controller's skills saw sporadic use in the wars since 2001, their occasional significant contributions brewed a desire for the close integration critical to a Forward Air Controllers' mission. As one author phrased it, "Relationships are better built face to face. A phone call or email can start a relationship, but nothing replaces actually meeting counterparts and seeing firsthand where they work."[95] As a result, in the summer of 2017 a quiet resurgence of Forward Air Controllers rose in Afghanistan. One F-16 squadron made a concerted effort to expand its FAC contingent and integrate them into ground operations. First, in the months leading up to the F-16 squadron's deployment, they brought the total number of Forward Air Controllers on their roster up to six. While they did not add a substantial number to the overall effort in Afghanistan, it allowed the squadron to cover major operations and particularly dangerous missions. Once the squadron arrived at its deployed operating base, they teamed with a special forces unit deployed to the same base to facilitate planning efforts and maximize their skills. Together with the special forces' leaders and Joint Tactical Air Controllers, they proactively planned, briefed, and debriefed together. Most importantly, although not explicitly tasked to do so by the Combined Air Operations Center, the squadron deliberately scheduled FACs to fly in support of the team's missions.

This effort was not universal, even among other F-16 or A-10 squadrons. In fact, the scenario resembled the individual efforts made during development of the Horseflies in World War II. Initially, senior Army and Air Force leaders were unaware of the cooperation, and the units only coordinated with their superiors after a close working relationship grew between the Forward Air Controllers and Joint Tactical Air Controllers. Instead, it was an effort by one squadron to train more FACs, persistently coordinate with the nearby special forces team, actively schedule qualified pilots to support their mission, and engage in a two-way educational process. The work paid dividends, enabling education and understanding of missions and capabilities to flow in both directions. As a result, the Forward Air Controllers often supported that team, other special forces teams, and other Coalition ground forces. As the squadron weapons officer recounted, "We utilized [Forward Air Controllers] for something like 15 terminal attack controls, probably 30 instances of [attack brief, situation update, target marking, and airspace control], and around 10 artillery [calls for fire]."[96] The Joint Tactical Air Controllers were faster, safer, and more lethal when paired with Forward Air Controllers.[97]

In one instance, the squadron's Forward Air Controllers saved a small contingent of special forces soldiers from being overrun. As the team returned at dusk from a mission, their small encampment came under attack by mortar and heavy machinegun fire. Two died in the initial attack, and twenty others were wounded as the fire continued from every direction. The pilot took control and began locating the enemy firing positions, coordinating with the team's wounded Joint Tactical Air Controller, other units on a nearby ridge, a second flight of F-16s, a B-52 bomber, and medevac helicopters. Bombs began raining down on enemy positions within minutes, dispersing the attack and allowing medevac operations to begin quickly. As the Joint Tactical Air Controller on the scene recounted, "When life was literally at its lowest point ever it was a [F-16 Forward Air Controller] who stepped up on the CAS Team and carried us through. It's really hard to convey in text, and you'd be hard pressed to get an Army guy to admit this but my crew owes our lives to a [Forward Air Controller]."[98]

Forward Air Controllers emerged to do just that; sort through the chaos to apply air power quickly and correctly for ground forces. Both halves of the air-ground team subconsciously need the Forward Air Controller to be the bridge between the two, particularly in the most critical situations. When planners failed to use Forward Air Controllers, as in Anaconda, integration suffered. When ground forces were in the most dangerous of situations and no Forward Air Controllers were available, soldiers and airmen created a substitute on the fly, as in the case of the Warden and Combat Outpost Keating. The dearth of available Forward Air Controllers since 2001 led the Air Force Special Operations Command to qualify their own, because those dangerous missions rely on the seamless integration of air and ground forces.[99] Despite the loss of aircraft, reduction in qualified pilots, technological advances enabling speed and accuracy, and increasingly centralized control, Forward Air Controllers have proven to be a critical resource for integration in the wars since September 11, 2001. It remains to be seen, however, if history will repeat itself and the Air Force will again cast aside Forward Air Controller for the air-centric missions of strategic attack and air superiority.

Conclusion

The Future of FACs

Since the decade following 2010 political and military leaders of the United States recognized a steady shift in the global strategic picture, as the rebirth of Sino-American strategic competition and a resurgent Russia brought peer rivals back to the center of American grand strategy. As the 2017 National Security Strategy states, "These competitions require the United States to rethink the policies of the past two decades—policies based on the assumption that engagement with rivals and their inclusion in international institutions and global commerce would turn them into benign actors and trustworthy partners. For the most part, this premise turned out to be false."[1] The changing global strategic balance prompted an examination of the assumptions behind the US defense strategy. Across the Department of Defense, large-scale combat operations against technologically peer adversaries became the focusing ideology after nearly two decades of counterinsurgencies, just as AirLand Battle served as the doctrinal driver of the US military's transformation after Vietnam. The Air Force's new focus included a reexamination of the role of Forward Air Controllers and their capabilities on the battlefield.

As part of this review, in September 2017 General Mike Holmes, Commander of the Air Force's Air Combat Command, released a memorandum chartering a new CAS Integration Group within the 57th Wing at Nellis Air Force Base, Nevada. General Holmes directed the CAS Integration Group "to advance joint CAS and joint fires culture through a mission-focused organization, build CAS and joint air-ground expertise through the Air Force and into the joint community, and is empowered to train CAS, joint fires, Killbox/SCAR, and maneuver experts to dominate joint operations through air-ground integration."[2]

Central to the CAS Integration Group's mission is the resurrected 24th Tactical Air Support Squadron. An outgrowth of the Jungle Jim program, the 24th operated as a composite squadron supporting special operations forces in the Panama Canal Zone from 1969 until 1991. In 1989 the squadron's A-37s saw action during Operation Just Cause. The Air Force shuttered the squadron during the force drawdown of the 1990s, along with most of the service's Forward Air Controller capabilities. In 2018 the squadron reactivated flying F-16s and, as the group's charter memo stated, was charged with the critical mission of training Air Force Forward Air Controllers and Sandy rescue mission commanders. As a result, the 24th assumed the role as the centralized training squadron for F-16 Forward Air Controller and Sandy pilots from the 310th Fighter Squadron. The Air Force also charged the group with supporting Joint Tactical Air Controller, Air Support Operations Center, and Air Liaison Officer training; running the Air-Ground Operations Course's descendant, the Joint Firepower Course; creating close air support doctrine, tactics, and procedures; and finally, to coordinate the focused close air support Green Flag series of exercises.[3]

The CAS Integration Group and the 24th Tactical Air Support Squadron were products of the realization within the Air Force that two decades of air power support to counterinsurgencies had, paradoxically, hindered air-ground cooperation. Unfortunately, the Air Force again shuttered the squadron in 2020. Despite the lessons of Operation Anaconda and Combat Outpost Keating, doctrinal structures and institutional knowledge critical to efficient air-ground coordination atrophied because of the way the wars in Iraq and Afghanistan were fought. Technological capabilities, heroic efforts by airmen and soldiers, temporary stopgaps, on-the-spot substitutes, and restrictive rules of engagement masked the absence of true integration. Lost in the last two decades were the critical systems and the knowledge for smooth coordination between air and land forces. As a 2009 Air Force lessons learned report observed, neglect in the Air Force close air support system "led to a state of 'creeping normalcy' whereby systemic atrophy became the accepted norm."[4] Since 2001, these tools also enabled the Air Force and Army to simultaneously leave Forward Air Controllers off the battlefield, defying history in the process. Looking back on the history of Air Force Forward Air Controllers illuminates the path to how and why this happened.

First, the Air Force and its predecessors always relied on ground-based liaisons to serve as the primary connection for air-ground cooperation. Beginning in World War I, individual Air Liaison Officers assisted Army divisions in planning and the use of air power to support their operations. During World War II, Rover Joes served as the first Tactical Air Control Parties, extending the division Air Liaison Officers' and Pineapple Tactical Air Control Centers' eyes and ears to the front. In Korea, Forward Air Controllers increasingly controlled airstrikes in place of Tactical Air Control Parties, and Forward Air Controllers' control of airstrikes reached its pinnacle in Vietnam. With the development of AirLand Battle and the expansion of advanced surface-to-air and air-to-air threats, reliance on Forward Air Controllers waned, returning the Tactical Air Control Party to its place of primacy. Since 2001, the shortage of Joint Tactical Air Controllers forced both the Army and the Air Force to rely on video, satellite communications, and datalink networks to expand the Tactical Air Control Party's reach. The simultaneous shortage of Air Force Forward Air Controllers tipped the scales in favor of the ground-based system as fewer on the air-ground team understood Forward Air Controllers' capabilities and purpose on the battlefield.

Second, technology and force reductions also shaped the Air Force's doctrinal focus in the decade between Desert Storm and Enduring Freedom. Desert Storm reopened the premise that air power alone could win wars, and the campaigns in the Balkans seemed to confirm that possibility. Stealth and precision-guided weapons dramatically upended the calculus of what air power could do. Blessed with massive technological advantages, fighting wars that seemed to confirm institutional bias, and experiencing a decade of budget, personnel, and aircraft losses, the Air Force prioritized core air-centric missions. In the decade after Desert Storm, air superiority and strategic attack won primacy over air-ground cooperative missions, structures, and doctrine.[5]

Third, despite Forward Air Controllers' stated doctrinal mission, the Air Force's predilection for air-centric missions, combined with muddled control over the battlespace straddling the bomb line and fire support coordination line, resulted in a historical bias by the Air Force to use Forward Air Controllers in interdiction missions rather than close air support. During World War I and II, radio technology often prevented reliable air-to-ground communications, making attacking enemy forces intermingled with friendly forces dangerous. Lacking reliable communications, cavalry reconnaissance patrols directed their attacks well ahead of the advancing infantry, and Horseflies regularly operated in the space beyond the reach of friendly artillery, near the bomb line. Similarly, when the Korean War stagnated near the 38th Parallel in mid-1951, the Mosquitoes regularly prowled behind the enemy front conducting reconnaissance and controlling interdiction strikes. The entire Out-Country FAC program in Vietnam, minus support to special forces raids and the brief US incursion into Cambodia, was a pure interdiction campaign by any doctrinal definition. With little ground force to speak of during the wars in the Balkans, only a handful of strikes fell into the category of close air support. Instead, the A-10 and F-16 FACs led a hunter-killer style interdiction campaign against Serbian forces. Finally, with a focus on reducing Iraqi land-based combat power before initiating the Coalition invasion, A-10 FACs along with F-16 Killer Scouts controlled interdiction strikes more often than close air support during Desert Storm.[6]

Fourth, the slow pace of air operations since 2001 built the illusion that Forward Air Controllers were not necessary; this contributed to the breakdown of Army-Air Force close air support integration, fostered the implementation of restrictive rules of engagement, and allowed the use of technological and Forward Air Controller-like substitutes. Historically FACs thrived on rapidly shifting, chaotic battlefields where land forces struggled to discern the enemy arrayed against them, and air forces struggled to maintain awareness of the ground situation. Indeed, they became necessary in such situations. Often FACs appear in the most chaotic situations: cavalry reconnaissance missions of World War I, the Horseflies in World War II, the Mosquitoes in Korea, and the Killer Scouts of Desert Storm.

Forward Air Controllers were also necessary on battlefields where command-and-control struggled. This struggle usually coincided with rapidly shifting and chaotic fronts, but also where technological limits and geographic factors necessitated Forward Air Controllers to help sort out the air-ground picture. As happened in World War II, Korea, Vietnam, and the Balkans, they provided the critical link between command-and-control, attack aircraft, and ground forces where no technological solution was readily available or suitable to overcome the distances and terrain of the battlefield. Their ability to persistently survey the battlefield from above in real-time gave Forward Air Controllers a perspective unavailable to others. After 2001, however, the air-ground command-and-control team instead used streaming video, satellite communications, datalink networks, and a reliance on remotely piloted aircraft almost constantly streaming video into headquarters.

Finally, the paltry number of close air support missions flown in US air campaigns since 2001 allowed for these developments. To this point, during the dual surge year of 2010, Operations Iraqi Freedom and Enduring Freedom reached a combined high-water mark of just over forty-one thousand close air support sorties for the entire year. For comparison, that equals the close air support and battlefield air interdiction total in just one month of Desert Storm. Comparing the current counterinsurgency operations to those in Vietnam, the totals of Iraq and Afghanistan represent only one-quarter the number of annual close air support and interdiction sorties flown at the peak of air operations in South Vietnam.[7] Further, close air support missions in Iraq and Afghanistan, at their high mark in 2010, used weapons just 12 percent of the time. By comparison, more than 60 percent of close air support and interdiction sorties in Desert Storm delivered weapons against Iraqi Army targets. That number also pales compared to World War II, where nearly 90 percent of all missions delivered weapons against their targets. In short, since 2001 the Air Force has, comparatively, flown far fewer close air support missions per year of conflict while using weapons at a dramatically reduced rate.[8]

In wars before 2001, when force structure changes, budget reductions, divergent doctrine, and battlefield conditions created gaps in air-ground integration, the Air Force repeatedly implemented Forward Air Controller programs to fill those gaps. Historically, the Air Force augmented Tactical Air Control Parties with a rapidly expanding Forward Air Controller force during conflicts, including a sustained training program and acquisition of new aircraft—as demonstrated by the creation of the Mosquitoes, and the explosion of programs in Vietnam. Even during the brief conflict of Desert Storm, the Air Force added F-16 Killer Scouts to provide FAC capabilities in areas unsafe for A-10s to operate. Eventually, F-16 Forward Air Controllers permanently joined the Air Force during the wars in the Balkans.

After 2001, however, a glacial operational tempo on the battlefield, technological substitutes, and specific rules of engagement that centralized execution to the highest levels of the air component, all simultaneously restricted and negated the use of Forward Air Controllers. Significantly, these substitutes were quickly overwhelmed when operations grew larger, or the urgency of the situation increased—as happened during Anaconda, the first month of Iraqi Freedom, the firefights with the Army of Heaven, at Combat Outpost Keating, and the case of the special forces team attacked in 2017. The post-2001 conflicts denied Forward Air Controllers their critical and historic role in dynamic and chaotic battlefield situations and created an unsustainable close air support operations framework. As Mike Benitez wrote, "this plan surely won't work in a high-intensity conflict requiring air support that demands increased integration."[9]

What is the possible future for the Air Force Forward Air Controller with the return to peer competition? The situation facing the Air Force in the decade after 2021 mirrors the period between 1965 and 1975: a budget-constrained force; still fighting insurgencies; in need of modernized aircraft, doctrine, and warfighting concepts; and facing modernizing threats in both peer adversaries and less-capable regional powers. Just as the threat realities of a NATO-Warsaw Pact war in Europe divided the Forward Air Controller community during the Cold War and diminished the role of Forward Air Controllers, the Air Force now seems similarly torn in the course Forward Air Controllers should adopt. Facing both an expanding threat of modern air defense technology and multiple simultaneous global counterinsurgencies and counterterrorism campaigns, Forward Air Controllers are caught in the turmoil of an on again-off again Light Attack program, as well as questions surrounding the role of the F-35, F-16, and the A-10 on the future battlefield. Predicting the future of warfare is a notoriously tricky business. However, examining historical constants within the Forward Air Controller mission will provide insights for the next decade or more.

First, the ground-based personnel of the Air Support Operation Squadrons will remain as the primary point of contact between the Army and the Air Force. By extension, the Tactical Air Control Party will remain as the primary air power conduit for friendly ground forces. Both history and current doctrine confirms this as the Air Force's preferred construct.[10] To that end, in 2009 the Air Force began increasing its pool of Air Liaison Officers and Joint Tactical Air Controllers with the implementation of new Air Force career fields. These new Tactical Air Control Party Officers will serve as Air Liaison Officers and Joint Tactical Air Controllers, replacing and augmenting the pool of aircrew who historically filled those positions.[11] Second, based on historical precedents, Forward Air Controllers will be used in interdiction as often as close air support, despite doctrinal insistence that Forward Air Controllers primarily facilitate close air support. Even the F-16 and A-10 FACs acknowledge this reality. To that end, they are developing new concepts to facilitate cooperation and increase airpower's efficiency in support of the land component, particularly in large land campaigns.

One such concept is the battlefield coordination line. An airspace control measure used by the US Marines, the battlefield coordination line aims to reduce the historic sanctuary provided to the enemy near the fire support coordination line. To do so, the battlefield coordination line serves as an intermediate point of coordination marked by the maximum range of Army artillery systems. Attacks in the area between the battlefield coordination line and the fire support coordination line blend elements of close air support and interdiction. As a result, successfully engaging such targets is historically difficult and contentious. Forward Air Controllers argue this is the critical reason they exist in the first place, to serve as the bridge between air and land power. In this space the FAC, who has a deeper understanding of the ground

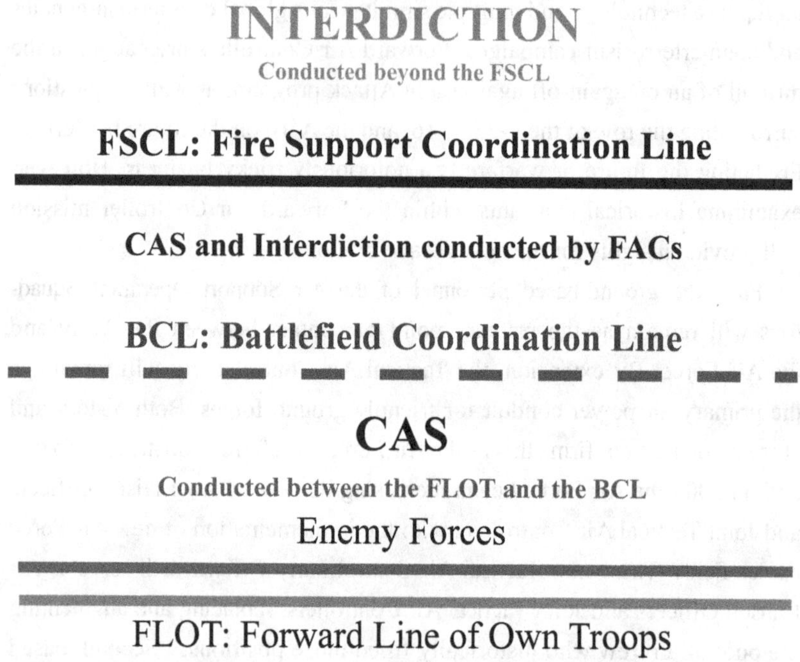

INTERDICTION
Conducted beyond the FSCL

FSCL: Fire Support Coordination Line

CAS and Interdiction conducted by FACs

BCL: Battlefield Coordination Line

CAS
Conducted between the FLOT and the BCL
Enemy Forces

FLOT: Forward Line of Own Troops

Figure 3 Battlefield Coordination Line with interdiction and CAS

scheme of maneuver than most aircraft, coordinates attacks against enemy forces to facilitate friendly advances or deflate enemy attacks, just as they have done since World War I.[12]

One reason for this change is because close air support is historically difficult to do well. It requires immense amounts of time, personnel, interservice-cooperation, and dedicated structures and processes. Even when done quickly, it is still slower than organic Army firepower.[13] Historically, the Army solved this problem with attack helicopters and, more recently, with improved artillery systems.[14] To that end, the Army will undoubtedly continue to pursue more advanced organic firepower systems, such as improved GPS guided artillery, improved attack helicopters, and unmanned aircraft of all sizes and capabilities, in order to provide the fastest fire support possible. As a result, in large-scale land combat scenarios close air support will revert to its historical role of augmenting Army firepower, rather than substituting for them as it has since 2001.

Finally, one type of Forward Air Controller aircraft is insufficient to meet the diverse battlefields of the future. While future Slow FACs in light attack airplanes will be networked, possess advanced sensors, and carry smart weapons, they are vulnerable to advanced surface-to-air threats, as was the case in Vietnam. In contrast, Fast FACs will need to be stealthy to survive on the advance battlefield of the future, but their ability to survive advanced threats makes them a poor choice for low threat and unconventional war zones. In between, Fourth Generation F-16s can operate in both scenarios but are optimized for neither. F-16s are more survivable, as Desert Storm and the Balkans proved, but still vulnerable to the latest threat technology. Similarly, they have repeatedly proven capable in the counter-insurgencies in Iraq and Afghanistan, but their high fuel consumption and maintenance requirements make them an expensive substitute for capable FAC aircraft in a low-tech conflict. Similarly, A-10s are less survivable in advanced air defense networks, as demonstrated in Desert Storm. The A-10 was designed based on close air support experiences in Vietnam, making them well suited for medium and low threat scenarios, but their aging airframes are increasingly expensive to maintain.[15] The current Air Force fleet is again caught at a crossroads, just as happened after every major conflict since World War I.

For the Air Force, however, fielding an expanded and diverse fleet of Forward Air Controllers to augment Tactical Air Control Parties and organic Army firepower is likely to become increasingly important in future conflicts, not less so. First, given the developmental pace and direction of defense technology, in future conflicts the strategic attack missions will likely be conducted by a combination of cyber-attack, space-based attack, and advanced hypersonic missiles, as well as manned and unmanned advanced stealth aircraft.[16] Further, in each conflict since World War II, the percentage of strategic attack and air superiority missions has decreased as a percentage of the total sorties. In contrast, interdiction and close air support missions have increased. These missions, historically, are the realm of Forward Air Controllers. To explore how future Forward Air Controllers might conduct close air support and interdiction missions, it is best to examine three broad potential conflict scenarios.[17]

The first of the three scenarios is a peer adversary war. Although the least likely, it is the most dangerous scenario, one modeled on the Cold War NATO-Warsaw Pact conflict in Europe. In this scenario, the US faces an adversary with advanced land, sea, and air forces, along with nuclear weapons. Potential enemy capabilities to oppose the Air Force include advanced surface-to-air missiles, advanced fourth and fifth-generation fighter aircraft, and advanced cyber and space capabilities. The second scenario is a conflict against a regional power. This conflict pits US forces against older but still capable threats. Iraq in 1991 fits as a good historical model for this scenario. Finally, US involvement in counterinsurgency and counterterror operations will likely continue for the foreseeable future. These scenarios match the campaigns in Afghanistan and Iraq after 2003 but also include hybrid threats such as the Islamic State. In these scenarios, the US enjoys significant technological advantages but faces an enemy meshed into the surrounding civilian population.

In the peer adversary war, it is easy to imagine stealthy and networked F-35 Forward Air Controllers prowling the deep battle. Using their sensors along with the other networked manned and unmanned platforms, the F-35s will direct the attacks of other aircraft, long-range artillery, artificial-intelligence-powered drone swarms, or cruise missiles via datalink messages and advanced communications. Against a regional power, a similar scenario will likely take place in the opening phases of the campaign. However, after degrading the enemy air defenses, Fast FAC Hunter-Killers operating on both sides of the fire support coordination line will use advanced networked technology and traditional tactics to mark targets and provide control for interdiction and close air support missions. Until their retirement, A-10s will support close air support missions nearest friendly forces, enabling rapid and efficient air power in support of the most critical areas of the land war. In these two scenarios, the use of the battlefield coordination line will allow both ground and airborne controllers even more flexibility to locate and engage enemy forces on a dynamically shifting battlefield.

Finally, as demonstrated by their use in Vietnam (and their absence since 2001), Forward Air Controllers serve a purpose in the third scenario of low-intensity, counterinsurgency, counterterrorism, and special forces operations. Using smaller propeller-driven aircraft, updated Slow FACs will feature the same advanced sensors, streaming video, and network capabilities as modern fighters at a fraction of the cost.[18] In future low-intensity conflicts, they will also be able to harness manned, unmanned, and organic autonomous drone platforms to assist ground teams in need of air power. Operating from forward bases similar to the Slow FACs in Vietnam, they will adopt abbreviated procedures to use firepower, allowing for the faster response of both air- and land-based firepower. As a recent article highlighted, "An MQ-9 Reaper can provide overwatch for a long time without refueling, but lacks the pilot-in-the-cockpit situational awareness of an F-16, and the kinetic capability of an AC-130. In the future, [Air Force Special Operations Command] may need to, in a time of limited budgets, be able to combine many of these capabilities into one airframe that can operate in austere areas."[19]

To match the operational and tactical demands for more Forward Air Controllers, the Air Force will need to expand its pool of current and qualified pilots, in addition to adding a Slow FAC capability. The F-16 and shrinking A-10 community represent the only available pools of Air Force Forward Air Controllers. Unfortunately, as of 2019, there is no plan to add a Forward Air Controller mission to F-35 squadrons. Given the Air Force plans to replace the F-16 with the F-35 as its primary multirole fighter, it seems imperative the Air Force begin qualifying F-35 pilots with the mission. The F-35 pilot community is a hodgepodge of pilots from the F-16, A-10, and F-15E. Therefore, the pool of former Forward Air Controllers already exists within the F-35 squadrons. The Air Force needs a Fast FAC platform to survive advanced threats, and the F-35 makes the most sense to fill that role in both capabilities and pilot knowledge.[20]

Similarly, the F-15E, the Air Force's only other multirole fourth-generation fighter besides the F-16, does not currently perform the mission. Although initially fielded as a strategic attack and interdiction platform, the F-15E may be the best suited to bridge the Forward Air Controller gap as the A-10 phases out. The aircraft features advanced sensor, communication, and network capabilities, it also carries a large and varied weapons payload and a large fuel capacity, allowing for long loiter times on the battlefield. Additionally, the airplane has a two-seat cockpit, a feature common to almost every Forward Air Controller aircraft in history. In flight, the two-person F-15E crew divides the combat duties between the pilot and weapons system operator, enabling the fighter to manage a higher workload than a single seat airplane. Despite its potential, the F-15E community has historically avoided adding the FAC mission to their already full agenda.

Beyond adding to the potential Forward Air Controller pool for future combat operations, increasing the number of aircraft increases the opportunities for joint training, which, in turn, increases familiarity and knowledge of air-ground integration within both the Army and the Air Force. The increased knowledge and familiarity at the tactical level will shape planning and doctrine from the tactical up to the strategic level—influencing long term force planning, acquisitions, and budgeting for both services.[21] A similar process occurred after Vietnam as a decade of improved and integrated air-ground operations led to new operational concepts, joint processes, and weapons systems via the 31 Initiatives. These new processes, concepts, and weapons systems led directly to the overwhelming success of Desert Storm.

Ultimately, it is the level of interservice cooperation that Forward Air Controllers both instill and facilitate in the Air Force and the Army that makes them critical to success on the battlefield. Future success depends on Forward Air Controllers not being neglected as they were after nearly every American conflict. The lack of a robust Forward Air Controller program historically reflected an inward turn by both the Air Force and Army, producing a focus on doctrine, tactics, and systems that rendered air power in a single dimension. These low tides of cooperation inevitably led to struggles on the battlefield, contentious service relationships, and negative press for the Air Force. As Farmer and Strumwasser put it in their analysis of the Mosquitoes' efforts in Korea and their broader effects on the Air Force, "Perhaps an even more important lesson than that of the effectiveness of the airborne Forward Air Controller is the course an innovation takes in Air Force attitudes and doctrine."[22]

The multidimensional version of air power, which is created by a force that embraces the level of air-ground cooperation where the

Air Force Forward Air Controller thrives, is what leads to success on the battlefield and highlights the accomplishments of air power. While the Air Force recently made positive steps in this direction, the question remains, will the Air Force learn from its mistakes? Will Air Force Forward Air Controllers again be a key enabler of air-ground cooperation, just as they have since World War I, or will the Air Force leave them off the battlefield again?

Endnotes

Notes for Introduction

1. Timothy Kline, *The Airborne FAC in the Korean War* (Washington, D.C.: Department of the Air Force, 1976), 4, Air Force Historical Research Agency (AFHRA) IRIS Num 01020874.
2. Kline, *The Airborne FAC*, 3.
3. Phillip S. Meilinger, ed., *The Paths of Heaven: The Evolution of Air power Theory* (Maxwell Air Force Base: Air University Press, 1997), 225, 357. Alfred F. Hurley, *Billy Mitchell: Crusader for Air Power* (Bloomington: Indiana University Press, 1975), 31. Samuel Hynes, *The Unsubstantial Air: American Fliers in the First World War* (Kindle ed., New York: Farrar, Straus and Giroux, 2014), 3167, 3192.
4. Richard P. Hallion, *Strike from the Sky: The History of Battlefield Air Attack, 1911–1945* (Washington, D.C.: Smithsonian Institute Press, 1989). Thomas Alexander Hughes, *Overlord: General Pete Quesada and the Triumph of Tactical Air Power in World War II* (New York: The Free Press, 1995). David N. Spires, *Air Power for Patton's Army: The XIX Tactical Air Command in the Second World War* (Washington, D.C.: Air Force History and Museums Program, 2002).
5. Robert F. Futrell, *The United States Air Force in Korea 1950–1953* (Washington, D.C.: Office of Air Force History, 1961). William Y'Blood, *Down in the Weeds—Close Air Support in Korea* (Washington, D.C.: Air Force History and Museums Program, 2002). W.M. Cleveland, *Mosquitoes in Korea* (Portsmouth: Peter R. Randall Publisher, 1991). Timothy Kline, *The Airborne FAC*. Gary Robert Lester, *Mosquitoes to Wolves: The Evolution of the Airborne Forward Air Controller* (Maxwell Air Force Base: Air University Press, 1997).
6. Bernard C. Nalty, *Air War over South Vietnam 1968–1975* (Washington, D.C.: Air Force History and Museums Program, 2000). Bernard C. Nalty, *The War against Trucks: Aerial Interdiction in Southern Laos 1968–1972* (Washington, D.C.: Air Force History and Museums Program, 2005). Donald J. Mrozek, *Air Power and the Ground War in Vietnam: Ideas and Actions* (Maxwell Air Force Base: Air University Press, 1988).
7. Mike Jackson and Tara Dixon-Engel, *Naked in Da Nang: A Forward Air Controller in Vietnam* (St. Paul: Zenith Press, 2004). Christopher Robbins, *The Ravens: The Men Who Flew in America's Secret War*

in Laos (New York: Crown, 1987). Rick Newman and Don Shepperd, *Bury Us Upside Down: The Misty Pilots and the Secret Battle for the Ho Chi Minh Trail* (New York: Ballantine Books, 2006).

8. Christopher E. Haave and Phil M. Haun, *A-10s Over Kosovo: The Victory of Airpower Over a Fielded Army as Told by the Airmen Who Fought in Operation Allied Force* (Maxwell Air Force Base: Air University Press, 2003).

9. Craig D. Wills, *Airpower, Afghanistan, and the Future of Warfare: An Alternative View* (Maxwell Air Force Base, AL: Air University Press, 2006). Dag Henriksen, ed., *Airpower in Afghanistan 2005–10: The Air Commanders' Perspectives* (Maxwell Air Force Base: Air University Press, 2014). Larry Goodson, *Afghanistan's Endless War: State Failure, Regional Politics and the Rise of the Taliban* (Seattle: University of Washington Press, 2001). For further reading see Lester Grau, "Wars in Afghanistan," *Oxford Bibliographies in Military History*, October 28, 2014, reviewed April 28, 2017. DOI: 10.1093/OBO/9780199791279-0128.

10. Wolf, *The United States Air Force: Basic Documents.*

11. Chairman of the Joint Chiefs of Staff, *DOD Dictionary of Military and Associated Terms* (Washington, D.C.: Chairman of the Joint Chiefs of Staff, 2019), 89 (hereafter CJCS).

12. Vincent Aiello and David Culpepper, "055: Forward Air Controllers." *The Fighter Pilot Podcast*, The Muscle Car Place Podcast Network, podcast. MP3 audio, 1:33:24, https://www.fighterpilotpodcast.com/episodes/055-forward-air-controllers/.

13. CJCS, *DOD Dictionary*, 211.

14. J. Farmer and J. M. Strumwasser, *The Evolution of the Airborne Forward Air Controller: An Analysis of Mosquito Operations in Korea* (Maxwell AFB, AL: Air University Press, 1967), 84.

15. Ibid., 84.

16. Kevin N. Lewis, *The U.S. Air Force Budget and Posture over Time* (Santa Monica, CA: RAND, 1990), 65–72. U.S. Air Force Financial Management and Comptroller, "Budget FY97 to FY21," *FM Resources*, https://www.saffm.hq.af.mil/FM-Resources/Budget/.

17. United States Air Force, "Definition," *Air Force Doctrine Annex 3–70: Strategic Attack* (Maxwell AFB, AL: Curtis E. LeMay Center for Doctrine, 2019), 5.

18. United States Air Force, "Introduction to Counterair Operations," *Air Force Doctrine Annex 3–01: Counterair Operations* (Maxwell AFB, AL: Curtis E. LeMay Center for Doctrine, 2019), 2–3.

19. United States Air Force, *Air Force Doctrine Annex 3–01.*

20. United States Air Force, "Role of Counterland Operations," *Air Force Doctrine Annex 3–03*, 3.

21. United States Air Force, "Air Interdiction," *Air Force Doctrine Annex 3–03*, 36.

22. United States Air Force, "Air Interdiction," *Air Force Doctrine Annex 3–03*. Burt Brown, interview with author, October 8, 2019.

23. United States Air Force, "Close Air Support," *Air Force Doctrine Annex 3–03*, 36.

24. For a broader discussion of air power strategy, see Colin S. Gray, *Air Power for Strategic Effect* (Maxwell Air Force Base: Air University Press, 2012); Phillip A. Meilinger, *Airmen and Air Theory: A Review of the Sources*, rev. ed (Maxwell Air Force Base, AL: Air University Press, 2001); and Meilinger, *The Paths of Heaven.*

25. Hearings before the House Committee on Armed Services, The National Defense Program—Unification and Strategy, 81st Cong., 1st sess., 1949, 193–200. John Schlight, *Help from Above: Air Force Close Air Support of the Army 1946–1973* (Washington, D.C.: Air Force History and Museums Program, 2003), 98.

Notes for Chapter 1

1. Hynes, *The Unsubstantial Air*, chapter 1 and 2. Allan Janus, "Lion Cubs? Yeah, We've Got Lion Cubs, Too." *National Air and Space Museum Archives Division, Online.* December 27, 2010, https://airandspace.si.edu/stories/editorial/lion-cubs-yeah-weve-got-lion-cubs-too.

2. Holley, *Ideas and Weapons*, 25–27.

3. David McCullough, *The Wright Brothers* (New York: Simon and Schuster, 2015), 100–101.

4. Bernard C. Nalty, *Winged Shield, Winged Sword: A History of the United States Air Force. Vol. 1, 1907–1950* (Washington, D.C.: Air Force History and Museums Program, 1997), 4.

5. McCullough, *The Wright Brothers*, 110–111.

6. Nalty, *Winged Shield Vol. 1*, 9–12.

7. Nalty, *Winged Shield Vol. 1*, 12–14. McCullough, *The Wright Brothers*, 131–137.

8. Walter J. Boyne, "Foulois," *Air Force Magazine* (June 20, 2008), 82.

9. Boyne, "Foulois," 82.

10. Shiner, *Foulois, and the U.S. Army Air Corps*, 1–2.

11. Foulois, *From the Wright Brothers to Astronauts*, 45–47.

12. Shiner, *Foulois, and the U.S. Army Air Corps*, 3.

13. Foulois, *From the Wright Brothers to the Astronauts*, 2.

14. Benjamin Foulois, "Chronology of Aviation," unpublished manuscript, USAFA, Benjamin Foulois Papers, Series Two, Box 6, Folder 4. Megan Cunningham, ed., *Logbook of Signal Corps No. 1: The U.S. Army's First Airplane* (Washington, D.C.: Air Force History and Museums Program, 2004).

15. Sherman, *Air Warfare*, vii–viii. Captain Chandler, "Field Order Number 1, March 5, 1913," Headquarters, First Aero Squadron, Texas City, Texas, AFHRA IRIS Number K-SQ-Bomb-1-H1.

16. Shiner, *Foulois, and the U.S. Army Air Corps*, 6.

17. Shiner, *Foulois, and the U.S. Army Air Corps*, 6. Alan Stephens, *The War in the Air 1914–1994*, American Edition (Maxwell Air Force Base: Air University Press, 2001), 11.

18. "1914," *WWI Aviation History Timeline*, http://www.worldwar1 centennial.org.

19. Geoffrey Wawro, *A Mad Catastrophe: The Outbreak of World War I and the Collapse of the Habsburg Empire* (New York: Basic Books, 2014), 140.

20. Stephens, *War in the Air*, 187.

21. Nalty, *Winged Shield Vol. 1*, 27. *WWI Aviation History Timeline*. http://www.worldwar1centennial.org. Tony Reichhardt, "The First Aerial Combat Victory," *Air and Space Magazine, Online* (October 4, 2014), https://www.airspacemag.com/daily-planet/first-aerial-combat-victory-180952933/.

22. "1914," *WWI Aviation History Timeline*, http://www.worldwar1 centennial.org. Miller, *A Preliminary to War*, 51. T. Dodson Stamps and Vincent J. Esposito, *A Short Military History of World War I: Atlas* (West Point: United States Military Academy, 1950), 12 and 22.

23. Hurst, *Pancho Villa and Black Jack Pershing*, 121. Benjamin Foulois, *The Operations of the U.S. Punitive Expeditionary Forces from May 11 to June 30, 1916*, unpublished manuscript dated October 10, 1916, USAFA Benjamin Foulois Papers, Series Two, Box 5, Folder 1.

24. Miller, *A Preliminary to War*, 15.

25. Shiner, *Foulois, and the U.S. Army Air Corps*, 8. Foulois, *The Operations of The U.S. Punitive Expeditionary Forces*.

26. Miller, *A Preliminary to War*, 43–46. Benjamin Foulois, *Air Service Lessons Learned during the War*, unpublished manuscript dated January 29, 1919, USAFA, Foulois Papers, Series V, Box 8, Folder 5.

27. Shiner, *Foulois, and the U.S. Army Air Corps*, 8.

28. Foulois, *From the Wright Brothers to the Astronauts*, 46. Hurley, *Billy Mitchell*, 20–26. Hynes, *The Unsubstantial Air*, 3167.

29. Levine, *Mitchell, Pioneer of Air Power*, 89–96. Stephens, *The War in the Air*, 11.

30. Levine, *Mitchell, Pioneer of Air Power*, 100. Hurley, *Billy Mitchell*, 20–26.

31. "Flying Yanks," *American Battlefield Monuments Commission*, accessed April 16, 2017, https://www.abmc.gov/sites/default/files/interactive/interactive_files/FY/.

32. Hynes, *The Unsubstantial Air*, 141.

33. James Norman Hall and Charles Bernard Nordhoff, *The Lafayette Flying Corps during the First World War*. Vol. 1 (New York: Leonaur Ltd., 2014), 16–18.

34. *"La Fayette Nous Voici!: L'entré en guerre des États-Unis Avril 1917"* Collection *"Mémoire er Citoyenneté"* No. 41. Paris: Ministrére de la Défense. Hynes, *The Unsubstantial Air*, 141–218, 3390.

35. Georges Thenault, *The Story of the LaFayette Escadrille, Told by its Commander Captain Georges Thenault*, trans. Walter Duranty (Boston: Small, Maynard, 1921), 12–13.

36. Levine, *Mitchell, Pioneer of Air Power*, 101.

37. Levine, *Mitchell, Pioneer of Air Power*, 101–102. Foulois. *Air Service Lessons Learned during the War*.

38. Thomas G. Bradbeer, *Battle for Air Supremacy Over the Somme: 1 June–30 November 1916* (Fort Leavenworth: U.S. Army Command and General Staff College, 2004), 15–16.

39. Ian Sumner, *German Air Forces 1914–1918* (New York: Osprey Publishing, 2006), 16–18.

40. Stephens, *The War in the Air*, 15.

41. Sumner, *German Air Force*, 13–16. Stephens, *The War in the Air*, 8.

42. Foulois, *From the Wright Brothers to the Astronauts*, 150–156.

43. Foulois, *From the Wright Brothers to the Astronauts*, 147. Maurer Maurer, ed., *The U.S. Air Service in World War I*, Vol. 4 (Washington, D.C.: Air Force Historical Research Agency, 1979), 17. Hynes, *The Unsubstantial Air*, 128.

44. Hurley, *Billy Mitchell*, 34–35.

45. Foulois, *From the Wright Brothers to the Astronauts*, 160–161. Levine, *Mitchell, Pioneer of Air Power*, 106. Hurley, *Billy Mitchell*, 1. Barney Sneiderman, *Warriors Seven: Seven American Commanders, Seven Wars, and the Irony of Battle* (Havertown: Savas Beatie, 2006), 201. Hurley, *Billy Mitchell*, 34–35.

46. Foulois, *From the Wright Brothers to the Astronauts*, 162.

47. Foulois, *From the Wright Brothers to the Astronauts*, 171.

48. Maurer, *U.S. Air Service in World War I*, Vol. 4, 29. Hynes, *The Unsubstantial Air*, 3412.

49. John Guttman, *The USAS 1st Pursuit Group* (New York: Osprey Publishing, 2008), 13.

50. Stamps and Esposito, *A Short Military History of World War I*, 62–65.

51. Maurer, *U.S. Air Service in World War I*, Vol. 4, 30–31.

52. Maurer, *U.S. Air Service in World War I*, Vol. 4, 31. Hynes, *The Unsubstantial Air*, 3305, 3800.

53. Mitchell, *Mitchell Diary*, 197. USAFA, Mitchell Papers, Box 5 Series Six, Microfilm Roll 3.

54. Frandsen, "America's First Air-Land Battle," 31–38.

55. Maurer, *U.S. Air Service in World War I*, Vol. 4, 197–220.

56. Foulois, *From the Wright Brothers to the Astronauts*, 175–176.

57. Brigadier General William Mitchell, "Battle Orders No. 1. September 11, 1918." *Mitchell Diary 1917–1919* (Headquarters Air Service, First Army, American Expeditionary Forces). USAFA, Mitchell Papers, Box 5 Series Six, Microfilm Roll 1. Maurer, *U.S. Air Service in World War I*, Vol. 4, 37.

58. Maurer, *U.S. Air Service in World War I*, Vol. 4, 39–40. Hurley, *Billy Mitchell*, 34–35. Hynes, *The Unsubstantial Air*, 3345–3353, 3682, 3800, 4133.

59. Hynes, *The Unsubstantial Air*, 3682–3710, 4133.

60. Geoffrey Wawro, *Sons of Freedom: The Forgotten American Soldiers Who Defeated Germany in World War I* (New York: Basic Books, 2018), 228–229. Hurley, *Billy Mitchell*, 34–35.

61. Hynes, *The Unsubstantial Air*, 2919–2935.

62. U.S. Army War College, ed., *The Means of Communication between Aeroplanes and the Ground* (Washington, D.C.: Government Printing Office, 1917). USAFA, Jefferson H. Davis Papers, Series 2, Box 1, Folder 2. Stephens, *The War in the Air*, 14. Hynes, *The Unsubstantial Air*, 2960.

63. U.S. Army War College, *Utilization and Role of Artillery Aviators in Trench Warfare* (Washington, D.C.: Government Printing Office, 1917). U.S. Army War College, *The Means of Communication between Airplanes and the Ground* (Washington, D.C.: Government Printing Office, 1917). General Headquarters American Expeditionary Forces, *Liaison for All Arms* (France, 1918). USAFA, Jefferson H. Davis Papers, Series 2, Box 1, Folder 2.

64. "Carte De France et Des Frontieres A Type 1912, Remonville" USAFA, Jefferson H. Davis Papers, Series 2, Box 1, Folder 2.

65. Brigadier General William Mitchell, *Supplementary Plan of Usement of Air Service Units, 1st American Army*. September 23, 1918 (Headquarters Air Service, First Army, American Expeditionary Forces,) 4. USAFA, Jefferson H. Davis Papers, Series 2, Box 1, Folder 6. Hynes, *The Unsubstantial Air*, 4174.

66. Mitchell. *Mitchell Diary*, 257.
    Maurer, *U.S. Air Service in World War I*, Vol. 4, 42.

67. Henry William Dwight, *Diary 30 June 1918–15 November 1918*. USAFA, Henry Dwight Papers, Series Two, Box 3, Folder 1. Hynes, *The Unsubstantial Air*, 3022–3030.

68. Dwight, "Monday, October 28, 1918," *Diary*. Hynes, *The Unsubstantial Air*, 3665.

69. Ibid., "Saturday, November 2, 1918."

70. Ibid., "Monday, November 4, 1918."

71. Ibid., "Tuesday, November 5, 1918." Hynes, *The Unsubstantial Air*, 4174.

72. Maurer, *U.S. Air Service in World War I*, Vol. 4, 17.

73. Stephens, *The War in the Air*, 12. Hynes, *The Unsubstantial Air*, 128, 4494.

74. Stephens, *The War in the Air*, 6 and 15.

75. Maurer, *U.S. Air Service in World War I*, Vol. 4, 43–44.

76. Mitchell, *Mitchell Diary*, 257, 290. Hynes, *The Unsubstantial Air*, 3451–3468, 3792, 3849, 4199.

77. Shiner, *Foulois and the U.S. Army Air Corps*, 15–16. *Hearing before the Subcommittee of the Committee on Military Affairs*, Reorganization of the Army, 66th Congress, 1st sess., October 16, 1919. USAFA, Benjamin Foulois Papers, Series 5, Box 7, Folder 12.

78. Foulois, *The Tactical and Strategical Value of Dirigible Balloons and Dynamical Flying Machines*. Shiner, *Foulois and the U.S. Army Air Corps*, 3, 8. Foulois, *From the Wright Brothers to the Astronauts*. James W. Hurst, *Pancho Villa and Black Jack Pershing: The Punitive Expedition in Mexico* (Westport: Praeger, 2008), 121. Roger G. Miller, *A Preliminary to War: The First Aero Squadron and the Mexican Punitive Expedition* (Washington, D.C.: Air Force History and Museums Program. 2003), 15.

79. Shiner, *Foulois and the U.S. Army Air Corps*, 19.

80. Hynes, *The Unsubstantial Air*, 3200.

81. Shiner, *Foulois and the U.S. Army Air Corps*, 17.

82. William Mitchell, *Winged Defense: The Development and Possibilities of Modern Air Power—Economic and Military* (London: G. P. Putnam's Sons, 1925).

83. Shiner, *Foulois and the U.S. Army Air Corps*, 43.

84. Ibid., 125–149.

85. Ibid., 171–192.

86. Shiner, *Foulois and the U.S. Army Air Corps*, 193–265. Haun, *Lectures of the Air Corps Tactical School*, 19–45. Meilinger, *Airmen and Air Theory*, 14. Hurley, *Billy Mitchell*, chapter 4.

87. Meilinger, *Airmen and Air Theory*. Meilinger, *The Paths of Heaven*. Haun, *Lectures of the Air Corps Tactical School*. Huges, *Overlord*, 52–53. Thomas Kessner, *The Flight of the Century: Charles Lindbergh and the Rise of American Aviation* (Oxford: Oxford University Press, 2010). David T. Courtwright, *Sky as Frontier: Adventure, Aviation, and Empire* (College Station: Texas A&M University Press, 2004).

88. Kessner, *The Flight of the Century*. Courtwright, *Sky as Frontier*.

89. Rebecca Hancock Cameron, *Training to Fly: Military Flight Training 1907–1945* (Washington, D.C.: Air Force History and Museums Program, 1999). Futrell, *Ideas Concepts, and Doctrine*. Nalty, *Winged Shield Vol. 1*. Tate. *The Army and Its Air Corps.*, 83–105. Brigadier General William Mitchell, "Testimony Before Congressional Committees. *United Air Service Hearing before a subcommittee*, 66th Congress, 2d Session December 4–5, 8–13, 15, 20 1919; February 3, 9, 12–13, 1920 (Washington, D.C.: Government Printing Office 1920). 1–2. USAFA, Mitchell Papers, Series Two, Box 4, Folder 1.

90. Hynes, *The Unsubstantial Air*, 2984, 3022–3030.

91. Mitchell, *Mitchell Diary*, 257.

92. Brigadier General William Mitchell, *Report of Inspection during Winter of 1923*. 122. USAFA, Mitchell Papers, Series One, Box 2, Folder 3. Cameron. *Training to Fly*, 259, 262, 264, 295.

93. Cameron, *Training to Fly*, 259, 262, 264, 295.

94. Ibid., 295.

95. Brigadier General William Mitchell, *Report of Inspection during Winter of 1923*. 66–67, 108–116. USAFA, Mitchell Papers, Series One, Box 2, Folder 3.

96. Mitchell, *Report of Inspection*, 49–63, 122–153.

97. Ibid., 126–127.

98. Haywood Hansell, "Course: Air Force." The Air Corps Tactical School, Maxwell Field, Alabama, 1932–1933, 4, USAFA, Haywood Hansell Papers, Addendum 2, Series Seven, Box 8, Folder 8.

99. Haywood Hansell, "Course: Air Force. Map Problem 2, Immediate Support of Ground Forces—The Strategic Defensive," The Air Corps Tactical School, Maxwell Field, Alabama, 1937–1938, 4, USAFA, Haywood Hansell Papers, Series Six, Box 5, Folder 4. Hughes, *Overlord*, 56.

100. Haywood Hansell, "Development of the U.S. Air Forces Philosophy of Air Warfare Prior to Our Entry into World War II," undated and unpublished, 3–4, Hansell Papers, USAFA, Series Three, Box 4, Folder 1.

101. Futrell, *Volume I, Ideas, Concepts, Doctrine*, 68, 92. Haun, *Lectures of the Air Corps Tactical School*, 21–23. Hughes, *Overlord*, 53–54.

102. Haun, *Lectures of the Air Corps Tactical School*, 24–26.

103. Haun, *Lectures of the Air Corps Tactical School*, 26.

104. Hughes, *Overlord*, 76–77.

105. Haun, *Lectures of the Air Corps Tactical School*, 31.

106. Futrell, *Volume I, Ideas, Concepts, Doctrine, 157*.

107. James P. Tate, *The Army and its Air Corps: Army Policy toward Aviation, 1919–1941* (Maxwell Air Force Base: Air University Press, 1998), chapter 6.

108. Tate, *The Army and its Air Corps*, 178.

Notes for Chapter 2

1. Futrell, *Volume I, Ideas, Concepts, Doctrine*, 101. Wesley Craven and James Cate, eds., *The Army Air Forces in World War II, Vol. 1* (1948; repr., Washington, D.C.: Office of Air Force History, 1983), 247.

2. War Department, *Basic Field Manual FM 31–35: Aviation in Support of Ground Forces* (Washington, D.C.: US Government Printing Office, 1942). Haun, *Lectures of the Air Corps Tactical School*, 8–18. Craven and Cate, *The Army Air Forces in World War II, Vol. 1*. 141–150, 157, 246–257. Bechthold, "A Question of Success," 821–851.

3. Christopher Shores, Giovanni Massimello, and Russell Guest, *A History of the Mediterranean Air War 1940–1945*, 4 Vols. (Philadelphia: Casemate, 2012–2018). Robert S. Ehlers Jr., *The Mediterranean Air War: Airpower and Allied Victory in World War II* (Lawrence: University Press of Kansas, 2015) 250–251, 255.

4. 12th Air Force, "12th Air Force History," *Fact Sheet*, (October 07, 2015), https://www.12af.acc.af.mil/About-Us/Fact-Sheets/Display/Article/667427/12th-air-force-afsouth-history/. Matthew St. Clair, *The Twelfth U.S. Air Force Tactical and Operational Innovations in the Mediterranean Theater of Operations 1943–1944* (Maxwell Air Force Base: Air University Press, 2007), 5–9. Bechthold, "A Question of Success," 821–851. Ehlers, *The Mediterranean Air War*, 250–251.

5. 12th Air Force, "12th Air Force History." St. Clair, *The Twelfth U.S. Air Force*, 5–9. T. Dodson Stamps and Vincent Esposito, *A Military History of World War II: Atlas* (West Point: United States Military Academy, 1953), 80–86. Shores, Massimello, and Guest, *A History of the Mediterranean Air War*, Vol. 2. Hughes, *Overlord*, 86–87. Ehlers, *The Mediterranean Air War*, 250–252. Rick Atkinson, *An Army at Dawn: The War in North Africa, 1942–1943* (New York: Henry Holt and Co, 2002).

6. United States War Department, *FM 1–10: Tactics and Techniques of Air Attack* (Washington, D.C.: US Government Printing Office, 1940). United States War Department, *FM 31–35: Basic Field Manual, Aviation in Support of Ground Forces* (Washington, D.C.: US Government Printing Office, 1942). United States War Department, *FM 1–15 Tactics and Techniques of Air Fighting* (Washington, D.C.: US Government Printing Office, 1942). Hughes, *Overlord*, 86–87. Hallion, *Strike from the Sky*, 163–165. Bechthold, "A Question of Success," 821–851.

7. United States War Department, *FM 31–35*, 10–15.

8. Ehlers, *The Mediterranean Air War*, 258.

9. St. Clair, *The Twelfth U.S. Air Force*, 5–9. Maurer Maurer, ed., *Combat Squadrons of the Air Force: World War II* (Washington, D.C.: Office of the Air Force History Headquarters, 1969). Spires, Air *Power for Patton's Army*, 440. Shores, Massimello, and Guest, *A History of the Mediterranean Air War*, Vol. 2.

10. Christopher R. Gabel, *The US Army GHQ Maneuvers of 1941* (Washington, D.C.: US Army Center of Military History, 1992). Robert Citino, "The Louisiana Maneuvers," The National WWII Museum, https://www.nationalww2museum.org/war/articles/louisiana-maneuvers.

11. United States War Department, *FM 31–35*, 3–7. Hughes, *Overlord*, 86–87. Hallion, *Strike from the Sky*, 157–162. Spires, Air *Power for Patton's Army*, 383–416, 440. Bechthold, "A Question of Success," 821–851. Ehlers, *The Mediterranean Air War*, 256–257.

12. Hallion, *Strike from the Sky*, 171. Emphasis original.

13. Daniel R. Mortensen, *A Pattern for Joint Operations: World War II Close Air Support North Africa* (Washington, D.C.: U.S. Army Center of Military History, 1987), 70. Shores, Massimello, and Guest, *A History of the Mediterranean Air War*, Vol. 3.

14. St. Clair, *The Twelfth U.S. Air Force Tactical*, 10–14. Stamps and Esposito, *World War II: Atlas*, 89–89. Spires, *Air Power for Patton's Army*, 446, 467. Bechthold, "A Question of Success," 821–851. Ehlers, *The Mediterranean Air War*, chapter 11.

15. St. Clair, *The Twelfth U.S. Air Force*, 24–29. Hallion, *Strike from the Sky*, 177–179. Hughes, *Overlord*, 94–97, 100. Ehlers, *The Mediterranean Air War*, 295–296, 303. Shores, Massimello, and Guest, *A History of the Mediterranean Air War*, Vol. 4. Rick Atkinson, *The Day of the Battle: The War in Sicily and Italy 1943–1944* (London: Little, Brown, 2007), 96–105.

16. Spires, *Air Power for Patton's Army*, 332. St. Clair, *The Twelfth U.S. Air Force*, 24–29. Michael C. McCarthy, *Air-to-Ground Battle for Italy. Second ed* (Maxwell Air Force Base: Air University Press, 2006), 12–24. Maurer, *Combat Squadrons of the Air Force*. Hughes, *Overlord*, 94–97. Hallion, *Strike from the Sky*, 224.

17. Maurer, *Combat Squadrons of the Air Force*. St. Clair, *The Twelfth U.S. Air Force Tactical Air Command*, 24–29. U.S. War Department War Film, "Film Communique #02: A Day with the A-36's (Sicilian Campaign); Report from Berlin." *WF 20–2*, U.S. Army Signal Corps, 1943. Penn State Special Collections Library, Penn State University, Eighth Air Force Archive, Vol. 140 (hearafter Penn State). "North American A-36A Mustang," *National Museum of the U.S. Air Force* (April 14, 2015), https://www.nationalmuseum.af.mil/Visit/Museum-Exhibits/Fact-Sheets/Display/Article/196283/north-american-a-36a-mustang/. Hallion, *Strike from the Sky*, 177–178.

18. Hardy D. Cannon, *Box Seat Over Hell: The True Story of America's Liaison Pilots and Their Light Planes in World War Two* (San Antonio: Alamo Liaison Squadron, 2007), 17–18.

19. Cannon, *Box Seat Over Hell*, 17–18, 103–135. Maurer, *Combat Squadrons of the Air Force*, 344.

20. Cannon, *Box Seat Over Hell*, 57–58.

21. Ibid., 58–60.

22. Cannon, *Box Seat Over Hell*, 62–69. Alfred W. Schultz with Kirk Neff, *Janey: A Little Plane in a Big War* (Middletown, CT: Southfarm Press, 1998). Ehlers, *The Mediterranean Air War*, 306.

23. "Brodie System," *Smithsonian's National Air and Space Museum*, https://airandspace.si.edu/multimedia-gallery/nasm-9a001183640jpg.

24. St. Clair, *The Twelfth U.S. Air Force*, 35–39. Stamps and Esposito, *World War II: Atlas*, 94–96. Atkinson, *The Day of the Battle*, 179–185, 197–201.

25. Mortensen, *A Pattern for Joint Operations*, 78–83.

26. United States War Department, *FM 100-20: Command and Usement of Airpower* (Washington, D.C.: US Government Printing Office, 1943). Hallion, *Strike from the Sky*, 177. Hughes, *Overlord*, 110. War Department, *FM 100–20*, 1.

27. St. Clair, *The Twelfth U.S. Air Force*, 35–39. Stamps and Esposito, *World War II: Atlas*, 94–96. Ehlers, *The Mediterranean Air War*, 299–300, 310–312.

28. CJCS, *DOD Dictionary of Military and Associated Terms*, 12. St. Clair, *The Twelfth U.S. Air Force*, 40–42. Ehlers, *The Mediterranean Air War*, 307. Lt. Col. John W. Hansborough, *Activities of Air Support Control Squadron, 19 February 1944*, Headquarters Fifth Army, Air Support Control, AFHRA IRIS Num 9480-39A.

29. St. Clair, *The Twelfth U.S. Air Force*, 41.

30. Stamps and Esposito, *World War II: Atlas*, 98–100. Atkinson, *The Day of the Battle*, 323–328, 351–357. Ehlers, *The Mediterranean Air War*, 324–328.

31. Hallion, *Strike from the Sky*, 181. Brigadier General Gordon P Saville, *Operation Memorandum No. 1: Pineapple Missions*, AFHRA IRIS Num 00246718. U.S. Air Force Oral History Interview, "Lt Gen Glenn O. Barcus, 10–13 August 1976" (Washington, D.C.: Office of Air Force History, 1976), 126–145, AFHRA IRIS Num 1032024. Hansborough, *Activities of Air Support Control Squadron, 8 July 1944*.

32. Ehlers, *The Mediterranean Air War*, 307. Quentin C. Aanenson, *A Fighter Pilot's Story*, Quentin C. Aanenson; in association with WETA-TV, 1999, 180 mins., https://quentinaanenson.com/. Hallion, *Strike from the Sky*, 181. "Lt Gen Glenn O. Barcus," 126–145. Hansborough, *Activities of Air Support Control Squadron, 8 July 1944*. Ian Gooderson, *Air Power at the Battlefront: Allied Close Air Support in Europe 1943–1945* (London: Frank Cass, 1998), 28.

33. XXIX Tactical Air Command, "The Effectiveness of Close in Air Support," AFHRA IRIS Num 239305. Saville, *Pineapple Missions*. Hallion, *Strike from the Sky*, 181.

34. "Technique 'Horsefly' in Ground Cooperation," *Air Operations Brief* vol. 4 (Feb. 12, 1945): 8–10, AFHRA IRIS Num 00156676. Hansborough, *Activities of Air Support Control Squadron*, 8 July 1944. Hallion, *Strike from the Sky*, 181. Atkinson, *The Day of the Battle*, 536–555. Ehlers, *The Mediterranean Air War*, 335–340.

35. Cannon, *Box Seat Over Hell*, 124–131. Department of the Navy, Bureau of Ships, *SC 3585S, Catalogue of Electronic Equipment: Vol. 3 of 3, S Series through Miscellaneous*, NavShips 900,116—Supplement No. 5, 1 October 1952, AR-19, http://www.radiomanual.info/schemi/Surplus_Handbooks/Catalogue_of_electronic_equipment_Part3_extract_NAVSHIPS_900-116_1952.pdf.

36. "Technique 'Horsefly' In Ground Cooperation." 8–10. Hansborough, *Activities of Air Support Control Squadron*, 8 July 1944. Hallion, *Strike from the Sky*, 181–182.

37. "Technique 'Horsefly' In Ground Cooperation," 10.

38. Ehlers, *The Mediterranean Air War*, 349–355.

39. Spires, *Air Power for Patton's Army*, 151.

40. Hughes, *Overlord*, 59.

41. Hughes, *Overlord*, 111–115. Spires, *Air Power for Patton's Army*, 679–733.

42. Hughes, *Overlord*, 111–115. Hallion, *Strike from the Sky*, 193.

43. Hughes, *Overlord*. Spires, *Air Power for Patton's Army*, 679–733. Hallion, *Strike from the Sky*, 193. Craven and Cate, *The Army Air Forces in World War II*, Vol. 2, 134–135.

44. Williamson Murray, *Strategy for Defeat: The Luftwaffe 1933–1945* (Maxwell Air Force Base: Air University Press, 1983). Craven and Cate, *The Army Air Forces in World War II, Vol. 1.*, 655. Craven and Cate, *The Army Air Forces in World War II*, Vol. 2, 63–66, 162–166. Haun, *Lectures of the Air Corps Tactical School*, 208–224.

45. "Eighth Air Force Intelligram, May 1945," Penn State, Box 123, Folder 22. Penn State, Joe Fulton Papers, Box 68, Folder 40.

46. Craven and Cate, *The Army Air Forces in World War II*, Vol. 2, 141.

47. Craven and Cate, *The Army Air Forces in World War II*, Vol. 2, 138–181. Spires, *Air Power for Patton's Army*, 794. Hughes, *Overlord*, 114, 126–134. Hallion, *Strike from the Sky*, 193.

48. William B. Reed et al., eds., *Condensed Analysis of the Ninth Air Force in the European Theater of Operations* (1946; repr. Washington, D.C.: Office of Air Force History, 1984), 20.

49. Reed et al., *Ninth Air Force Operations*, 22. Craven and Cate, *The Army Air Forces in World War II*, Vol. 2, 138–181. Spires, *Air Power for Patton's Army*, 794. Hughes, *Overlord*, 114. Hallion, *Strike from the Sky*, 193.

50. Aanenson, *A Fighter Pilot's Story*. Spires, *Air Power for Patton's Army*, 862–867. Hallion, *Strike from the Sky*, 203. Hughes, *Overlord*, 141. Reed et al., *Ninth Air Force Operations*, 21.

51. Reed et al., *Ninth Air Force Operations*, 21. Hughes, *Overlord*, 144–154, 156–169.

52. Reed et al., *Ninth Air Force Operations*, 23.

53. Gooderson, *Air Power at the Battlefront*, 149–151, 251. Reed et al., *Ninth Air Force Operations*, 26. Hughes, *Overlord*, 200–219. Hallion, *Strike from the Sky*, 206–214. Stamps and Esposito, *World War II: Atlas*, 61–62.

54. Craven and Cate, *The Army Air Forces in World War II*, Vol. 2, 230. Hughes, *Overlord*, 200–219. Hallion, *Strike from the Sky*, 206–214.

55. Craven and Cate, *The Army Air Forces in World War II*, Vol. 2., 231–234. Gooderson, *Air Power at the Battlefront*, 149–151, 251. Reed et al., *Ninth Air Force Operations*, 26. Hughes, *Overlord*, 200–219. Hallion, *Strike from the Sky*, 206–214. Stamps and Esposito, *World War II: Atlas*, 61–62.

56. Gooderson, *Air Power at the Battlefront*, 149–151, 251. Reed et al., *Ninth Air Force Operations*, 26. Hughes, *Overlord*, 200–219. Hallion, *Strike from the Sky*, 206–214. Stamps and Esposito, *World War II: Atlas*, 61–62.

57. Craven and Cate, *The Army Air Forces in World War II*, Vol. 2, 234–236. Gooderson, *Air Power at the Battlefront*, 149–151, 251. Reed et al., *Ninth Air Force Operations*, 26–30. Hughes, *Overlord*, 228–231. Hallion, *Strike from the Sky*, 226–227. Spires, *Air Power for Patton's Army*, 1395, 1485–1502.

58. "U.S. Tactical Air Power in Europe, A Special Issue." *Impact* 3, no. 5 (May 1945), 29. Craven and Cate, *The Army Air Forces in World War II*, Vol. 2, 239–241. Gooderson, *Air Power at the Battlefront*, 149–151, 251. Reed et al., *Ninth Air Force Operations*, 26–30. Hughes, *Overlord*, 228–231. Hallion, *Strike from the Sky*, 226–227. Spires, *Air Power for Patton's Army*, 1395, 1485–1502.

59. XXIX Tactical Air Command, "The Effectiveness of Close in Air Support," AFHRA IRIS NUM 239305. Gooderson, *Air Power at the Battlefront*, 149–151, 251. Reed et al., *Ninth Air Force Operations*, 26–30.

Hughes, *Overlord*, 228–231. Hallion, *Strike from the Sky*, 226–227. Spires, *Air Power for Patton's Army*, 1395, 1485–1502.

60. Aanenson, *A Fighter Pilot's Story*.

61. Ibid. Maurer, *Combat Squadrons of the Air Force*, 250–252. Stamps and Esposito, *World War II: Atlas*, 64. Hughes, *Overlord*, 240–241. Hallion, *Strike from the Sky*, 218–219.

62. Aanenson, *A Fighter Pilot's Story*.

63. Aanenson, *A Fighter Pilot's Story*. Hughes, *Overlord*, 274–293. Hallion, *Strike from the Sky*, 106–107, 198.

64. Aanenson, *A Fighter Pilot's Story*. Maurer, *Combat Squadrons of the Air Force*, 250–252. Stamps and Esposito, *World War II: Atlas*, 64.

65. Edward J. Krasnoborski et al, "WWII Major Operations of the Asian and Pacific Theater," United States Military Academy, West Point. Ronald H. Spector, *Eagle against the Sun: The American War with Japan* (New York: Free Press, 1985). Ian W. Toll, *The Conquering Tide: War in the Pacific Islands, 1942–1944* (New York: W. W. Norton, 2015). William Manchester, *American Caesar: Douglas MacArthur, 1880–1964* (Boston: Little, Brown, 1978).

66. Craven and Cate, *The Army Air Forces in World War II*, Vol. 4. 6–16, 44. Kit C. Carter and Robert Mueller, *The Army Air Forces in World War II: Combat Chronology 1941–1945* (1973; repr., Washington, D.C.: Center for Air Force History, 1990), 8, 12.

67. Stamps and Esposito, *World War II: Atlas*, 130. Carter and Mueller, *Combat Chronology*, 8. Maurer, *Combat Squadrons of the Air Force*, 9. Craven and Cate, *The Army Air Forces in World War II*, Vol. 4., 592, 690–693. Benjamin Cooling, ed., *Case Studies in the Development of Close Air Support* (Washington, D.C.: Office of Air Force History, 1990), 295.

68. Craven and Cate, *The Army Air Forces in World War II*, Vol. 4 and Vol. 5. George C. Kenney, *Air War in the Pacific: The Journal of General George Kenney, Commander of the Fifth U.S. Air Force* (Los Angeles: P-47 Press, 2018). William N. Hess, *Pacific Sweep: The V and XIII Fighter Commands in World War II* (Garden City, NY: Doubleday, 1974). Krasnoborski et al, "WWII Major Operations of the Asian and Pacific Theater."

69. Joe G. Taylor, *Close Air Support in the War against Japan* (Washington, D.C.: U.S. Air Force Historical Division, 1945), 52, 241.

70. Cooling, *Case Studies in the Development of Close Air Support*, 295–311. Taylor, *Close Air Support in the War against Japan*, 6–10, 14, 32, 45, 52.

71. Cooling, *Case Studies in the Development of Close Air Support*, 311, 328.

72. Kenney, *Air War in the Pacific*, 340–430, 2588. Craven and Cate, *The Army Air Forces in World War II*, Vol. 4. 626.

73. Taylor, *Close Air Support in the War against Japan*, 341, 345. Craven and Cate, *The Army Air Forces in World War II*, Vol. 4., 504. Maurer, *Combat Squadrons of the Air Force*, 19, 25, 29.

74. Kenney, *Air War in the Pacific*. Hess, *Pacific Sweep*. Craven and Cate, *The Army Air Forces in World War II*, Vol. 4. 126, 626, 690–693. Cooling, *Case Studies in the Development of Close Air Support*, 325–333. Carl A. Post, "Forward Air Control: A Royal Australian Air Force Innovation" *Air Power History* 53, no. 4 (Winter 2006): 4–11. Taylor, *Close Air Support in the War against Japan*, 52, 131–133.

75. Cooling, *Case Studies in the Development of Close Air Support*, 329.

76. Taylor, *Close Air Support in the War against Japan*, 138–150, 155, 158.

77. Kenney, *Air War in the Pacific*. Hess, *Pacific Sweep*. Craven and Cate, *The Army Air Forces in World War II*, Vol. 4., 126, 626, 690–693. Taylor, *Close Air Support in the War against Japan*, 222, 254, 261, 305–307, 340.

78. Taylor, *Close Air Support in the War against Japan*, 222, 254, 261, 305–307, 340.

79. Ibid., 222, 254, 261, 305–307, 340.

80. Taylor, *Close Air Support in the War against Japan*, 176–178. Office of Statistical Control, *Army Air Forces Statistical Digest: World War II* (Washington, D.C.: Comptroller of the Air Force), 230. Craven and Cate, *The Army Air Forces in World War II*, Vol. 4, 690–693.

81. Taylor, *Close Air Support in the War against Japan*, 210.

82. Ibid., 216.

83. Taylor, *Close Air Support in the War against Japan*, 342–345. Craven and Cate, *The Army Air Forces in World War II*, Vol. 5, 532, chapter 18, chapter 20. Paul W. Tibbets, *Return of the Enola Gay* (Columbus: Mid Coast Marketing, 1998), 216. US Air Force, "Twentieth Air Force Fact Sheet," *Fact Sheets* (May 1, 2019), https://www.20af.af.mil/About-Us/Fact-Sheets/Display/Article/825610/twentieth-air-force-fact-sheet/.

84. Hughes, *Overlord*. Spires, *Air Power for Patton's Army*. Kenney, *Air War in the Pacific*. Hess, *Pacific Sweep*.

85. Spires, *Air Power for Patton's Army*, 5754.

86. Reed et al., *Ninth Air Force Operations*, 93. Krasnoborski et al., "WWII Major Operations of the Asian and Pacific Theater." Spector,

Eagle against the Sun. Toll, *The Conquering Tide.* Manchester, *American Caesar.*

87. XXIX Tactical Air Command, "The Effectiveness of Close in Air Support," AFHRA IRIS NUM 239305. Reed et al., *Ninth Air Force Operations,* 92, 123–125. Craven and Cate, *The Army Air Forces in World War II,* Vol. 2, 272. *History of the XXII Tactical Air Command: 20 September 1944–31 December 1944,* AFHRA IRIS Num 6205-1. 35–43. Spires, Air *Power for Patton's Army,* chapter 7. Robert F. Futrell, *Volume I, Ideas, Concepts, Doctrine,* 173–178.

88. George L. Murphy, *History of 162nd Liaison Squadron: Lafayette Air Field, Lafayette, Louisiana. 3rd Tactical Air Command (1 August–2 September 1945),* 4, AFHRA IRIS Num SQ-LIA-162-HI.

89. Murphy, *History of 162nd Liaison Squadron,* AFHRA IRIS Num SQ-LIA-162-HI, Historical Branch, AC/S-2, Air Proving Ground Command, Eglin Field, Florida. *A History of Air Proving Ground Command. AAF Board: Its Divisions and Their Work, 1943–1946,* Vol. 2, 340–343, AFHRA ISRISNUM 3956–2.

90. Wolf, *Basic Documents on Roles and Missions,* 63.

91. Comptroller of the Air Force, "United States Air Force Statistical Digest 1947." *U.S. Air Force Statistical Digest,* 109–116, 250–253. Comptroller of the Air Force, "United States Air Force Statistical Digest 1948, part 1." *U.S. Air Force Statistical Digest,* 276–277. Comptroller of the Air Force, "United States Air Force Statistical Digest 1948, part 2." *U.S. Air Force Statistical Digest,* 12–15, 50–61.

92. Futrell, *Volume I, Ideas, Concepts, Doctrine,* 173–179, 212–219, 232–235, 240–245. Hughes, *Overlord,* chapter 11.

Notes for Chapter 3

1. For more on the Korean War see Xiaobing Li, *China's Battle for Korea: The 1951 Spring Offensive* (Bloomington: Indiana University Press, 2014); T. R. Fehrenbach, *This Kind of War: A Study in Unpreparedness* (New York: Macmillan, 1963); Shu Guang Zhang, *Mao's Military Romanticism: China and the Korean War, 1950–1953* (Lawrence: University Press of Kansas, 1995); and Bruce Cumings, *The Korean War: A History* (New York: Modern Library, 2010).

2. For more on Soviet, Chinese, and North Korean efforts during the air war in Korea, see Xiaoming Zhang, *Red Wings Over the Yalu: China, the Soviet Union, and the Air War in Korea* (College Station: Texas A&M University Press, 2002).

3. Futrell, *The USAF in Korea*, 690. For more on UN and USAF air campaigns in Korea, see Robert F. Futrell, *The United States Air Force in Korea 1950–1953* (Washington, D.C.: Office of Air Force History, 1961) and Conrad Crane, *American Airpower Strategy in Korea, 1950–1953* (Lawrence, KS: University Press of Kansas, 2000).

4. 6147th Tactical Control Squadron, *Historical Report for July 1950*, 6147th Tactical Control Squadron (Airborne), Fifth Air Force, 1–3, AFHRA IRIS Number K-SQ-AW-6147. Ehlers, *The Mediterranean Air War*, 331–335.

5. Schlight, *Help from Above*, 115.

6. Futrell, *The USAF in Korea*, 33–35. Schlight, *Help from Above*, 120.

7. Futrell, *The USAF in Korea*, 9–15. Crane, *American Airpower Strategy in Korea*, 24–26.

8. Farmer and Strumwasser, *The Evolution of the Airborne Forward Air Controller*, vi.

9. Henry D. Gower and Kenneth W. Heist, *History of the Ninth Air Force 1 January 1948–1 January 1950* (Headquarters Ninth Air Force, Langley AFG Virginia), 304–306, Archives Ninth Air Force/CENTAF Office of the Historian (hereafter Ninth AF/HO). Schlight, *Help from Above*, 98–111.

10. Schlight, *Help from Above*, 98–111.

11. 6147th Tactical Control Squadron, *Historical Report for July 1950*, 1–3.

12. Ibid., 1–3.

13. "493 Fighter Squadron (USAFE)," *Organizational Records, Squadrons and Flights* (October 16, 2016), https://www.afhra.af.mil/Desktop Modules/ArticleCS/Print.aspx?PortalId=16&ModuleId=2404& Article=433505. "ORD Changes to General Type Base; New Organizational Heads Take Over," *The Rotator, 106 AAFBU, Greensboro, NC*, vol. 5, no. 48, July 12, 1946, Greensboro Historical Newspapers, Online Archive, UNC Greensboro, Greensboro, NC, http://libcdm1.uncg.edu/cdm/ref/collection/GSOPatriot/id/37423. U.S. Air Force, "Lt. Gen. John R. Murphy," *Biography*, https://www.af.mil/DesktopModules/ArticleCS/Print.aspx?PortalId=1&ModuleId=858&Article=106128.

14. U.S. Air Force, "General Earl E. Partridge," *Biography*, https://www.af.mil/DesktopModules/ArticleCS/Print.aspx?PortalId=1&ModuleId=858&Article=106052. U.S. Air Force, "General Otto P. Weyland," *Biography*, https://www.af.mil/DesktopModules/ArticleCS/Print.aspx?PortalId=1&ModuleId=858&Article=105233.

15. 6147th Tactical Control Squadron, *Historical Report for July 1950*, 2.
16. 6147th Tactical Control Squadron, *Historical Report for July 1950*, 2. Crane, *American Airpower Strategy in Korea*, 30–34.
17. 6147th Tactical Control Squadron, *Historical Report for July 1950*, 3.
18. Crane, *American Airpower Strategy in Korea*, 29.
19. 6147th Tactical Control Squadron, *Historical Report for July 1950*, 3.
20. Ibid., 4.
21. Ibid., 4.
22. Ibid., 13.
23. Ibid., 4.
24. Ibid., 13.
25. Ibid., 15.
26. Ibid., 13.
27. 6147th Tactical Control Squadron, *Historical Report for August 1950*, 6147th Tactical Control Squadron (Airborne), Fifth Air Force, AFHRA IRIS Number K-SQ-AW-6147, 1–2, 7–8.
28. 6147th Tactical Control Squadron, *Historical Report for March 1951*, Appendix III.
29. 6147th Tactical Control Squadron, *Historical Report for August 1950*, 3. Defense POW/MIA Accounting Agency, "Korean War Airloss Database—Sorted by Name," *Korean War POW/MIA List*, updated April 26, 2018, accessed October 23, 2019, https://www.dpaa.mil/portals/85/Documents/KoreaAccounting/korwald_name.pdf.
30. 6147th Tactical Control Squadron, *Historical Report for August 1950*, 3–4.
31. Ibid., 11.
32. Ibid., 11.
33. 6147th Tactical Control Squadron, *Historical Report for July 1950*, 4.
34. 6147th Tactical Control Squadron, *Historical Report for August 1950*, 5–7.
35. Cleveland, *Mosquitoes in Korea*, 198–199, 201.
36. Major Merrill H. Carlton, *SOP Number: 5, 22 September 1950*, Headquarters 6147th Tactical Control Squadron (Airborne), Fifth Air Force, AFHRA IRIS Number K-SQ-AW-6147.
37. 6147th Tactical Control Squadron, *Historical Report for August 1950*, 9.
38. Lt. John E. Thompson, "Thompson Report," in Cleveland, *Mosquitoes in Korea*, 268.
39. Cleveland, *Mosquitoes in Korea*, 258. 6147th Tactical Control Squadron, *Historical Report for November 1950*, Appendix II.

Lt. John E. Thompson, "Thompson Report," in Cleveland, *Mosquitoes in Korea*, 266–272.

40. Ibid.
41. Lt. John E. Thompson, "Thompson Report," in Cleveland, *Mosquitoes in Korea*, 268.
42. Futrell, *The USAF in Korea*, 274, 350.
43. Brigadier General E. J. Timberlake, *General Orders Number 258, 7 April 1951*, Headquarters Fifth Air Force, AFHRA IRIS Number K-SQ-AW-6150-HI.
44. Cleveland, *Mosquitoes in Korea*, 234.
45. Judy G. Endicott, "USAF Organizations in Korea 1950–1953," *U.S. Air Force Historical Research Agency*, https://www.afhra.af.mil/Portals/16/documents/Timelines/Korea/USAFOrganizationsinKorea.pdf?ver=2016-08-30-151054-960.
46. 6147th Tactical Control Squadron, *Historical Report for March 1951*, Appendix III. Kline, *The Airborne FAC in the Korean War*, 17.
47. 6147th Tactical Control Squadron, *Historical Report for November 1950* "Appendix IV: 'Mosquito' Tunes." Cleveland, *Mosquitoes in Korea*, 294–298.
48. Cleveland, *Mosquitoes in Korea*, 259.
49. Cleveland, *Mosquitoes in Korea*, 294. Farmer and Strumwasser, *The Evolution of the Airborne Forward Air Controller*, 25, 29, Appendix C. Schlight, *Help from Above*, 134.
50. Cleveland, *Mosquitoes in Korea*, 294. Farmer and Strumwasser, *The Evolution of the Airborne Forward Air Controller*, 25, 29, Appendix C. Schlight, *Help from Above*, 134. Jesse Breau, interview with author, October 9, 2019.
51. Schlight, *Help from Above*, 140.
52. 6147th Tactical Control Group, *History of the 6147 Tactical Control Group 1 July 1955–31 December 1955*, USAFHRA IRIS Num 445960.
53. Bernard Katz, *History of the Ninth Air Force (Tactical Air Command) 1 January–30 June 1954* (Headquarters Ninth Air Force, Pope AFB NC,) 151. Ninth AF/HO.
54. Katz, *History of the Ninth Air Force 1 January–30 June 1954*, 142–152.
55. Ibid., 156, 158.
56. Ibid., 161.
57. Ibid., 143–154.
58. Katz, *History of the Ninth Air Force 1 January–30 June 1954*, 142–161. Linn, *Elvis's Army*, chapter 3, 4, 7, 8.

59. Kenneth W. Heist, *History of the Ninth Air Force 1 January–31 June 1958* (Headquarters Ninth Air Force), 122. Ninth AF/HO.

60. Heist, *History of the Ninth Air Force 1 January–31 June 1958*, 3. Linn, *Elvis's Army*.

61. Howard W. Kayner, *History of Ninth Air Force, Tactical Air Command, USAF, 1 January–30 June 1962. Vol. 1, Narrative* (Shaw Air Force Base NC: Historical Division, Headquarters Ninth Air Force), 128–132. Ninth AF/HO.

62. Kayner, *History of Ninth Air Force, 1 January–30 June 1962*, 128–132.

63. Farmer and Strumwasser, *The Evolution of the Airborne Forward Air Controller*, v.

64. Ben F. Mariska, *History of the Ninth Air Force 1 January–30 June 1953, Volume 1*. Katz, *History of the Ninth Air Force 1 January–30 June 1954*. Katz, *History of the Ninth Air Force (Tactical Air Command) 1 July–31 December 1954, Vol. 1*. Heist, *History of the Ninth Air Force 1 January–31 June 1958*. Kayner, *History of Ninth Air Force, 1 January–30 June* 1962, 128–132. Linn, *Elvis's Army*, chapter 3.

65. Katz, *1 July–31 December 1954*, i. U.S. Air Force, "Major General Homer L. Sanders," *Biographies*, https://www.af.mil/About-Us/Biographies/Display/Article/105718/major-general-homer-l-sanders/.

Notes for Chapter 4

1. Vietnam Task Force, "Part II: U.S. Involvement in the Franco-Viet Minh War, 1950–1954" *Report of the Office of the Secretary of Defense Vietnam Task Force*. NARA ID 5890486. Vietnam Task Force, "Part IV. B. 3: Evolution of the War. Counterinsurgency: The Advisory Build-up, 1961–67," *Report of the Office of the Secretary of Defense Vietnam Task Force*, 1–9, NARA ID 5890495, https://www.archives.gov/research/pentagon-papers. Daniel J. Lawler and Erin R. Mahan, eds., *Foreign Relations of the United States, 1949, The Far East and Australasia, Volume VII, Part 2* (Washington, D.C.: Government Printing Office, 2010), Document 387. Graham A. Cosmas, *United States Army in Vietnam: MACV The Joint Command in the Years of Escalation, 1962–1967* (Washington, D.C.: Center of Military History United States Army, 2006), 3–50. Stanley Karnow, *Vietnam* (New York: Viking Press. 1983).

Mark A. Lawrence, *The Vietnam War: A Concise International History* (New York: Oxford University Press, 2008). Tonkin Gulf Resolution; Public Law 88–408, 88th Congress, August 7, 1964; General Records of the United States Government; Record Group 11; National Archives. Nalty, *Winged Shield Vol. 2*, 245–255.

2. Jeff Michalke, "Air Commandos History Revived in 1960s," *16th SOW History Office, Air Force Special Operations Command* (October 13, 2006), https://www.afsoc.af.mil/News/Article-Display/Article/163454/air-commandos-history-revived-in-1960s/. Capt. E. Vallentiny, "VNAF FAC Operations in SVN," *Project CHECO Report #242. 28 January 1969*, 1, USAFA Call Number K717.0413-49. Nalty, *Winged Shield Vol. 2*, 245–255.

3. Vallentiny, *Project CHECO Report #242*, 1–7.

4. Vallentiny, *Project CHECO Report #242*, 1–7. Nalty, *Winged Shield Vol. 2*, 245–255.

5. Vallentiny, *Project CHECO Report #242*, 9–10. Darrel Whitcomb, "Farm Gate," *Air Force Magazine* (December 2005): 82–87. Nalty, *Winged Shield Vol. 2*, 245–255.

6. Fact Sheet, "North American T-28B Trojan," *National Museum of the United States Air Force* (May 18, 2015), accessed November 4, 2019, https://www.nationalmuseum.af.mil/Visit/Museum-Exhibits/Fact-Sheets/Display/Article/196029/north-american-t-28b-trojan/. Fact Sheet, "Douglas A-1E Skyraider," *National Museum of the United States Air Force* (May 18, 2015), https://www.nationalmuseum.af.mil/Visit/Museum-Exhibits/Fact-Sheets/Display/Article/195821/douglas-a-1e-skyraider/. Nalty, *Winged Shield Vol. 2*, 245–255.

7. Andi Biancur, Phil Caine, Lindy Royer, Jim Titus, and Jim Gaston, interview with author, September 25, 2019. Fact Sheet, "Douglas B-26K (A-26) Counter Invader," *National Museum of the United States Air Force* (May 15, 2015), https://www.nationalmuseum.af.mil/Visit/Museum-Exhibits/Fact-Sheets/Display/Article/196070/douglas-b-26k-a-26-counter-invader/. Andrew R. Walton, *The History of the Airborne Forward Air Controller in Vietnam*, thesis (U.S. Army Command and General Staff College, Fort Leavenworth KS, 2004), 25–28.

8. Whitcomb, "Farm Gate," 87. Christopher Michael Hobson, *Vietnam Air Losses* (Hinkley, England: Midland Publishing, 2001), 5.

9. Fact Sheet, "Cessna O-1G Bird Dog," *National Museum of the United States Air Force* (May 18, 2015), https://www.nationalmuseum.af.mil/Visit/Museum-Exhibits/Fact-Sheets/Display/Article/196068/cessna-o-1g-bird-dog/. Nalty, *Winged Shield Vol. 2*, 245–255.

10. Sharon Heilbrunn, "U.S.S. *Midway* Air Boss Remembers Heroic Birddog Airplane Rescue," *KPBS* (April 25, 2010), https://www.kpbs.org/news/2010/apr/25/uss-midway-air-boss-remembers-heroic-bird-dog-airp/.

11. Hale Burr, "FAC Survey," email to author. December 10, 2018. "The 19th TASS" Forward Air Controllers Association. http://www.fac-assoc.org/19%20TASS/19thTASS.htm. Fact Sheet, "FAC in SEA: South Vietnam—'In-Country'" *National Museum of the United States Air Force*. May 18, 2015. Maurer Maurer, ed., *Air Force Combat Units of World War II* (Air Force Historical Research Agency, 1963), 132. Nalty, *Winged Shield Vol. 2*, 245–255.

12. Hobson, *Vietnam Air Losses*, 9. Richard D. Gabbert and Gary B. Streets, *A Comparative Analysis of USAF Fixed-Wing Aircraft Losses in Southeast Asia Combat* (Wright-Patterson Air Force Base: Air Force Flight Dynamics Laboratory, 1977), 25, 32, 39, 44, 51, 57, 59, 61.

13. Hobson, *Vietnam Air Losses*, 9.

14. James B. Overton, "FAC Operations in Close Air Support Role in SVN," *Project CHECO Report #103. 31 January 1969*, 1–7, USAFA Call Number K717.0413-45. History Department, "The Vietnam War," *The United States Military Academy at West Point*, https://westpoint.edu/academics/academic-departments/history/vietnam-war. Harold G. Moore and Joseph L. Galloway, *We Were Soldiers Once—and Young: Ia Drang, the Battle That Changed the War in Vietnam* (New York: Harper Perennial, 1993).

15. James H. Wilbanks, *The Tet Offensive: A Concise History* (New York: Columbia University Press, 2007).

16. Overton, *Project CHECO Report #103*, *The United States Military Academy at West Point*. "The Vietnam War."

17. For more on the air war in North Vietnam see Jack Broughton, *Thud Ridge* (New York: Popular Library, 1969); Ed Cobleigh, *War for the Hell of It* (New York: Berkley Caliber, 2005); Marshall L. Michel III, *Clashes: Air Combat over North Vietnam 1965–1972* (Annapolis: Naval Institute Press, 1997); and Marshall L. Michel III, *The 11 Days of Christmas: America's Last Vietnam Battle* (New York: Encounter Books, 2002).

18. Overton, *Project CHECO Report #103*.

19. Overton, *Project CHECO Report #103*, 8–10.

20. Overton, *Project CHECO Report #103*, 9. For more on the In-Country and interdiction air-to-ground war in Vietnam see Nalty, *Air War over South Vietnam*; Nalty, *The War against Trucks*; and Mrozek, *Air Power and the Ground War in Vietnam*.

21. Overton, *Project CHECO Report #103*. Melvin F. Porter, "Control of Airstrikes: January 1967–December 1968," *Project CHECO Report #79, 30 June 1969*, 28–32, USAFA Call Number K717.0414-74.

22. Overton, *Project CHECO Report #103*. Porter, *Project CHECO Report #79*. Lt. Col. John Schlight, "Jet Forward Air Controllers in SEASIA," *Project CHECO Report #104, 15 October 1969*, USAFA Call Number K717.0915-70. Rick Newman and Don Shepperd, *Bury Us Upside Down: The Misty Pilots and the Secret Battle for the Ho Chi Minh Trail* (New York: Ballantine Books, 2006).

23. Overton, *Project CHECO Report #103*. Porter, *Project CHECO Report #79*. Warren A. Trest, "Control of Air Strikes in SEA 1961–1968," *Project CHECO Report #78*, USAFA Call Number K717.0414-4.

24. Overton, *Project CHECO Report #103*. Porter, *Project CHECO Report #79*. Trest, *Project CHECO Report #78*. Porter, *Project CHECO Report #11*.

25. Overton, *Project CHECO Report #103*. Porter, *Project CHECO Report #79*. Trest, *Project CHECO Report #78*. Porter, *Project CHECO Report #11*. CJCS, *Joint Publication 3-09.3: Close Air Support* (Washington, D.C.: Chairman of the Joint Chiefs of Staff, 2019), V17–V24.

26. Overton, *Project CHECO Report #103*. Porter, *Project CHECO Report #79*. Trest, *Project CHECO Report #78*.

27. Hale Burr, interview with author, December 11, 2019. Clyde Edgerton, interview with author, October 18, 2017. Jesse Breau, interview with author, October 9, 2019. Burt Brown, interview with author, October 8, 2019. Bobcat Baker, interview with author, October 8, 2019. Scout Randolph, interview with author, October 9, 2019. Owen Birkett, email to author, October 3, 2017. Owen Birkett, interview with author, October 8, 2019.

28. Ibid.

29. Ibid.

30. Broughton, *Thud Ridge*. Cobleigh, *War for the Hell of It*. Michel III, *Clashes*.

31. Overton, *Project CHECO Report #103*, 12–14. Walton, *Airborne Forward Air Controller in Vietnam*, 43.

32. Overton, *Project CHECO Report #103*, 12–14. Walton, *Airborne Forward Air Controller in Vietnam*, 43. Mike Jackson and Tara Dixon-Engel, *Naked in Da Nang: A Forward Air Controller in Vietnam* (St. Paul: Zenith Press, 2004), 91–110.

33. Hale Burr, interview with author, December 11, 2019. Clyde Edgerton, interview with author, October 18, 2017.

34. Ibid. Technically Fighter Navigators, Weapons System Operators do not navigate but instead use the myriad of weapons and sensors on fighter aircraft. During the Vietnam era, this included operating the air-to-air radar, electronic warfare systems, and after 1972 the laser designator pod for laser-guided bombs.

35. Hale Burr, interview with author, December 11, 2019. Clyde Edgerton, interview with author, October 18, 2017.

36. Hale Burr, unpublished memoirs, private archive. Hale Burr, interview with author, August 21, 2019, and December 11, 2019. Hale Burr, email to author, November 3, 2019, and November 11, 2013. For more on Operation Rolling Thunder see James C. Thompson, *Rolling Thunder: Understanding Policy and Program Failure* (Chapel Hill: University of North Carolina Press, 1980); Earl H. Tilford, *Setup: What the Air Force Did in Vietnam and Why* (Maxwell Air Force Base: Air University Press, 1991); Broughton, *Thud Ridge*; Cobleigh, *War for the Hell of It*; and Michel III, *Clashes*.

37. Hale Burr, unpublished memoirs, private archive.

38. Ibid. Hale Burr, interview with author, August 21, 2019, and December 11, 2019. Hale Burr, email to author, November 3, 2019, and November 11, 2013.

39. Ibid. Staff Writer, "From the Archives: 82nd Airborne in Vietnam," *The Fayetteville Observer* (March 29, 2019), https://www.fayobserver.com/photogallery/NC/20190329/NEWS/328009988/PH/1.

40. Ibld.

41. Fact Sheet, "Cessna O-2A Skymaster," *National Museum of the United States Air Force* (May 18, 2015), https://www.nationalmuseum.af.mil/Visit/Museum-Exhibits/Fact-Sheets/Display/Article/196063/cessna-o-2a-skymaster/. Fact Sheet, "North American Rockwell OV-10A Bronco," *National Museum of the United States Air Force* (May 18, 2015), https://www.nationalmuseum.af.mil/Visit/Museum-Exhibits/Fact-Sheets/Display/Article/196062/north-american-rockwell-ov-10a-bronco/.

42. Ben Petty, interview with author, October 18, 2019. Hale Burr, unpublished memoirs, private archive. Hale Burr, interview with author, August 21, 2019, and December 11, 2019. Hale Burr, email to author, November 3, 2019, and November 11, 2013.

43. Ibid.

44. Joseph V. Potter, "OV-10 Operations in SEA Jul 1968-Jun 1969." *Project CHECO Report #156, 15 September 1969*, USAFA Call Number K717.0413-60.

45. Potter, *Project CHECO Report #156*, 16–18.

46. Potter, *Project CHECO Report #156*. 17. Nalty, *Air War over South Vietnam*, 88–89. Nalty, *The War against Trucks*, 129. Mrozek, *Air Power and the Ground War in Vietnam*, 82, 91, 107, 125.

47. Potter, *Project CHECO Report #156*, 14.

48. Potter, *Project CHECO Report #156*.

49. Ben Petty, Mike Peters, and Dan Keyes, interview with author, October 18, 2019. Clyde Edgerton, interview with author, October 18, 2017. Potter, *Project CHECO Report #156*, 7–9. Captain Andy Walton, "Bronco 12 Cleared Hot," *U.S. Navy Institute Proceedings* 142, no. 6 (June 2016): 1369, https://www.usni.org/magazines/proceedings/2016/june/bronco-12-cleared-hot.

50. Potter, *Project CHECO Report #156*, 8.

51. Correll, *The Air Force in Vietnam*, 21. Bruce P. Layton, "Commando Hunt VI," *Project CHECO Report #76, 7 July 1972*, USAFA Call Number K717.0414-28. For an account of Commando Hunt operations see Bernard Nalty, *The War against the Trucks*.

52. Thomas D. Wade, "Seventh Air Force Tactical Air Control Center Operations," *Project CHECO Report #217, 15 October 1968*, USAFA Call Number K717.0413-40.

53. Darrel Whitcomb, email to author, November 9, 2019. Loel Tibbitts, email to author, November 9, 2019. Hale Burr, interview with author, August 21, 2019, and December 11, 2019. Hale Burr, email to author, November 3, 2019, and November 11, 2013.

54. For more on the Raven program see Craig W. Duehring, *The Lair of the Raven* (self-published, Craig W. Duehring, 2014); Christopher Robbins, *The Ravens: The Men Who Flew in America's Secret War in Laos* (New York: Crown, 1987); and Jim Roper, *Quoth the Raven* (Baltimore: America House, 2001).

55. Ralph Wetterhahn, "Ravens of Long Tieng," *Air and Space Magazine* (November 1998), https://www.airspacemag.com/military-aviation/ravens-of-long-tieng-284722/?all. "The Fall of Lima Site 85," *The Central Intelligence Agency* (April 14, 2007), https://www.cia.gov/library/center-for-the-study-of-intelligence/csi-publications/csi-studies/studies/95unclass/Linder.html#ft1. John T. Correll, "The Fall of Lima Site 85," *Air Force Magazine* (April 2006).

56. The "G" in "high G" refers to the force of gravity. The normal force of gravity on earth is 1 G, or 1 times the force of gravity. On a roller coaster, a rider might experience 3 or 4 Gs. Vietnam era fighter aircraft were capable of 7 to 9 Gs. Misty pilots made turns at 4 or 5 Gs, every few seconds for the entire duration of their patrol. This means a two-hundred-pound pilot would experience eight hundred to one thousand pounds of force, every five seconds, for nearly two hours.

57. Schlight, *Project CHECO Report #104*. Ron Schuh, interview with author, October 17, 2017. Hale Burr, interview with author, August 21, 2019, and December 11, 2019. Hale Burr, email to author, November 3, 2019, and November 11, 2013.

58. Phil M. Haun, *Airpower Versus a Fielded Force: Misty FACs of Vietnam and A-10 FACs of Kosovo, a Comparative Analysis* (Maxwell AFB, AL: Maxwell Air Force Base: Air University Press, 2004), 17.

59. Haun, *Misty FACs of Vietnam*, 19

60. Walton, *The Airborne Forward Air Controller in Vietnam*, 65.

61. Nalty, *The War against Trucks*, 129. Don Bell and The Tiger FACs, *The Tiger FACs: A Dance with the Devil* (Denver: Outskirts Press, 2014). Ross C. Detwiler, *The Great Muckrock and Rosie* (Bloomington: Abbott Press, 2013). Barry J. Howard, *Wow! And they Even Paid Me, by Colonel Lucky* (self-published, North Charleston, SC: CreateSpace, 2013). Newman and Shepperd, *Bury Us Upside Down.*

62. Schlight, *Project CHECO Report #104*, 29–30.

63. Hale Burr, interview with author, August 21, 2019, and December 11, 2019. Hale Burr, email to author, November 3, 2019, and November 11, 2013. Bell, *The Tiger FACs*. Detwiler, *The Great Muckrock and Rosie.* Newman and Shepperd, *Bury Us Upside Down.*

64. Hale Burr, unpublished memoirs, private archive. Hale Burr, interview with author, August 21, 2019, and December 11, 2019. Hale Burr, email to author, November 3, 2019, and November 11, 2013.

65. Hale Burr, unpublished memoirs, private archive. Hale Burr, interview with author, August 21, 2019, and December 11, 2019. Hale Burr, email to author, November 3, 2019, and November 11, 2013.

66. Hale Burr, unpublished memoirs, private archive. Hale Burr, interview with author, August 21, 2019, and December 11, 2019. Hale Burr, email to author, November 3, 2019, and November 11, 2013.

67. Hale Burr, unpublished memoirs, private archive. Hale Burr, interview with author, August 21, 2019, and December 11, 2019. Hale Burr, email to author, November 3, 2019, and November 11, 2013.

68. Hale Burr, unpublished memoirs, private archive. Hale Burr, interview with author, August 21, 2019, and December 11, 2019. Hale Burr, email to author, November 3, 2019, and November 11, 2013.

69. Hale Burr, unpublished memoirs, private archive. Hale Burr, interview with author, August 21, 2019, and December 11, 2019. Hale Burr, email to author, November 3, 2019, and November 11, 2013.

70. Broughton, *Thud Ridge*. Cobleigh, *War for the Hell of It*. Michel III, *Clashes*. Hale Burr, interview with author, August 21, 2019, and December 11, 2019. Hale Burr, email to author, November 3, 2019, and November 11, 2013. Bell, *The Tiger FACs*. Detwiler, *The Great Muckrock and Rosie*. Newman and Shepperd, *Bury Us Upside Down*.

71. Hale Burr, unpublished memoirs, private archive. Hale Burr, interview with author, August 21, 2019, and December 11, 2019. Hale Burr, email to author, November 3, 2019, and November 11, 2013. Knots, or Nautical Miles Per Hour, is a measurement of airspeed. An aircraft measures speed via pressure, known as Indicated Airspeed. This is the speed a pilot sees in the cockpit. Since pressure changes with altitude, air temperature, and an increase in speed, his True Air Speed (how fast he is traveling through the air) at five hundred knots indicated is about 575 Knots. As 1 Knot equals 1.15 Miles Per Hour, Burr's speed over the ground would be about 660 MPH. For reference, the speed of sound is roughly 750 MPH in the same conditions. Burr's "135-degree slicing turn" means Burr would have rolled his aircraft to 135 degrees of bank, or nearly completely upside down, before executing a maximum G turn back in the opposite direction. This method allowed Burr the ability to maintain sight of the SAM and turn his F-4 around as quickly as possible. This maneuver enabled him to immediately fire his marking rockets, but subjected Burr and his GIB to about 7 Gs of force during the maneuver.

72. Hale Burr, unpublished memoirs, private archive. Hale Burr, interview with author, August 21, 2019, and December 11, 2019. Hale Burr, email to author, November 3, 2019, and November 11, 2013.

73. Hale Burr, unpublished memoirs, private archive. Hale Burr, interview with author, August 21, 2019, and December 11, 2019. Hale Burr, email to author, November 3, 2019, and November 11, 2013.

74. Hale Burr, unpublished memoirs, private archive. Hale Burr, interview with author, August 21, 2019, and December 11, 2019. Hale Burr, email to author, November 3, 2019, and November 11, 2013. Biography "Major General H. Hale Burr Jr.," *U.S. Air Force*, accessed November 14, 2019,

https://www.af.mil/About-Us/Biographies/Display/Article/107539/
major-general-h-hale-burr-jr/.

75. Gabbert and Streets, *USAF Fixed-Wing Aircraft Losses*, 25, 32, 39, 44, 51, 57, 59, 61.

76. Hobson, "11 May 1972." *Fixed Wing Air Losses, Vietnam*, https://vietnamairlosses.com/loss.php?id=3120. "The Tomb of the Unknown Soldier," *Arlington National Cemetery*, https://www.arlingtoncemetery.mil/Explore/Tomb-of-the-Unknown-Soldier.

77. Hobson, "16 June 1973," *Fixed Wing Air Losses, Vietnam*, https://vietnamairlosses.com/loss.php?id=3120.

78. *The Raven Songbook*, The Edgar Allen Poe Literary Society of Texas, undated.

79. Various Authors, *Project CHECO Reports #196-201*, 24 October 1966–27 November 1974. USAF Search and Rescue in SEA 1961–1973, USAFA Call Number K717.0414-1. Gabbert and Streets, *USAF Fixed-Wing Aircraft Losses*, 25, 32, 39, 44, 51, 57, 59, 61.

80. Forward Air Controller's Association, "History Archive," *History*, https://www.fac-assoc.org/history/history01.html.

Notes for Chapter 5

1. Wolf, *The United States Air Force: Basic Documents*. Alfred Goldberg and Donald Smith, *Army Air Force Relations: The Close Air Support Issue* (Santa Monica, CA: RAND, 1971), 6–16.

2. Goldberg and Smith, *Army Air Force Relations*, 6–16. Nalty, *Winged Shield Vol. 2*, chapter 17. For more on the Army between the Korean and Vietnam Wars, see Maxwell Taylor, *Uncertain Trumpet*, and Brian McAllister Linn, *Elvis's Army*.

3. Wolf, *The United States Air Force: Basic Documents*, 251–324. Goldberg and Smith, *Army Air Force Relations*, 17–27. Goldberg and Smith, *Army Air Force Relations*, 17–27.

4. Wolf, *The United States Air Force: Basic Documents*, 251–324. Goldberg and Smith, *Army Air Force Relations*, 17–27. Goldberg and Smith, *Army Air Force Relations*, 17–27.

5. Goldberg and Smith, *Army Air Force Relations*, 17–27. David R. Jacques and Dennis Strouble, *A-10 Thunderbolt II (Warthog) Systems Engineering Case Study* (Wright-Patterson Air Force Base: Air Force Center for Systems Engineering, Air Force Institute of Technology, 2010), 13–15.

6. Goldberg and Smith, *Army Air Force Relations*, 28–36. Jacques and Strouble, *A-10 Thunderbolt*, 13–15. Trest, *Project CHECO Report #78*, 11–30. Melvin F. Porter, "Control of Airstrikes, Jan 1967–Dec 1968," *Project CHECO Report #79*, 30 June 1969, USAFA Call Number K717.0414-4. Overton, *Project CHECO Report #103*.
7. Ibid.
8. Ibid.
9. *Air Force and Space Digest* 49, No. 3 (March 1966).
10. Matt Schudel, "Otis G. Pike, Maverick N.Y. Congressman, Dies at 92," *The Washington Post* (January 20, 2014).
11. Cong. Rec., 89th Cong. 2nd Sess, 1966, Vol. 112, Part 7, 8486.
12. Goldberg and Smith, *Army Air Force Relations*, 28–36. Jacques and Strouble, *A-10 Thunderbolt*, 13–15. Trest, *Project CHECO Report #78*, 11–30. Porter, *Project CHECO Report #79*. Overton, *Project CHECO Report #103*. JR Lind, K. Harris, and S. G. Spring, *Fast-Val: A Study of Close Air Support (Briefing Summarizing the Comparisons of Model with Combat Results and Illustrating the Influence of Supporting Arms on Fire-fight Outcomes)* (Santa Monica, CA: RAND, 1971).
13. Cong. Rec., 89th Cong. 2nd Sess, 1966, Vol. 112, Part 7, 8486.
14. Wolf, *The United States Air Force: Basic Documents*, 379–384.
15. Goldberg and Smith, *Army Air Force Relations*, 32
16. Wolf, *The United States Air Force: Basic Documents*, 379–384. Lind, Harris, and Spring, *Fast-Val*. Goldberg and Smith, *Army Air Force Relations*, 32.
17. Nalty, *Air War over South Vietnam*, 26–27, 329, 364, 396. Michel III, *The 11 Days of Christmas*, 13, 19–20. Wolf, *The United States Air Force: Basic Documents*, 401–405. Goldberg and Smith, *Army Air Force Relations*, 32.
18. Wolf, *The United States Air Force: Basic Documents*, 401–405. Goldberg and Smith, *Army Air Force Relations*, 32. Anderegg, *Sierra Hotel: Flying Air Force Fighters in the Decade After Vietnam* (Washington, D.C.: Air Force History and Museums Program, 2001), chapter 16.
19. Wolf, *The United States Air Force: Basic Documents*, 401–405. Goldberg and Smith, *Army Air Force Relations*, 32.
20. Jacques and Strouble, *A-10 Thunderbolt*, 13–14. Goldberg and Smith, *Army Air Force Relations*, 28–38.
21. Cong. Rec., 89th Cong. 2nd Sess, 1966, Vol. 112, Part 7, 8486. Jacques and Strouble, *A-10 Thunderbolt*, 13–14. Goldberg and Smith, *Army Air Force Relations*, 28–38.

22. Jacques and Strouble, *A-10 Thunderbolt*, 16–31. Goldberg and Smith, *Army Air Force Relations*, 31–38. Anderegg, *Sierra Hotel*, chapter 16. Nalty, *Winged Shield Vol. 2*, 358.
23. Goldberg and Smith, *Army Air Force Relations*, 32. Jacques and Strouble, *A-10 Thunderbolt*, 38–40. Anderegg, *Sierra Hotel*, chapter 16. Nalty, *Winged Shield Vol. 2*, 358.
24. Michel III, "The Revolt of the Majors," 126–132. Jacques and Strouble, *A-10 Thunderbolt*, 41–42, 45. Anderegg, *Sierra Hotel*, chapter 16. Nalty, *Winged Shield Vol. 2*, chapter 21.
25. Jacques and Strouble, *A-10 Thunderbolt*, 42. Original text note added for clarity: "The Piper Enforcer, which had a strong resemblance to the WWII era P-51 Mustang, was designed and originally flown by David Lindsay, owner of Cavalier Aircraft. Lindsay sold the program to Piper Aircraft in 1970 but remained involved in the project after the closure of Cavalier in 1971. The original prototype Piper Enforcers were heavily modified P-51's with a single turboprop engine. The Enforcer was evaluated, but rejected, as a counterinsurgency aircraft by the Air Force in the early 1970s."
26. Wolf, *The United States Air Force: Basic Documents*, 405.
27. Stamps and Esposito, *World War II: Atlas*, 12–14. United States Military Academy and the United States Air Force Academy, "Kuwait and Vicinity, 1991—The Liberation of Kuwait, 24–26 February," *The First Gulf War*, https://westpoint.edu/sites/default/files/inline-images/academics/academic_departments/history/Gulf%20War%20and%20Iraq/GulfWarFirst03.pdf.
28. U.S. Army, *Field Manual 100–5: Operations* (Washington, D.C.: Headquarters, Department of The Army, 1976), 8–1.
29. Ibid.
30. U.S. Army, *Field Manual 100–5*, 8–1. Emphasis in the original.
31. For more on the Yom Kippur War see Abraham Rabinovich, *The Yom Kippur War: The Epic Encounter That Transformed the Middle East* (New York: Schocken, 2005).
32. Robert A. Doughty, *The Evolution of U.S. Army Tactical Doctrine, 1946–76* (Fort Leavenworth: U.S. Army Command and General Staff College, 1979), 41.
33. For more on the US Army post-Vietnam transformation see John L. Romjue, *From Active Defense to Airland Battle: The Development of Army Doctrine, 1973–1982* (Fort Monroe: Historical Office, US Army Training and Doctrine Command, 1984).

34. U.S. Army, *Field Manual 100–5*, 8–2—8–5.
35. U.S. Army, *Field Manual 100–5*, 8–5. Emphasis in the original.
36. Douglas W. Skinner, *Airland Battle Doctrine* (Alexandria: Center for Naval Analysis, 1988), 5. Doughty, *The Evolution of U.S. Army Tactical Doctrine*, 43.
37. U.S. Army, *Field Manual 100–5: Operations* (Washington, D.C.: Headquarters, Department of The Army, 1986), 16.
38. Robert S. Dotson, "Tactical Air Power and Environmental Imperatives," *Air University Review* 28, no. 5 (July–August 1977): 27–35. Terrance J. McCaffrey, *What Happened to Battlefield Air Interdiction? Army and Air Force Battlefield Doctrine Development from Pre–Desert Storm to 2001* (Maxwell Air Force Base: Air University Press, 2004).
39. Futrell, *The United States Air Force in Korea 1950–1953*, chapter 10, chapter 15. Crane, *American Airpower Strategy in Korea*, chapter 7. Nalty, *The War against Trucks*. Ehlers, *The Mediterranean Air War*, 331–335.
40. McCaffrey, *What Happened to Battlefield Air Interdiction*, 18–20. St. Clair, *The Twelfth U.S. Air Force*, 40–42. Hansborough, *Activities of Air Support Control Squadron*. Albert F. Simpson, *Air Phase of the Italian Campaign to 1 January 1944* (Washington, D.C.: Air Force Historical Research Agency, 1946), IRIS Num 467699. Futrell, *The USAF in Korea*, 475–504. Nalty, *The War against Trucks*. Layton, *Project CHECO Report #76*.
41. Dotson, "Tactical Air Power and Environmental Imperatives," 29.
42. McCaffrey, *What Happened to Battlefield Air Interdiction*, 19.
43. Wolf, *The United States Air Force: Basic Documents*, 415–423. Richard G. Davis, *Air Staff Historical Study. The 31 Initiatives: A Study in Air Force-Army Cooperation* (Washington, D.C.: Office of Air Force History, 1987).
44. Ibid.
45. U.S. Army, *Field Manual 100–5*, 8–5.
46. Dotson, "Tactical Air Power and Environmental Imperatives," 35.
47. Mike Peters and Dan Keyes, interview with author, October 18, 2019.
48. Mike Peters and Dan Keyes, interview with author, October 18, 2019.
49. Dotson, "Tactical Air Power and Environmental Imperatives," 30.
50. *Report to the Chairman, Committee on Armed Services, House of Representatives. Close Air Support: Status of Air Force's Efforts to Replace the A-10 Aircraft*, by Harry R. Finley (Washington, D.C.:

Government Accountability Office, National Security and International Affairs Division, 1988).

51. Mike Mangus, interview with author, December 5, 2019. Mike Peters and Dan Keyes, interview with author, October 18, 2019. Stephen Renner, interview with author, July 26, 2018. Finley, *Air Force's Efforts to Replace the A-10 Aircraft*. Jeff Hagen et al., *Needs, Effectiveness, and Gap Assessment for Key A-10C Missions: An Overview of Findings* (Santa Monica, CA: RAND, 2016).

52. Finley, *Air Force's Efforts to Replace the A-10 Aircraft*, 15. Hagen et al., *Needs, Effectiveness, and Gap Assessment*. John Matsumura, John Gordon IV, and Randall Steeb, *Defining and Approach for Future Close Air Support Capability* (Santa Monica, CA: RAND, 2017), 5–8.

53. U.S. Air Force, *Amended FY 1990/91 Biennial Budget* (Washington, D.C.: Financial Management and Comptroller, 1989), D-6.

54. Finley, *Air Force's Efforts to Replace the A-10 Aircraft*, 17. Mike Mangus, interview with author, December 5, 2019. Mike Peters and Dan Keyes, interview with author, October 18, 2019. Stephen Renner, interview with author, July 26, 2018.

55. U.S. Army and U.S. Air Force, *TRADOC Training Text 17-50-3: Joint Air Attack Team* (Fort Rucker, AL: Commander USAAVNC, 1983), iii.

56. Finley, *Air Force's Efforts to Replace the A-10 Aircraft*, 17. Mike Mangus, interview with author, December 5, 2019. Mike Peters and Dan Keyes, interview with author, October 18, 2019. Stephen Renner, interview with author, July 26, 2018. Jacques and Strouble, *A-10 Thunderbolt*, 43–45.

57. Jacques and Strouble, *A-10 Thunderbolt*, 49. Anderegg, *Sierra Hotel*, chapter 16.

58. Ibid.

59. Mike Mangus, interview with author, December 5, 2019. Mike Peters and Dan Keyes, interview with author, October 18, 2019. Stephen Renner, interview with author, July 26, 2018.

60. Finley, *Air Force's Efforts to Replace the A-10 Aircraft*, 12. Jacques and Strouble, *A-10 Thunderbolt*, 51.

61. Finley, *Air Force's Efforts to Replace the A-10 Aircraft*, 12.

62. Ibid.

63. Ibid.

Notes for Chapter 6

1. Thomas A. Keaney and Eliot A. Cohen, *Gulf War Air Power Survey, Unclassified,* vol. 1, part 1 (Washington, D.C.: Department of the Air Force, 1993), 57–58. Geoffrey Wawro, *Quicksand: America's Pursuit of Power in the Middle East* (New York: Penguin Press, 2010), 406–411. Nalty, *Winged Shield Vol. 2,* 442–448.

2. U.S. President, National Security Directive 26, "U.S. Policy toward the Persian Gulf," October 2, 1989, 2.

3. U.S. Army, *CMH Pub 70-30-1. Whirlwind War: The United States Army in Operations Desert Shield and Desert Storm* (Washington, D.C.: Center of Military History, 1995), 48. Norman H. Schwarzkopf with Peter Petre, *It Doesn't Take a Hero* (New York: Linda Gray, 1992), 282–298. Wawro, *Quicksand,* 411–415. Nalty, *Winged Shield Vol. 2,* 442–448.

4. Diane T. Putney, *Airpower Advantage: Panning the Gulf War Air Campaign 1989–1991* (Washington, D.C.: Air Force History and Museums Program, 2004), 35–40. Nalty, *Winged Shield Vol. 2,* 442–448.

5. Putney, *Airpower Advantage,* 35–40. Hale Burr, interview with author, December 11, 2019. Keaney and Cohen, *Gulf War Air Power Survey, Unclassified,* vol. 1, part 1, 46. Wawro, *Quicksand,* 422–424.

6. Putney, *Airpower Advantage,* 35–40. Hale Burr, interview with author, December 11, 2019. Keaney and Cohen, *Gulf War Air Power Survey, Unclassified,* vol. 1, part 1, 46. Wawro, *Quicksand,* 422–424. Schwarzkopf, *It Doesn't Take a Hero,* 318–321. Nalty, *Winged Shield Vol. 2,* 442–448.

7. Keaney and Cohen, *Gulf War Air Power Survey, Unclassified,* vol. 1, part 1, 127. Hale Burr, interview with author, December 11, 2019. Nalty, *Winged Shield Vol. 2,* 442–448.

8. Keaney and Cohen, *Gulf War Air Power Survey, Unclassified,* vol. 1, part 1, 127. Hale Burr, interview with author, December 11, 2019. Tom Clancy, and Chuck Horner, *Every Man a Tiger* (New York: G.P. Putnam's, 1999), 179–219.

9. Keaney and Cohen, *Gulf War Air Power Survey, Unclassified,* vol. 1, part 1, 112–113.

10. Keaney and Cohen, *Gulf War Air Power Survey, Unclassified,* vol. 1, part 1, 169. Putney, *Airpower Advantage,* chapter 2. Hale Burr, interview with author, December 11, 2019.

11. Charles J. Quilter II, *U.S. Marines in the Persian Gulf, 1990–1991: With the I Marine Expeditionary Force in Desert Shield and Desert Storm* (Washington, D.C.: History and Museums Division, Head Quarters U.S. Marine Corps, 1993), Appendix B. Keaney and Cohen, *Gulf War Air Power Survey, Unclassified*, vol. 5, part 1, 20–45. U.S. Army, *CMH Pub 70-117-1. War in the Persian Gulf: Operations Desert Shield and Desert Storm August 1990–March 1991* (Washington, D.C.: Center of Military History, 2010), 157.

12. Keaney and Cohen, *Gulf War Air Power Survey, Unclassified*, vol. 1, part 1, 207. James A. Winnefeld, Preston Niblack, and Dana J. Johnson, *A League of Airmen: U.S. Air Power in the Gulf War* (Santa Monica, CA: RAND, 1994), 290. Clancy and Horner, *Every Man a Tiger*, 220–263.

13. Winnefeld, Niblack, and Johnson, *A League of Airmen*, 172. Keaney and Cohen, *Gulf War Air Power Survey, Unclassified*, vol. 2, part 1, 77–81. Wawro, *Quicksand*, 424.

14. Jeffrey A. Engel, *When the World Seemed New: George H.W. Bush and the End of the Cold War* (Boston: Houghton Mifflin Harcourt, 2017), 402–403. Keaney and Cohen, *Gulf War Air Power Survey, Unclassified*, vol. 1, part 1, 91.

15. Keaney and Cohen, *Gulf War Air Power Survey, Unclassified*, vol. 1, part 1, 179. Schwarzkopf, *It Doesn't Take a Hero*, 413–478. Nalty, *Winged Shield Vol. 2*, 456–458.

16. Keaney and Cohen, *Gulf War Air Power Survey, Unclassified*, vol. 2, part 1, 249.

17. Keaney and Cohen, *Gulf War Air Power Survey, Unclassified*, vol. 2, part 1, 255–256. Wawro, *Quicksand*, 421. Schwarzkopf, *It Doesn't Take a Hero*, 386–387. Clancy and Horner, *Every Man a Tiger*, 440–449. Nalty, *Winged Shield Vol. 2*, 456–458.

18. John Andreas Olsen, ed., *A History of Air Warfare* (Washington, D.C.: Potomac, 2010), 185–191.

19. Keaney and Cohen, *Gulf War Air Power Survey, Unclassified*, vol. 2, part 1, 258. United States Military Academy and the United States Air Force Academy, "Kuwait and Vicinity, 1991—The Liberation of Kuwait, 24–26 February," *The First Gulf War*.

20. Keaney and Cohen, *Gulf War Air Power Survey, Unclassified*, vol. 2, part 1, 261–291. Winnefeld, Niblack, and Johnson, *A League of Airmen*, 148–153. Clancy and Horner, *Every Man a Tiger*, 440–449.

21. Keaney and Cohen, *Gulf War Air Power Survey, Unclassified*, vol. 2, part 1, 219.

22. Keaney and Cohen, *Gulf War Air Power Survey, Unclassified*, vol. 2, part 1, 261–291. Winnefeld, Niblack, and Johnson, *A League of Airmen*, 148–153. Clancy and Horner, *Every Man a Tiger*, 440–442. Nalty, *Winged Shield Vol. 2*, 476–479.

23. Keaney and Cohen, *Gulf War Air Power Survey, Unclassified*, vol. 2, part 1, 310.

24. Keaney and Cohen, *Gulf War Air Power Survey, Unclassified*, vol. 1, part 1, 301–302. Winnefeld, Niblack, and Johnson, *A League of Airmen*, 174.

25. Keaney and Cohen, *Gulf War Air Power Survey, Unclassified*, vol. 2, part 1, 254–255.

26. Warden published his Five Rings study in 1995, continuing his line of argument from *The Air Campaign* and the efforts of Checkmate in creating Instant Thunder.

27. Winnefeld, Niblack, and Johnson, *A League of Airmen*, 157. Fred Frostic, *Air Campaign against the Iraqi Army in the Kuwaiti Theater of Operations* (Santa Monica, CA: RAND, 1994), 14. Keaney and Cohen, *Gulf War Air Power Survey, Unclassified*, vol. 2, part 2, 148. In the data, FAC and Combat Search and Rescue missions are lumped into a single category and total 2,655 sorties. For the following calculations, an assumed 1,500 FAC missions is used.

28. Winnefeld, Niblack, and Johnson, *A League of Airmen*, 157. Frostic, *Air Campaign against the Iraqi Army*, 14. Keaney and Cohen, *Gulf War Air Power Survey, Unclassified*, vol. 2, part 2, 148. The data listed in the Gulf War Air Power Survey is annotated as incomplete; to compensate for the missing data, a 15 percent increase was added to all the data provided based on other accounts of Desert Storm and the notes in the Gulf War Air Power Survey.

29. Winnefeld, Niblack, and Johnson, *A League of Airmen*, 157. Frostic, *Air Campaign against the Iraqi Army*, 14. Keaney and Cohen, *Gulf War Air Power Survey, Unclassified*, vol. 2, part 2, 148. Spires, *Air Power for Patton's Army*, 54–57. Futrell, *The United States Air Force in Korea*, 690. Correll, *Air Force in Vietnam*, 14–16. For calculating the percentages, the data does not differentiate strategic attacks from interdiction attacks in North Vietnam. The 79 percent calculation allows for 50 percent of all attacks in North Vietnam as interdiction. Allowing all strike missions in North Vietnam counted as interdiction yields 86 percent CAS and interdiction missions to all missions in Vietnam. Disregarding all air-to-ground missions in North Vietnam,

the percentage of CAS and interdiction missions to all missions in Vietnam falls to 73 percent.

30. Frostic, *Air Campaign against the Iraqi Army*, 34–38. Keaney and Cohen, *Gulf War Air Power Survey, Unclassified*, vol. 2, part 1, 313–315.

31. Quilter II, *U.S. Marines in the Persian Gulf*, 53, 88, 107.

32. Frostic, *Air Campaign against the Iraqi Army*, 48. Keaney and Cohen, *Gulf War Air Power Survey, Unclassified*, vol. 2, part 1, 279–281.

33. Frostic, *Air Campaign against the Iraqi Army*, 15–16.

34. Mark A. Welsh, "Day of the Killer Scouts," *Air Force Magazine* (September 3, 2008).

35. Frostic, *Air Campaign against the Iraqi Army*, 34–38. Paul Hibbeln and Nathan Simmons, "Vipers of '91: Hill's F-16s at War," *Air Combat Command, News* (January 15, 2016), http://www.acc.af.mil/News/Article-Display/Article/660228/vipers-of-91-hills-f-16s-at-war/. Welsh, "Day of the Killer Scouts."

36. Frostic, *Air Campaign against the Iraqi Army*, 15–16, 48. Keaney and Cohen, *Gulf War Air Power Survey, Unclassified*, vol. 2, part 1, 279–281.

37. Keaney and Cohen, *Gulf War Air Power Survey, Unclassified*, vol. 2, part 1, 313–315. Rebecca Grant, "The Great Escape," *Air Force Magazine* (June 18, 2008). Rebecca Grant, "The Great Escape," *Air Force Magazine* (June 18, 2008). Frostic, *Air Campaign against the Iraqi Army*, 45.

38. Ibid.

39. Ibid.

40. Keaney and Cohen, *Gulf War Air Power Survey, Unclassified*, vol. 2, part 1, 313–315. Grant, "The Great Escape." Frostic, *Air Campaign against the Iraqi Army*, 45. Wawro, *Quicksand*, 445–450.

41. Keaney and Cohen, *Gulf War Air Power Survey, Unclassified*, vol. 2, part 1, 326.

42. U.S. Army, *Field Manual 100–5*, 8–1.

43. "Defense Department News Briefing," given by Gen. Merrill McPeak on CSPAN, March 15, 1991, 55:20, https://www.c-span.org/video/?17250-1/defense-department-news-briefing.

44. Ibid., 53:10.

45. Ibid., 56:20.

46. Richard H. Shultz Jr. and Robert L. Pfaltzgraff Jr., *The Future of Air Power in the Aftermath of the Gulf War*, 8th ed., (Maxwell Air Force Base: Air University Press, 2005), 3–7, 9–38, 57–82.

47. Raphael S. Cohen, *Air Force Strategic Planning: Past, Present, and Future* (Santa Monica, CA: RAND, 2017), 22. Richard P. Hallion, *Storm over Iraq: Air Power and the Gulf War* (Washington, D.C.: Smithsonian Institution Press, 1992), 118.

48. Ibid.

49. Donald B. Rice, *The Air Force and U.S. National Security: Global Reach—Global Power* (Washington, D.C.: Office of the Secretary of the Air Force, 1990), 5. Formatting original. Nalty, *Winged Shield Vol. 2*, 512–515.

50. Derek Chollet and James Goldgeier, *America between the Wars: From 11/9 to 9/11. The Misunderstood Years between the Fall of the Berlin Wall and the Start of the War on Terror* (New York: Public Affairs, 2008), 48–52, 89. Mark Guzinger, *Shaping America's Future Military toward a New Force Planning Construct* (Washington, D.C.: Center for Strategic and Budgetary Assessments, 2013), 2.

51. Chollet and Goldgeier, *America between the Wars*, 126–127, 147. George C. Herring, *From Colony to Superpower: U.S. Foreign Relations since 1776. The Oxford History of the United States* (New York: Oxford University Press, 2008), chapter 20. Hal Brands, *Making the Unipolar Moment: U.S. Foreign Policy and the Rise of the Post-Cold War Order* (Ithaca: Cornell University Press, 2016). Warren I. Cohen, *The New Cambridge History of Foreign Relations, Vol. 4: Challenges to American Supremacy, 1945 to the Present* (New York: Cambridge University Press, 2013). Lloyd C. Gardner, *The Long Road to Baghdad: A History of U.S. Foreign Policy from the 1970s to the Present* (New York: The New Press, 2008). Michael H. Hunt, *The American Ascendancy: How the United States Gained and Wielded Global Dominance* (Chapel Hill: University of North Carolina Press, 2007), chapter 8.

52. David E. Jeremiah, "What's Ahead for the Armed Forces," *Joint Forces Quarterly* 1, no. 1 (Summer 1993): 25–35.

53. Guzinger, *Shaping America's Future Military*, 3.

54. NPR Staff, "What a Downed Black Hawk in Somalia Taught America," *All Things Considered, NPR*, October 5, 2013. For more on the Gothic Serpent raid see Mark Bowden, *Black Hawk Down: A Story of Modern War* (New York: Signet, 2001). Nalty, *Winged Shield Vol. 2*, 492–496.

55. Chollet and Goldgeier, *America between the Wars*, 72–76. Herring, *From Colony to Superpower*, chapter 20. Brands, *Making the Unipolar Moment*. Cohen, *The New Cambridge History of Foreign Relations*,

Vol. 4. Gardner, *The Long Road to Baghdad.* Hunt, *The American Ascendancy*, chapter 8.

56. Robert C. Owen, *Deliberate Force: A Case Study in Effective Air Campaigning. Final Report of the Air University Balkans Air Campaign Study* (Maxwell Air Force Base: Air University Press, 2000), chapter 1. Nalty, *Winged Shield Vol. 2*, 496–512.

57. Owen, *Deliberate Force*, 247. Nalty, *Winged Shield Vol. 2*, 496–512.

58. Owen, *Deliberate Force*, 318. Nalty, *Winged Shield Vol. 2*, 496–512.

59. Owen, *Deliberate Force*, 318–320. Nalty, *Winged Shield Vol. 2*, 496–512.

60. Owen, *Deliberate Force*, 258. Nalty, *Winged Shield Vol. 2*, 496–512.

61. Owen, *Deliberate Force*, chapter 3, 4, 16. Nalty, *Winged Shield Vol. 2*, 496–512.

62. Benjamin S. Lambeth, *NATO's Air War for Kosovo* (Santa Monica, CA: RAND, 2001), 6–7. Chollet and Goldgeier, *America between the Wars*, chapter 8. Ivo H. Daalder and Michael E. O'Hanlon, *Winning Ugly: NATO's War to Save Kosovo* (Washington, D.C: Brookings Institution Press, 2000), chapter 2.

63. Chollet and Goldgeier, *America between the Wars*, 212. Daalder and O'Hanlon, *Winning Ugly*, 65.

64. Chollet and Goldgeier, *America between the Wars*, chapter 8. Lambeth, *NATO's Air War for Kosovo*, xx, 43–47. Daalder and O'Hanlon, *Winning Ugly*, 69–72.

65. Lambeth, *NATO's Air War for Kosovo*, 17–42.

66. "Defense Department News Briefing," given by Gen. Merrill McPeak on CSPAN, March 15, 1991.

67. Chollet and Goldgeier, *America between the Wars*, 210. Lambeth, *NATO's Air War for Kosovo*, 17–42. Christopher E. Haave and Phil M. Haun, *A-10s Over Kosovo: The Victory of Airpower Over a Fielded Army as Told by the Airmen Who Fought in Operation Allied Force* (Maxwell Air Force Base: Air University Press, 2003), xxix. Daalder and O'Hanlon, *Winning Ugly*, 103–108, 117–125, 136.

68. Lambeth, *NATO's Air War for Kosovo*, 43. Original text note added for clarity: "On that account, Clark later acknowledged that his air commanders were no happier than he was with the absence of a ground threat, noting that it was 'sort of an unnatural act for airmen to fight a ground war without a ground component.'" Michael Ignatieff, "Issue: The Virtual Commander," Annals of Diplomacy, *The New Yorker* (August 2, 1999): 33. Daalder and O'Hanlon, *Winning Ugly*, 130–136.

69. Michael W. Lamb, *Operation Allied Force: Golden Nuggets for Future Campaigns* (Maxwell Air Force Base: Air University Press, 2002), 3. Chollet and Goldgeier, *America between the Wars*, 211–217. Daalder and O'Hanlon, *Winning Ugly*, 117–125.

70. Haave and Haun, *A-10s Over Kosovo*, 15–22. Congressional Research Service, *Instances of Use of United States Armed Forces Abroad, 1798–2020*, by Barbara Salazar Torreon (updated January 13, 2020), 14–18.

71. Haave and Haun, *A-10s Over Kosovo*, xxix, 15–22.

72. Greg Masters and Stuart Johnson, *Operation Allied Force Tactical Lessons Learned*, Secret (Ramstein Air Force Base: USAFE, 2000), 4–1. Ninth AF/HO, Non USAFCENT History Office Papers, Operation Allied Force. Haave and Haun, *A-10s Over Kosovo*, 23–36. Phil M. Haun, *Airpower Versus a Fielded Force: Misty FACs of Vietnam and A-10 FACs of Kosovo, a Comparative Analysis* (Maxwell Air Force Base: Air University Press, 2004), 57–72.

73. Daalder and O'Hanlon, *Winning Ugly*, 150–155, 198–210. Lambeth, *NATO's Air War for Kosovo*, 82–86.

74. Daalder and O'Hanlon, *Winning Ugly*, 144–147, 198–210. Lambeth, *NATO's Air War for Kosovo*, 83–85.

75. Daalder and O'Hanlon, *Winning Ugly*, 137–142. Lambeth, *NATO's Air War for Kosovo*, 67–70.

76. Lambeth, *NATO's Air War for Kosovo*, 86.

77. Daalder and O'Hanlon, *Winning Ugly*, 165–181. Lambeth, *NATO's Air War for Kosovo*, 60.

78. Bruce R. Nardulli, Walter L. Perry, Bruce Pirnie, John Gordon IV, and John G. McGinn, *Disjointed War: Military Operations in Kosovo, 1999* (Santa Monica, CA: RAND, 2002), 48–54. Daalder and O'Hanlon, *Winning Ugly*, 198–210. Lambeth, *NATO's Air War for Kosovo*, 198–210.

79. U.S. Department of Defense, "Joint Statement on the Kosovo After Action Review," *News Release No: 478–99* (October 14, 1999), https://archive.defense.gov/Releases/Release.aspx?ReleaseID=2220. Daalder and O'Hanlon, *Winning Ugly*, 210–226.

80. Rick Griset, *56th Fighter Wing and Luke Field/Air Force Base Heritage Pamphlet 1940–2015* (Luke Air Force Base: 56th Fighter Wing History Office, 2015). Larry Schneck, "Luke-trained Warriors Direct Fight While in Flight," *56th Fighter Wing Public Affairs* (August 12, 2011), https://www.luke.af.mil/News/Article-Display/

Article/358891/luke-trained-warriors-direct-fight-while-in-flight/, AFHRA. "310 Fighter Squadron (AETC)," *Fact Sheets* (April 6, 2010), https://www.afhra.af.mil/About-Us/Fact-Sheets/Display/Article/ 432812/310-fighter-squadron-aetc/.

81. Lamb, *Operation Allied Force*, 1–2.

82. Ibid., 1. U.S. Department of Defense, "Joint Statement on the Kosovo After Action Review."

83. Ibid.

Notes for Chapter 7

1. Joshua Rovner, "The Primacy Problem: Explaining America's Afghan Purgatory," *War on the Rocks* (January 8, 2020), https://warontherocks. com/2020/01/the-primacy-problem-explaining-americas-afghan-purga- tory/. Wawro, *Quicksand*, chapter 16.

2. Gregory Ball, "2001—Operation Enduring Freedom," *AFHRA* (August 23, 2011), http://www.afhistory.af.mil/FAQs/Fact-Sheets/ Article/458975/2001-operation-enduring-freedom/. Wawro, *Quick- sand*, chapter 16.

3. Benjamin S. Lambeth, *Air Power against Terror: America's Conduct of Operation Enduring Freedom* (Santa Monica, CA: RAND, 2001), 45–46. Ball, "2001—Operation Enduring Freedom." Wawro, *Quick- sand*, 495, 503.

4. Benjamin S. Lambeth, *Air Power against Terror: America's Conduct of Operation Enduring Freedom* (Santa Monica, CA: RAND, 2001), 135–153. Ball, "2001—Operation Enduring Free- dom." Wawro, *Quicksand*, 495, 498–502. For more on Operation Enduring Freedom and the recent history of Afghanistan see Steve Coll, *Ghost Wars: The Secret History of the CIA, Afghanistan and Bin Laden, from the Soviet Invasion to September 10, 2001* (New York: Penguin, 2004); Seth Jones, *In the Graveyard of Empires: America's War in Afghanistan* (New York: W. W. Norton, 2010); Larry Goodson, *Afghanistan's Endless War: State Failure, Regional Politics and the Rise of the Taliban* (Seattle: University of Washington Press, 2001); and Lester Grau, ed., *The Bear Went over the Mountain: Soviet Combat Tactics in Afghanistan* (Washington, D.C.: National Defense University Press, 1996).

5. "OEF Sorites by Mission Type," *Secret, Excel File*, Ninth AF/HO, Reports/Papers/Sorties. "OEF TAC CAS," *Secret, Excel File*, Ninth AF/HO, Reports/Papers/Sorties.

6. Lambeth, *Air Power against Terror*, 158–161.

7. Ibid., 161.

8. Ibid., 158. Wawro, *Quicksand*, 498–502, 503–504. U.S. Congress, Senate, *A Report to Members of the Committee on Foreign Relations of the United States, Tora Bora Revisited: How We Failed to Get Bin Laden and Why It Matters Today*, 111th Cong., 1st sess., 2009, S. Prt. 111–35.

9. Lambeth, *Air Power against Terror*, 258–262.

10. By the time Operations in Afghanistan began, the term Joint Terminal Attack Controller, or JTAC, came into use in place of TACP. Even though a JTAC is technically a part of the Tactical Air Control Party, JTAC supplanted TACP as the term of art used by both soldiers and airmen to mean ground-based air controllers. CJCS, *Joint Publication 3-09.3: Joint Close Air Support* (Washington, D.C.: Chairman of the Joint Chiefs of Staff, 2019), I–2 (hereafter JP 3-09.3). Joint Fire Support Executive Steering Committee, *Action Plan Memorandum of Agreement 2004–02 Joint Forward Air Controller (Airborne) (FAC (A))*, May 1, 2015, 6 (hereafter JMOA).

11. Lambeth, *Air Power against Terror*, 177.

12. Ibid.

13. U.S. Air Force, *Operation Anaconda: An Air Power Perspective (Unclassified)*, Headquarters Air Force AF/XOL (February 7, 2005), 44, 46, 49.

14. Ibid., 46–47

15. U.S. Air Force, *Operation Anaconda*, 46–47.

16. Nicholas Mellow, interview with author, October 8, 2019. United States Air Force, "Theater Air Ground System," *Air Force Doctrine Annex 3–03*, CJCS, *Joint Publication 3-30: Joint Air Operations* (Washington, D.C.: Chairman of the Joint Chiefs of Staff, 2019), II–11.

17. Ibid.

18. U.S. Air Force, *Operation Anaconda*, 50–53. Michael Longoria, "The Way We Were: Forward Air Controller War Stories at the Beginning of the Two Longest Wars in Our History," presentation, Texas Tech University, 2nd Annual Symposium on Modern Warfare Agenda "Insurgency and Counterinsurgency in the Modern Era of Warfare," Lubbock, TX, October 6, 2017.

19. U.S. Air Force, *Operation Anaconda*, 34–39.

20. U.S. Air Force, *Operation Anaconda*, 34–39. Lambeth, *Air Power against Terror*, 202.

21. "Joint Fires Review Panel Chronology," Secret, Ninth AF/HO, Non-AFCENT Papers, OEF. U.S. Air Force, *Operation Anaconda*, 37.

22. U.S. Air Force, *Operation Anaconda*, 66.

23. Richard L. Kulger, Michael Baranick, and Hans Binnendijk, *Operation Anaconda: Lessons for Joint Operations* (Washington, D.C.: Center for Technology and National Security Policy, 2009), 25.

24. U.S. Air Force, *Operation Anaconda*, 66.

25. U.S. Air Force, *Operation Anaconda*, 66, 79. Lambeth, *Air Power against Terror*, 192–200.

26. Kulger, Baranick, and Binnendijk, *Operation Anaconda*, 50.

27. Lambeth, *Air Power against Terror*, 197.

28. U.S. Air Force, *Operation Anaconda*, 78–79.

29. U.S. Air Force, *Operation Anaconda*, chapter 5. Lambeth, *Air Power against Terror*, 221–231.

30. U.S. Air Force, *Operation Anaconda*, 112–115, 121. Lambeth, *Air Power against Terror*, 200–231. Kulger, Baranick, and Binnendijk, *Operation Anaconda*, x–xiv. Joint Fire Support Executive Steering Committee, *JMOA*, 7–25.

31. Lambeth, *Air Power against Terror*, 158. Wawro, *Quicksand*, 507–508.

32. Walter L. Perry, Richard E. Darilek, Laurinda L. Rohn, and Jerry M. Sollinger, eds. *Operation Iraqi Freedom: Decisive War, Elusive Peace* (Santa Monica, CA: RAND, 2015), 4–8. Wawro, *Quicksand*, 552.

33. Perry, Darilek, Rohn, and Sollinger, *Operation Iraqi Freedom*, 47, 56. Wawro, *Quicksand*, 515, 542–543, 545.

34. Perry, Darilek, Rohn, and Sollinger, *Operation Iraqi Freedom*, 163, 165–166.

35. Perry, Darilek, Rohn, and Sollinger, *Operation Iraqi Freedom*, 163, 165–166. Porter and Overton, *Project CHECO Report #103*, 2.

36. Perry, Darilek, Rohn, and Sollinger, *Operation Iraqi Freedom*, 47, 56, 163, 165–166. Wawro, *Quicksand*, 515, 542–545.

37. Perry, Darilek, Rohn, and Sollinger, *Operation Iraqi Freedom*, 149–174.

38. Ibid., 149–175.

39. "OIF Sorite Trends," *Secret, Excel File*, Ninth AF/HO, Reports/Papers/ Sorties. Perry, Darilek, Rohn, and Sollinger, *Operation Iraqi Freedom*, 152.

40. Perry, Darilek, Rohn, and Sollinger, *Operation Iraqi Freedom*, 149–174. Sortie generation rates are based on the following assumptions: Total A-10 and F-16 aircraft in theater, a 2–1 aircrew to aircraft ratio, two

missions a day for aircrew, 1 FAC qualified pilot per 10 pilots in F-16 squadrons, and 1 FAC qualified pilot per 4.5 pilots in A-10 squadrons. One mission per FAC per day is used to calculate the reduced sortie amount.

41. Lambeth, *Airpower against* Terror, 149. Perry, Darilek, Rohn, and Sollinger, *Operation Iraqi Freedom*, 4–8. Wawro, *Quicksand*, 565–568, 572–573. For more on Operation Iraqi Freedom see Thomas E. Ricks, *Fiasco: The American Military Adventure in Iraq* (New York: Penguin Books, 2007) and John R. Ballard, David W. Lamm, and John K. Wood, *From Kabul to Baghdad and Back: The US at War in Afghanistan and Iraq* (Annapolis: Naval Institute Press, 2012).

42. Nicholas Mellow, interview with author, October 8, 2019. Paul Birch, interview with author, May 29, 2019. Howard D. Belote, "Counterinsurgency Airpower: Air-Ground Integration for the Long War," *Air and Space Power Journal* 20, no. 3 (Fall 2006): 55–68.

43. Headquarters U.S. Air Force, "HAC-D Request for Future CAS Day Discussion," undated, unpublished Powerpoint presentation, 10. Mike Benitez, "How Afghanistan Distorted Close Air Support and Why It Matters," *War on the Rocks* (June 29, 2016). Wawro, *Quicksand*, 584–592.

44. Belote, "Counterinsurgency Airpower," 55–68.

45. Darren Halford, interview with author, October 18, 2017. Jeff Allison, interview with author, October 3, 2017. Caleb Campbell, interview with author, May 10, 2017.

46. Ibid.

47. Ibid.

48. 455th Expeditionary Operations Group, "FCIF13-19B," Secret, Ninth AF/HO, USAFCENT HO Papers, Operation Enduring Freedom.

49. 361st Expeditionary Reconnaissance Squadron, "Air Warden Duties," Secret, undated Powerpoint presentation, Ninth AF/HO, USAFCENT HO Papers, Operation Enduring Freedom.

50. Christopher Koontz, *Valley of Fire: Air Power at the Battle of COP Keating*, Secret (Air Force Historical Support Division, 2018), 21, Ninth AF/HO, USAFCENT HO Papers, Operation Enduring Freedom.

51. Koontz, *Valley of Fire*, 21. David Faggard, "Airman Comments About Firefight," *455th Air Expeditionary Wing Public Affairs* (October 15, 2009), https://www.af.mil/News/Article-Display/Article/118844/airman-comments-about-firefight/.

52. Ibid.

53. Ibid. "Dude 25, 3 Oct 2009," *Mission Report*, Secret, Ninth AF/HO, USAFCENT HO Papers, Operation Enduring Freedom. "Dude 01, 4 Oct 2009," *Mission Report*, Secret, Ninth AF/HO, USAFCENT HO Papers, Operation Enduring Freedom.

54. Joint Fire Support Executive Steering Committee, *JMOA*, 1.

55. Government Accountability Office, National Security and International Affairs Division, *Report to the Chairman, Committee on Armed Services, House of Representatives. Close Air Support: Status of Air Force's Efforts to Replace the A-10 Aircraft*, by Harry R. Finley, September 2, 1988. Comptroller of the Air Force, "Statistical Digest FY 2001," *U.S. Air Force Statistical Digest*, https://www.afhistory.af.mil/Portals/64/Statistics/2001%20USAF%20STATISTICAL%20DIGEST.pdf?ver=2017-04-28-100500-117. Comptroller of the Air Force, "Statistical Digest FY 1991," *U.S. Air Force Statistical Digest*, https://media.defense.gov/2011/Apr/19/2001330028/-1/-1/0/AFD-110419-007.pdf.

56. Michel III, "The Revolt of the Majors," U.S. Air Force, *Air Force Instruction 11-2F-16 Volume 1: Flying Operations, F-16 Aircrew Training* (Joint Base Langley-Eustis, Virginia: Headquarters Air Combat Command, 2019). U.S. Air Force, *Air Force Instruction 1-2A-10C Volume 1: Flying Operations, A-10 Aircrew Training* (Joint Base Langley-Eustis, Virginia: Headquarters, Air Combat Command, 2019). Brian D. Laslie, *The Air Force Way of War: U.S. Tactics and Training after Vietnam* (Lexington: University Press of Kentucky, 2015). Anderegg, *Sierra Hotel*. US Air Force, *Air Force Manual 11–202 Volume 1: Flying Operations, Aircrew Training* (Washington, D.C.: Headquarters Air Force, 2019). Matthew Walsh, William W. Taylor, and John A. Ausink, *Independent Review and Assessment of the Air Force Ready Aircrew Program: A Description of the Model Used for Sensitivity Analysis* (Santa Monica, CA: RAND, 2019). Government Accountability Office, *Report to Congressional Committees, Ready Aircrew Program. Air Force Actions to Address Congressionally Mandated Study on Combat Aircrew Proficiency*. February 2020. Nalty, *Winged Shield Vol. 2*, chapter 21.

57. Robert G. Angevine, "Time to Revive Joint Concept Development and Experimentation," *War on the Rocks* (January 23, 2020), https://warontherocks.com/2020/01/time-to-revive-joint-concept-development-and-experimentation/.

58. Joint Fire Support Executive Steering Committee, *JMOA*, 8–9.

59. US Air Force, *Air Force Manual 11–202 Volume 1*, 6. Walsh, Taylor, and Ausink, *Independent Review and Assessment of the Air Force Ready Aircrew Program.* Government Accountability Office. *Report to Congressional Committees, Ready Aircrew Program.* U.S. Air Force, *Air Force Instruction 1-2A-10C Volume 1.* U.S. Air Force, *Air Force Instruction 11-2F-16 Volume 1.*

60. Joint Fire Support Executive Steering Committee, *JMOA*, 56.

61. Joint Fire Support Executive Steering Committee, *JMOA*, 56. U.S. Air Force, *Air Force Instruction 1-2A-10C Volume 1*, 42. U.S. Air Force, *Air Force Instruction 11-2F-16 Volume 1*, 42.

62. Ibid.

63. Ibid.

64. Jesse Breau, interview with author, October 9, 2019. Burt Brown, interview with author, October 8, 2019. Bobcat Baker, interview with author, October 8, 2019. Scout Randolph, interview with author, October 9, 2019.

65. U.S. Air Force, *Air Force Instruction 11-2F-16 Volume 1*, 17–19. Jesse Breau, interview with author, October 9, 2019. Burt Brown, interview with author, October 8, 2019. Bobcat Baker, interview with author, October 8, 2019. Scout Randolph, interview with author, October 9, 2019.

66. Albert A. Robbert, et al. *Reducing Air Force Fighter Pilot Shortages* (Santa Monica, CA: RAND, 2015). Mike Benitez, "Air Force in Crisis, Part II: How Did We Get Here?" *War on the Rocks*, March 8, 2018.

67. Some aircraft operated by US Air Force Special Operations Command (AFSOC) also qualify as FACs but only perform the mission in support of special operations missions. Further, AFSOC and other Special Operations units have their JTAC capabilities integral to their units. While significant within the discussion of Special Operations missions, the discussion of AFSOC JTAC/FAC(A) capabilities is beyond the scope of this discussion.

68. Comptroller of the Air Force, "Statistical Digest Fiscal Year 2010," *U.S. Air Force Statistical Digest*, https://www.afhistory.af.mil/Portals/64/Statistics/2010%20USAF%20Statistical%20Digest%20FY10.pdf?ver=2017-04-25-123540-013&timestamp=1493138487191. This number derives from the author's experience in theater and discussions with the USAF 549 Combat Training Squadron, which conducts training and evaluation of USAF close air support programs, and the 24th TASS.

Further examination of the JMOA, Joint Close Air Support Publications, and the USAF Ready Air Crew Program inform the calculations. These numbers represent an estimation made based on FY 2010 and FY2018 information and were not exact for every year of OIF or OEF. The Air Force does not maintain a record of individual squadron FAC qualification, nor did the Air Force maintain an equal distribution of F-16 and A-10 squadrons deployed throughout either campaign. Therefore, arriving at an exact number for any given year of either campaign would not be possible.

69. U.S. Air Forces Central Command Assessment and Analysis Division, *Operation Iraqi Freedom—By the Numbers, Unclassified.* by Michael Moseley, April 30, 2003, Ninth AF/HO. U.S. Air Forces Central Command Public Affairs, "Combined Forces Air Component Commander 2010–2014 Airpower Statistics, as of 31 Jan 2014," *U.S. Air Forces Central Command Air Power Summaries.* U.S. Air Forces Central Command Public Affairs, "Combined Forces Air Component Commander 2013–2019 Airpower Statistics, as of 31 Jan 2020," *U.S. Air Forces Central Command Air Power Summaries.*

70. "Kahuna 61, 17 April 2003," *Mission Report*, Secret, Ninth AF/HO, USAFCENT HO Papers, Operation Iraqi Freedom.

71. Dan Keller, "FAC(A) Research," email to Jesse Breau, December 20, 2018.

72. Dana J. H. Pittard and Wes J. Bryant, *Hunting the Caliphate* (New York: Post Hill Press, 2019), 31.

73. Charles H. Briscoe, "Fighting through the 'Fog of War': The Battle of An Najaf, 8–9 January 2007—Part I," *Veritas* 4, (2008): 2. Chaps Leaderhouse, interview with author, February 28, 2020.

74. Briscoe, "Fighting through the 'Fog of War,'" 2.

75. Ibid, 2. Charles H. Briscoe, "Fighting through the 'Fog of War': The Battle of An Najaf, 8–9 January 2007—Part II," *Veritas* 4, no. 3 (2008), 2.

76. Chaps Leaderhouse, interview with author, February 28, 2020. Mission Reports January 28, 2007: Tusk 05, Tusk 03, Tusk 07, Thud 27, Buzzard 53, Buzzard 57, Secret, Ninth AF/HO, USAFCENT HO Papers, Operation Iraqi Freedom. Dan Byrd, *"The Mother of All TICS": The Air Force Response at An Najaf, January 29–29, 2007*, Secret (Washington, D.C.: Air Force Historical Support Division, 2018), Ninth AF/HO, USAFCENT HO Papers, Operation Iraqi Freedom.

77. Briscoe, "Fighting through the 'Fog of War'—Part I," 2. Briscoe, "Fighting through the 'Fog of War'—Part II," 1. Chaps Leaderhouse, interview with author, February 28, 2020.

78. U.S. Air Force, "MQ-1B Predator," *Fact Sheets* (September 23, 2015), https://www.af.mil/About-Us/Fact-Sheets/Display/Article/104469/ mq-1b-predator/. James Thompson, "Sunsetting the MQ-1 Predator: A History of Innovation," *432nd Air Expeditionary Wing Public Affairs* (February 20, 2018), https://www.af.mil/News/Article-Display/Article/ 1445531/sunsetting-the-mq-1-predator-a-history-of-innovation/. U.S. Army, "Hellfire Family of Missiles," *USAASC*, https://asc.army. mil/web/portfolio-item/hellfire-family-of-missiles/.

79. Ibid. USAF, "MQ-9 Reaper," *Fact Sheet* (September 23, 2015), https:// www.af.mil/About-Us/Fact-Sheets/Display/Article/104470/mq-9- reaper/. USAF, "RQ-11B Raven," *Fact Sheets* (October 31, 2007), https://www.af.mil/About-Us/Fact-Sheets/Display/Article/104533/ rq-11b-raven/.

80. Thompson, "Sunsetting the MQ-1 Predator."

81. Ibid. USAF, "Sniper Pod," *Fact Sheets* (September 22, 2015), https://www. af.mil/About-Us/Fact-Sheets/Display/Article/104527/sniper-pod/. USAF, "Litening Advance Targeting," *Fact Sheets* (August 18, 2015), https://www.af.mil/About-Us/Fact-Sheets/Display/Article/104571/ litening-advance-targeting/.

82. Frank Anderson, interview with author, October 18, 2019. Jesse D. Breau, *Forward Air Controller (Airborne): Training to Win Large Scale Combat Operations*, thesis (School of Advanced Military Studies, Fort Leavenworth, 2019), 35. Myron Hura et al., *Interoperability: A Continuing Challenge in Coalition Air Operations* (Santa Monica, CA: RAND, 2000), chapter 9.

83. William Hague, "Statement from the Conference Chair Foreign Secretary William Hague Following the London Conference on Libya," London Conference on Libya, London, England, March 29, 2011. William Hague, "Chair's Statement," Fourth Meeting of the Libya Contact Group, Istanbul, Turkey, July 15, 2011. CJTF-OIR, "About CJTF-OIR," *Operation Inherent Resolve*, https://www.inherentresolve. mil/About-CJTF-OIR/.

84. Matthew C. Isler, interview with author, January 9, 2019. Scott Kindsvater, interview with author, March 29, 2019. Todd R. Phinney, "Reflections on Operation Unified Protector," *Joint Force Quarterly* 73 (2nd Quarter 2014): 73–86. Florence Gaub, *The North Atlantic*

Treaty Organization and Libya: Reviewing Operation Unified Protector (Carlisle: Strategic Studies Institute and U.S. Army War College Press, 2013). ACC Office of History, *A North African Odyssey: ACC and the 2011 War in Libya*, Secret (Langley Air Force Base: Air Combat Command, 2015). Ninth AF/HO, Non AFCENT Papers, Operation Odyssey Dawn. Office of History, Headquarters USAFE, *The United States Air Forces in Europe in Operation Odyssey Dawn* (Ramstein AFB: USAFE, 2012), Ninth AF/HO, Non AFCENT Papers, Operation Odyssey Dawn. The author also served as a NATO staff officer at the Combined Air Operations Center, Poggio Renatico, Italy, during Operation Odyssey Dawn/Unified Protector.

85. CJCS, *DOD Dictionary*, 218.

86. Office of the Secretary of Defense Joint Warfighters Joint Test and Evaluation, *Commander's Handbook for Joint Time-Sensitive Targeting* (Washington, D.C.: National Defense University Institute for National Strategic Studies, 2002). Mark A. Hewitt, *Time Sensitive Targeting: Overcoming the Intelligence Gap in Interagency Operations*, thesis, Naval War College, Newport, RI, 2003.

87. Pittard and Bryant, *Hunting the Caliphate*, xiii.

88. Ibid., xiii, 31, 104–105, 189, 204, 241.

89. Matthew C. Isler, interview with author, January 9, 2019. Scott Kindsvater, interview with author, March 29, 2019. Paul Birch, interview with author, May 29, 2019. Pittard and Bryant, *Hunting the Caliphate*, xiii, 104–105, 189, 204, 241.

90. Matthew C. Isler, interview with author, January 9, 2019. Scott Kindsvater, interview with author, March 29, 2019. Megan Friedl, "Crushing ISIS: Air Power Operations in a Complex Battle Space," *Defense Media Activity* (September 19, 2017), https://www.af.mil/News/Article-Display/Article/1315995/crushing-isis-air-power-operations-in-a-complex-battle-space/.

91. ACC Office of History, *A North African Odyssey*, Office of History, Headquarters USAFE, *Operation Odyssey Dawn*. Joe Quartararo Sr., Michael Rovenolt, and Randy White, *Libya's Operation Odyssey Dawn: Command and Control* (Washington, D.C.: National Defense University Press, 2012). Karl P. Mueller, *Precision and Purpose: Airpower in the Libyan Civil War* (Santa Monica, CA: RAND, 2015).

92. Matthew C. Isler, interview with author, January 9, 2019. Scott Kindsvater, interview with author, March 29, 2019. Dave Bickerstaff, interview with author, September 25, 2019. Mitch Walker, email to

author, October 3, 2017. Owen Birkett, email to author, October 3, 2017. Pittard and Bryant, *Hunting the Caliphate*, xiii, 104–105, 189, 204, 241.

93. John J. Schaefer, "Responsive Close Air Support," *Joint Forces Quarterly* 67 (Fourth Quarter 2012): 91–96. Overton, *Project CHECO Report #103*. Porter, *Project CHECO Report #79*. Trest, *Project CHECO Report #78*. Porter, *Project CHECO Report #11*.

94. Benitez, "How Afghanistan Distorted Close Air Support."

95. Schaefer, "Responsive Close Air Support," 92.

96. Smack Balzhiser, "FAC(A) Research," email to Jesse Breau, December 18, 2018.

97. Dan Keller, "FAC(A) Research," email to Jesse Breau, December 20, 2018.

98. "NY Freedoms Sentinel MISREP VIPER 31 A3247 CASTGT," *Mission Report Viper 31*, 12 October 2017, Secret, Ninth AF/HO. Dan Keller, "FAC(A) Research," email to Jesse Breau, December 20, 2018.

99. Aaron French, interview with author, September 22, 2017. Darren Halford, interview with author, October 18, 2017. Joint Fire Support Executive Steering Committee, *JMOA*, 7–25. Director Joint Staff, *Joint Publication 3–09 "Joint Fire Support,"* II–21.

Notes for Conclusion

1. U.S. President, "The National Security Strategy of The United States, 2017," 3.

2. James Holmes, "57th Operations Group's CAS Integration Group Charter," Air Combat Command, September 26, 2017, 1.

3. Jesse Breau, interview with author, October 9, 2019. James Holmes, "57th Operations Group's CAS Integration Group Charter," *Air Combat Command* (September 26, 2017), 2. USAF, "24 Tactical Air Support Squadron," *Fact Sheet*, https://www.afhra.af.mil/About-Us/Fact-Sheets/Display/Article/1702297/24-tactical-air-support-squadron-acc/. Nalty, *Winged Shield Vol. 2*, 439.

4. Office of Air Force Lessons Learned, *Total Force Integration Air Support Operations Center-Tactical Air Control Party (ASOC-TACP) Efficacy*, by Jacqueline R. Henningsen, 2009, 1.

5. Shultz Jr. and Pfaltzgraff Jr., *The Future of Air Power*, 3–7, 9–38, 57–82.

6. Keaney and Cohen, *Gulf War Air Power Survey, Unclassified*, vol. 2, part 2, 148.

7. U.S. Air Forces Central Command Public Affairs, "Combined Forces Air Component Commander 2010–2014 Airpower Statistics, as of 31 Jan 2014," *U.S. Air Forces Central Command Air Power Summaries*. U.S. Air Forces Central Command Public Affairs, "Combined Forces Air Component Commander 2013–2019 Airpower Statistics, as of 31 Jan 2020," *U.S. Air Forces Central Command Air Power Summaries*. Correll, *The Air Force in Vietnam*, 4. Winnefeld, Niblack, and Johnson, *A League of Airmen*, 157. Keaney and Cohen, *Gulf War Air Power Survey, Unclassified*, vol. 2, part 2, 148. Frostic, *Air Campaign against the Iraqi Army*, 14.

8. Winnefeld, Niblack, and Johnson, *A League of Airmen*, 157. Frostic, *Air Campaign against the Iraqi Army*, 14. Keaney and Cohen, *Gulf War Air Power Survey, Unclassified*, vol. 2, part 2, 148. Office of Statistical Control, *Army Air Forces Statistical Digest*, 220–230.

9. Benitez, "How Afghanistan Distorted Close Air Support."

10. United States Air Force, "Theater Air Ground System," *Air Force Doctrine Annex 3–03*.

11. Headquarters U.S. Air Force, "HAC-D Request for Future CAS Day Discussion," 10. U.S. Air Force, *CFETP 13LXX, Parts I and II: AFSC 13 LXX Air Liaison Officer Career Field Education and Training Plan*, Washington, D.C.: Headquarters, U.S. Air Force, 2013. Henningsen, *Total Force Integration*, Office of Air Force Lessons Learned. *CSAF Priority: Modernize Our Air Space and Cyber Inventories, Organizations, and Training. Collection: Air Liaison Officer (13LX) Career Field Effectiveness*, by Jacqueline R. Henningsen, 2013.

12. Burt Brown, interview with author, October 8, 2019. Jesse Breau, interview with author, October 9, 2019. Rebecca Grant, "Marine Air in the Mainstream," *Air Force Magazine* (June 8, 2008).

13. Ibid.

14. Todd South, "Return of Fires: How the Army is Getting Back to Its Big Guns as It Prepares for the Near-Peer Fight," *Army Time* (August 27, 2018), https://www.armytimes.com/news/your-army/2018/08/27/return-of-fires-how-the-army-is-getting-back-to-its-big-guns-as-it-prepares-for-the-near-peer-fight/.

15. Joel Bier, "Playing Moneyball: The Scouting Report on Light Attack Aircraft," *War on the Rocks* (April 21, 2017), warontherocks.com/2017/04/playing-moneyball-the-scouting-report-on-light-attack-aircraft/. Valerie Insinna, "U.S. Air Force to Keep A-10 off the Chopping Block in Next Budget Request," *Defense News* (September 4, 2019),

https://www.defensenews.com/smr/defense-news-conference/2019/09/04/
air-force-not-planning-to-cut-a-10-under-budget-reorg/.

16. Alan Cummings, "Hypersonic Weapons: Tactical Uses and Strategic
    Goals," *War on the Rocks* (November 12, 2019), https://warontherocks.
    com/2019/11/hypersonic-weapons-tactical-uses-and-strategic-goals/.
    Loren Thompson, "Ten Things We Know for Sure about the Air Force's
    Secret B-21 Bomber," *Forbes* (November 8, 2019), https://www.
    forbes.com/sites/lorenthompson/2019/11/08/ten-things-we-know-for-
    sure-about-the-air-forces-secret-b-21-bomber/#185127b31caf.

17. Office of Statistical Control, *Army Air Forces Statistical Digest*,
    220–230. Futrell, *The USAF in Korea*, 690. Correll, *The Air Force in
    Vietnam*, 14–16. Winnefeld, Niblack, and Johnson, *A League of Airmen*,
    157. Frostic, *Air Campaign against the Iraqi Army*, 14. Keaney and
    Cohen, *Gulf War Air Power Survey, Unclassified*, vol. 2, part 2, 148.

18. Bier, "Playing Moneyball."

19. Brian W. Everstine, "How AFSOC Plans to Use its Light Attack
    Aircraft," *Air Force Magazine* (November 18, 2019). Aaron Mehta,
    "U.S. Air Force Officially Buying Light-Attack Planes," *Defense News*
    (October 25, 2019), https://www.defensenews.com/air/2019/10/25/
    air-force-officially-buying-light-attack-planes/.

20. Jesse Breau, interview with author, October 9, 2019.

21. Dan Keller, "FAC(A) Research," email to Jesse Breau, December 20,
    2018.

22. Farmer and Strumwasser, *The Evolution of the Airborne Forward Air
    Controller*, 83.

Glossary

AAF	Army Air Forces
AAGS	Army Air Ground System
ACC	Air Combat Command
ACTS	Air Corps Tactical School
ADC	Air Defense Command
AEF	American Expeditionary Force
AFB	Air Force Base
AFCENT	US Air Forces Central Command
AFSOC	Air Force Special Operations Command
AGOW	Air Ground Operations Wing
ASOC	Air Support Operations Center
AWPD	Air War Plans Division
CAS	Close Air Support
CENTAF	US Central Command Air Forces
CJCS	Combined Joint Chiefs of Staff
CJTF	Combined Joint Task Force
DASC	Direct Air Support Center
DMZ	Demilitarized Zone
DOC	Designed Operational Capability Statement
DOD	Department of Defense
FAC	Forward Air Controller
FAC(A)	Forward Air Controller (Airborne)
FEBA	Forward Edge of the Battle Area
FM	Field Manual
FSCL	Fire Support Coordination Line
GHQ	General Headquarters Air Force
GIB	Guy in Back
HVI	High Value Individual
JFC	Joint Force Commander
J-MOA	Joint Forward Air Controller (Airborne) Memorandum of Agreement
JTAC	Joint Terminal Attack Controller
LANTIRN	Low Altitude Navigation and Targeting Infrared for Night
MACV	Military Assistance Command, Vietnam
MANPAD	Man Portable Air Defense System

NATO	North Atlantic Treaty Organization
OSD	Office of the Secretary of Defense
RAF	Royal Air Force
RFC	Royal Flying Corps
ROVER	Remotely Operated Video Enhanced Receiver
SAC	Strategic Air Command
SAM	Surface-to-Air Missile
SCAR	Strike Coordination and Reconnaissance
SEA	Southeast Asia
SOP	Standard Operating Procedure
TAC	Tactical Air Command
TACP	Tactical Air Control Party
TACS	Tactical Air Control System/Theater Air Control System
TASS	Tactical Air Support Squadron
USAF	US Air Forces
USAFE	US Air Forces in Europe
VNAF	Republic of Vietnam Air Force

Bibliography

Archives and Archival Documents

Ninth Air Force / US Air Forces Central Command, Command Historian Archive, Shaw Air Force Base, SC.

366 AEW at War. Command Historian Digital Archive, Non USAFCENT History Office Papers.

Ninth Air Force/AFCENT History. Command Historian Digital Archive, Reports and Papers.

Desert Shield-Storm. Command Historian Digital Archive, Reports and Papers.

Desert Shield-Storm. Command Historian Digital Archive, Non USAFCENT History Office Papers.

Operation Allied Force. Command Historian Digital Archive, Non USAFCENT History Office Papers.

Operation Enduring Freedom. Command Historian Digital Archive, Reports and Papers.

Operation Enduring Freedom. Command Historian Digital Archive, Non USAFCENT History Office Papers.

Operation Iraqi Freedom. Command Historian Digital Archive, Reports and Papers.

Operation Iraqi Freedom. Command Historian Digital Archive, Non USAFCENT History Office Papers.

Operation Odyssey Dawn. Command Historian Digital Archive, Non USAFCENT History Office Papers.

Southeast Asia. Command Historian Digital Archive, Non USAFCENT History Office Papers.

Theater History of Operations Reports Database. Command Historian Digital Archive, Non USAFCENT History Office Papers.

Urban CAS. Command Historian Digital Archive, Non USAFCENT History Office Papers.

US Air Forces Central Command. Command Historian Digital Archive, Non USAFCENT History Office Papers.

US Air Forces in Europe. Command Historian Digital Archive, Non USAFCENT History Office Papers.

World War II. Command Historian Digital Archive, Non USAFCENT History Office Papers.

Penn State Special Collections Library, Penn State University, State College, PA.

The Eighth Air Force Archive. Histories, Papers, Diaries, Manuscripts. Call Num 1538.

US Air Force Academy McDermott Library, Clark Special Collections, US Air Force Academy, Colorado Springs, CO.

Beardslee, Lily V. Papers of PFC Erwin W. Beardslee. SMS-16.

Davis, Jefferson Hayes. Papers. SMS-57.

Directorate, Tactical Evaluation, CHECO Division. *Project Contemporary Historical Evaluation of Combat Operations (CHECO) Reports*. 254 vols. Hickam Air Force Base: Headquarters Pacific Air Forces, 1961–1975.

Dwight, Henry William. Papers, 1917–1961. MS-5.

Foulois, Benjamin Delahauf. Papers, 1880–1968. MS-17.

Gimbel, Richard. Aeronautical Collection. SMS-337.

Green, Murry. "Hap" Arnold Collection, 1891–1988. MS-33.

Hansell, Haywood Shepherd. Papers, 1929–1964. MS-6.

Kuter, Laurence Sherman. Papers, 1905–1979. MS-18.

L' Escadrille LaFayette Flying Corps Association. Papers and Records, 1927–1960. MS-19.

Mitchell, William. Papers, 1917–1958. MS-14.

Pruden, Russell Godine. *The Wartime Diary of Russell Godine Pruden, Captain, Air Service, US Army, Adjutant, 27th Aero Squadron, A.E.F. July 16, 1917–March 31, 1919*. SMS-705.

US Air Force Historical Research Agency, Maxwell Air Force Base, AL.

Fifth Army. "Operation Rover Joe Photographs." December 1, 1944. IRIS Num 00249262.

Fifth Army. "Air Operations Bulletin." November 20, 1944. IRIS Num 00249264.

Fifth Army. XXII Support to US Fifth Army. IRIS Num 00249268 to 00249274, 00249277, 00249278.

VIII Air Support Command. "Air Operations in Support of Ground Forces in Northwest Africa." 2 vols. IRIS Num 00232189 and 00232190.

Ninth Air Force. Operational History of the Ninth Air Force. IRIS Nums 232241 to 232245.

Ninth Air Force. Operations and Reports. Multiple records within call series 533.01.

IX Tactical Air Command. Employment of Air Power in the Breakthrough at Saint Lo. IRIS Num 00467887.

IX Tactical Air Command. History of IX Tactical Air Command European Theater of Operations. IRIS Num 237284 to 237394, 237403, 237405, 237407, 237410, 237603, 237910, 238100 to 238118, 238130 to 238184.

XIX Tactical Air Command. XIX Tactical Air Command Operations in Europe. IRIS Num 238200 to 238214.

XXII Tactical Air Command. "History of the XXII Tactical Air Command." September 20, 1944 to December 31, 1944. IRIS Num 00246956 to 00246958.

XXIX Tactical Air Command. "Effectiveness of Close in Air Cooperation." January 1, 1943 to January 1, 1945. IRIS Num 00239305.

XXIX Tactical Air Command. Operations, Orders, and Reports. IRIS num. 00238787, 00238808, 00238993, 00239239, 00239299 to 00239301, 00239302, 00239307 to 00239313.

162nd Air Liaison Squadron. History of 162nd Air Liaison Squadron. IRIS Num 63401 to 63422.

324th Fighter Group. 324th Fighter Group Operational History and War Diaries. IRIS Num 83641 to 83657.

324th Fighter Group. "Odyssey of the 324th Fighter Group." IRIS Num 123756.

6147 Tactical Control Group. History and Operations of the 6147 Tactical Control Group. Multiple records within call series K-GP-TACT-6147-HI, K110.36147-1, K-SQ-AW-6147-SU-RE, K-GP-6147-SU-PE, K239.0512-1335 C.1, K-GP-TACT-6147-SU-RE, K-GP-TACT-6147-SU-RO, K-GP-TACT-6147-NE, K-GP-TACT-6147-SU-NE, K720.059-41.

6147 Tactical Control Squadron. History and Operations of the 6147 Tactical Control Squadron. Multiple records within call series K-SQ-AW-6147-HI.

6148 Tactical Control Squadron. History and Operations of the 6148 Tactical Control Squadron. Multiple records within call series K-SQ-AW-6148-HI.

6149 Tactical Control Squadron. History and Operations of the 6149 Tactical Control Squadron. Multiple records within call series K-SQ-AW-6149-HI.

6150 Tactical Control Squadron. History and Operations of the 6150 Tactical Control Squadron. Multiple records within call series K-SQ-AW-6150-HI.

Air Corps Tactical School. History and Records. IRIS Num 159930 to 159988.

Assistant Chief of Air Staff, Intelligence. Battle for Tunisia, Historical Division, Air Power. April 1, 1943 to May 1, 1943. IRIS Number 0467738.

Evaluation Division, Air University, "The Air University Evaluation Division Quarterly Project Report," 15 June 1948, 19–20; Maj. Gen. David M. Schlatter to chief of staff, US Air Force, letter, 22, October 1947. IRIS Num 00155290.

Grime, Lieutenant Colonel. *Summary Report on "Chiefs 31 Joint Force Development Initiatives."* May 6, 1985. Iris Number: 01125972.

Mediterranean Allied Air Forces. "Military Attaché Report, Great Britain." June 1, 1944. IRIS Num 00243328.

Mediterranean Allied Air Forces. "Tactical Bulletin Number 54, Air Support to 8 Army in Final Assault in Italy." April 23, 1945–May 22, 1945. IRIS Num 00243329.

Mediterranean Allied Air Forces. "Operational and Tactical Information, Bombing Accuracy." August 1, 1944 to January 1, 1945. IRIS Num 00243334.

Mediterranean Allied Tactical Air Forces. Mediterranean Allied Tactical Air Forces Microfilm Project. 24 vols. IRIS Num 20000646 to 20000669.

Mediterranean Theater of Operations. "Rover Joe Chart/Maps of Florence Italy." IRIS Num 00244119.

Mediterranean Theater of Operations. "Index to Air Operations, Translation of Code Names." IRIS Num 00244123.

Mediterranean Theater of Operations. "Situation Maps: Italy." IRIS Num 00244178.

South East Asia Air Command. Tactical Bulletins No. 1 through 19. IRIS Num 00265066.

Weyland, Otto P. Extracts from the Diary of Major General OP Weyland, September 15, 1944 to May 18, 1945. IRIS Num 00238252.

Other Archives

Quentin C. Aanenson. Unpublished personal papers, photographs, and manuscripts. Private collection.

Burr Jr., Hale H. Unpublished personal papers, photographs, memoir, emails, interviews, and manuscripts. Private collection.

Forward Air Controller Association and OV-10 Bronco Association. Museum, Manuscripts, and Collections. Fort Worth Aviation Museum, Fort Worth, TX.

George W. Bush Presidential Library and Museum, Dallas, TX.

George H. W. Bush Presidential Library Center, College Station, TX.
Richard Nixon Presidential Library and Museum, Yorba Linda, CA.
Ronald Reagan Presidential Library and Museum, Simi Valley, CA.
The University of Texas at Dallas, Eugene McDermott Library Special Collections and Archives, Richardson, TX.
The US National Archives and Records Administration, College Park, MD.
The US Department of State, Office of the Historian. Washington, D.C.

Primary Sources

"ORD Changes to General Type Base; New Organizational Heads Take Over." *The Rotator, 106 AAFBU, Greensboro, NC*, vol. 5, no. 48. July 12, 1946. Greensboro Historical Newspapers. Online Archive. UNC Greensboro, Greensboro, NC. http://libcdm1.uncg.edu/cdm/ref/collection/GSOPatriot/id/37423.

"Occupation Personals." *The Pacific Stars and Stripes*. November 11, 1948. https://newspaperarchive.com/pacific-stars-and-stripes-nov-11-1948-p-5/

Aanenson, Quentin C. *A Fighter Pilot's Story*. Quentin C. Aanenson; in association with WETA-TV, 1999, 180 mins. https://quentinaanenson.com/.

Boyd, John R. *Aerial Attack Study. Boyd*. Nellis AFB, Nevada, 1964. Air University Muir S. Fairchild Research Information Center, Fairchild Documents Catalog Number M-U 43947-5.

———. *A Discourse on Winning and Losing*. Maxwell AFB: Air University Press, 2018.

The Central Intelligence Agency. *The Pros and Cons of Bombing North Vietnam*. Undated. https://www.cia.gov/library/readingroom/docs/CIA-RDP78S02149R000100130005-1.pdf.

———. *SNIE 14.3-70: The Outlook from Hanoi: Factors Affecting North Vietnam's Policy on the War in Vietnam*, by Director of Central Intelligence, Concurred in by the US Intelligence Board. February 5, 1970. https://www.cia.gov/library/readingroom/docs/DOC_0001166466.pdf.

The Central Intelligence Agency and The Defense Intelligence Agency. *An Appraisal of the Bombing of North Vietnam: Through 31 December 1967*. https://www.cia.gov/library/readingroom/docs/CIA-RDP82S00205R000100060002-6.pdf.

Congressional Research Service. *Operation Enduring Freedom: Potential Air Power Questions for Congress*, by Christopher Bolkcom. RS21020, Updated November 21, 2001. https://digital.library.unt.edu/ark:/67531/metacrs7569/.

_____. *Troop Levels in the Afghan and Iraq Wars, FY2001-FY2012: Cost and Other Potential Issues*, by Amy Belasco. R40682, July 2, 2009.   https://digital.library.unt.edu/ark:/67531/metadc26175/m1/1/high_res_d/R40682_2009Jul02.pdf.

_____. *Operation Odyssey Dawn (Libya): Background and Issues for Congress*, by Jeremiah Gertler. R41725, March 30, 2011. https://digital.library.unt.edu/ark:/67531/metadc99034/.

_____. *Proposed Retirement of A-10 Aircraft: Background Brief*, by Jeremiah Gertler. R43843 Version 4 New, January 5, 2015. https://crsreports.congress.gov/product/pdf/R/R43843.

_____. *Instances of Use of United States Armed Forces Abroad, 1798–2020*, by Barbara Salazar Torreon. Updated January 13, 2020. https://crsreports.congress.gov/product/pdf/R/R42738.

CSPAN, "Defense Department News Briefing" given by Gen. Merrill McPeak. Aired March 15, 1991 on CSPAN. https://www.c-span.org/video/?17250-1/defense-department-news-briefing.

Cunningham, Megan, ed. *Logbook of Signal Corps No. 1: The US Army's First Airplane*. Washington, D.C.: Air Force History and Museums Program, 2004.

Department of Defense. "Joint Statement on the Kosovo After Action Review." *News Release No: 478–99*. October 14, 1999.

_____. *Joint Test and Evaluation Program, 1999 Annual Report*, by George R. Schneiter, 2000. https://apps.dtic.mil/dtic/tr/fulltext/u2/a387709.pdf.

_____. *Directive: Functions of the Department of Defense and Its Major* Components, by Robert M. Gates. DoDD 5001.01, December 21, 2010.

Department of the Navy, Bureau of Ships. *SC 3585S, Catalogue of Electronic Equipment: Vol. 3 of 3, S Series Through Miscellaneous*. NavShips 900,116—Supplement No. 5 1 October 1952. http://www.radiomanual.info/schemi/Surplus_Handbooks/Catalogue_of_electronic_equipment_Part3_extract_NAVSHIPS_900-116_1952.pdf.

Dulles, John Foster. "The Evolution of Foreign Policy," Before the Council of Foreign Relations, New York, N.Y., Department of State, Press Release No. 81, January 12, 1954. http://www.airforcemag.com/MagazineArchive/Documents/2013/September%202013/0913keeperfull.pdf.

Evans, Ryan. "Interview of General David L. Goldfein." *Podcast: War on the Rocks*, August 29, 2018. https://warontherocks.com/2018/08/wotr-podcast-a-chat-with-the-chief-gen-david-goldfein-on-the-people-and-future-of-the-u-s-air-force/2/2

Forward Air Controllers Association. "MACV Tam #32, Gia Nghia, Quang Duc Province, Vietnam 1970." Unpublished home video recordings.

Gabbert, Richard D., and Gary B. Streets. *A Comparative Analysis of USAF Fixed-Wing Aircraft Losses in Southeast Asia Combat.* Wright-Patterson Air Force Base: Air Force Flight Dynamics Laboratory, 1977. https://apps.dtic.mil/dtic/tr/fulltext/u2/c016682.pdf.

Government Accountability Office. *Report to Congressional Committees, Ready Aircrew Program. Air Force Actions to Address Congressionally Mandated Study on Combat Aircrew Proficiency.* February 2020. GAO-20-91. https://www.gao.gov/resources/710/704996.pdf.

Government Accountability Office, National Security and International Affairs Division. *Comptroller General's Report to the Chairmen, Appropriations and Armed Services Committees, Congress of the United States. The Close Air Support: Principal Aircraft Choices, Department of Defense.* December 8, 1971.

_____. *Report to the Chairman, Committee on Armed Services, House of Representatives. Close Air Support: Status of Air Force's Efforts to Replace the A-10 Aircraft,* by Harry R. Finley. September 2, 1988. GAO/NSIAD-88-211. https://www.gao.gov/resources/nsiad-88-211.pdf.

_____. *Report to the Chairman, Committee on Armed Services, House of Representatives. Close Air Support: Airborne Controllers in High-Threat Areas May Not Be Needed,* by Nancy R. Kingsbury. April 1990. GAO/NSIAD-90-116. https://www.gao.gov/resources/nsiad-90-116.pdf.

Goldwater—Nichols Department of Defense Reorganization Act of 1986. Public Law 99–433, 99th Congress. October 1, 1986.

Hague, William. "Statement from the Conference Chair Foreign Secretary William Hague Following the London Conference on Libya." London Conference on Libya, London, England. March 29, 2011. https://www.nato.int/nato_static_fl2014/resources/pdf/pdf_2011_03/20110927_110329_-London-Conference-Libya.pdf.

_____. "Chair's Statement." Fourth Meeting of the Libya Contact Group, Istanbul, Turkey. July 15, 2011. https://www.nato.int/nato_static_fl2014/resources/pdf/pdf_2011_07/20110926_110715-Libya-Contact-Group-Istanbul.pdf.

Headquarters US Air Force. "HAC-D Request for Future CAS Day Discussion." Undated, unpublished Powerpoint presentation.

Headquarters US Air Force. *USAF Strategic Master Plan,* by Deborah Lee James and Mark A. Welsh, III. May 2015. https://apps.dtic.mil/dtic/tr/fulltext/u2/a618021.pdf.

Headquarters US Military Assistance Command, Vietnam. *MACV Combat Experiences 4–69*, by T.B. Mancinelli. November 3, 1969.

Joint Deployable Analysis Team (JDAT), Deputy Director, Command, Control, Communications, and Computers (DDC4), Joint Staff J-8. *Joint Close Air Support Action Plan Assessment Report*, by Jabari J. Reneau and Leonard J Gordon. JDAT-TR-11-09, August 2011.

Joint Fire Support Executive Steering Committee. *Chairman of the Joint Chiefs of Staff Instruction CJCSI 5127.01*. May 23, 2014. https://archive.org/details/Joint-Fire-Support-Executive-Steering-Committee-Governance-and-Management-23-May-2014/mode/2up.

_____. *Action Plan Memorandum of Agreement 2004–02 Joint Forward Air Controller (Airborne) (Forward Air Controller(A))*. May 1, 2015.

Jumper, John P. "The US Air Force: 'Our Mission is to Fly and Fight.'" Remarks to the Air Force Association Air Warfare Symposium, Orlando, Fla., Feb. 8, 2007. https://www.af.mil/About-Us/Speeches-Archive/Display/Article/143965/the-us-air-force-our-mission-is-to-fly-and-fight/

Lawler, Daniel J., and Erin R. Mahan, eds. *Foreign Relations of The United States*. Washington: Government Printing Office, 2010.

McConnell, Joseph P. "The Role of Airpower in Viet-Nam: Address by Gen. J. P. McConnell, Chief of Staff, US Air Force," before the Dallas Council on World Affairs, Dallas, Texas, 16 September, 1965. Washington, D.C.: The United States Air Force, 1965.

Office of Air Force Lessons Learned. *Enduring Airpower Lessons From OEF/OIF UAS Predator/Reaper Surge Operations*, by Jacqueline R. Henningsen. 2009. https://www.jllis.mil/USAF/.

_____. *Total Force Integration Air Support Operations Center-Tactical Air Control Party (ASOC-TACP) Efficacy*, by Jacqueline R. Henningsen. 2009. https://www.jllis.mil/USAF/.

_____. *Enduring Lessons From OEF/OIF: Adapting to Evolving Combat Realities*, by Jacqueline R. Henningsen. 2012. https://www.jllis.mil/USAF/.

_____. *CSAF Priority: Modernize Our Air Space and Cyber Inventories, Organizations, and Training. Collection: Air Liaison Officer (13LX) Career Field Effectiveness*, by Jacqueline R. Henningsen. 2013. https://www.jllis.mil/USAF/.

The Raven Songbook. The Edgar Allen Poe Literary Society of Texas. Undated, unpublished. Mike Leonard personal archive.

Secretary of the Air Force. *The Air Force and US National Security: Global Reach—Global Power, a White Paper*, by Donald B. Rice. June 1990.

Shaud, John A. *Air Force Strategy Study 2020–2030*. Maxwell Air Force Base: Air Force Research Institute, Air University Press, 2011.

Tonkin Gulf Resolution; Public Law 88–408, 88th Congress, August 7, 1964; General Records of the United States Government; Record Group 11; National Archives.

US Air Force. *Amended FY 1990/91 Biennial Budget*, by Deputy Assistant Secretary (Cost and Economics) Assistant Secretary of the Air Force (Financial Management and Comptroller). December 15, 1989. https://media.defense.gov/2011/Apr/12/2001330025/-1/-1/0/AFD-110412-038.pdf.

――――. *The Air Force and US National Security: Global Reach—Global Power, a White Paper*, by Donald B. Rice. June 1990. https://secure.afa.org/EdOp/2012/GRGP_Rice_1990.pdf.

US Air Force. *Air Superiority 2030 Flight* Plan, by Enterprise Capability Collaboration Team. May 2016. https://www.af.mil/Portals/1/documents/airpower/Air%20Superiority%202030%20Flight%20Plan.pdf.

US Air Force, Headquarters Air Combat Command. *57th Operations Group's CAS Integration Group Charter*, by Gen. James M Holmes. September 26, 2017. Jesse Breau personal archive.

US Air Force Financial Management and Comptroller. "Budget FY97 to FY21." *FM Resources*. https://www.saffm.hq.af.mil/FM-Resources/Budget/.

US Air Forces Central Command Public Affairs. "Combined Forces Air Component Commander 2007–2012 Airpower Statistics, as of 31 Jan 2012." *US Air Forces Central Command Air Power Summaries*. https://www.afcent.af.mil/Portals/82/Documents/Airpower%20summary/AFD-120208-021.pdf.

――――. "Combined Forces Air Component Commander 2010–2014 Airpower Statistics, as of 31 Jan 2014." *US Air Forces Central Command Air Power Summaries*. https://www.afcent.af.mil/Portals/82/Documents/Airpower%20summary/AFD-140219-003.pdf?ver=2016-01-13-143748-070.

――――. "Combined Forces Air Component Commander 2013–2019 Airpower Statistics, as of 31 Jan 2020." *US Air Forces Central Command Air Power Summaries*. https://www.afcent.af.mil/Portals/82/Documents/Airpower%20summary/Jan%202020%20Airpower%20Summary.pdf?ver=2020-02-13-032911-670.

US Congress. "The National Security Act of 1947." Public Law 253, 80th Congress; Chapter 343, 1st Session; S. 758. July 26, 1947.

_____. *Congressional Record.* 2nd sess., 1966. Vol. 112, pt. 7.

US Congress, House. *Roles and Missions of Close Air Support. Hearing Before the Investigations Subcommittee of The Committee on Armed Services, House of Representatives.* 101st Cong., 1st sess. September 27, 1990.

US Congress, Senate. *A Report to Members of the Committee on Foreign Relations of the United States, Tora Bora Revisited: How We Failed to Get Bin Laden and Why It Matters Today.* 111th Cong., 1st sess., 2009. S. Prt. 111–35. https://www.govinfo.gov/content/pkg/CPRT-111SPRT53709/html/CPRT-111SPRT53709.htm.

_____. *Department of the Air Force. Presentation to the Senate Armed Services Committee, Subcommittee on Airland Forces. Fiscal Year 2017 Air Force, Force Structure and Modernization Programs.* 114th Cong, 2nd sess. March 8, 2016.

US President. National Security Directive 26, "US Policy toward the Persian Gulf." October 2, 1989. https://bush41library.tamu.edu/files/nsd/nsd26.pdf.

_____. National Security Directive 45, "US Policy in Response to the Iraqi Invasion of Kuwait." August 20, 1990. https://bush41library.tamu.edu/files/nsd/nsd45.pdf.

_____. National Security Directive 54, "Responding to Iraqi Aggression in the Gulf." January 15, 1991. https://bush41library.tamu.edu/files/nsd/nsd54.pdf.

_____. "The National Security Strategy of The United States, 2017." https://www.whitehouse.gov/wp-content/uploads/2017/12/NSS-Final-12-18-2017-0905.pdf.

US Secretary of War. *United States Strategic Bombing Surveys. Summary Report: European War, Pacific War*, by Franklin D'Olier, Chairman, United States Strategic Bombing Survey. 1945. Reprint, Maxwell Air Force Base: Air University Press, 1987.

DOD Doctrine Publications and Training Manuals

93 AGOW/A3T. *TACP Officer Assessment and Selection. Application Process.* Moody Air Force Base, 93 Air Ground Operations Wing, 2019.

Air Combat Command. *ACC Syllabus Forward Air Controller(A): F-16 Forward Airborne Controller (Airborne) Training Course.* Joint Base Langley-Eustis, Virginia: Headquarters Air Combat Command, 2018.

Chairman of the Joint Chiefs of Staff. *Joint Publication 3-03: Joint Interdiction*. Washington, D.C.: Chairman of the Joint Chiefs of Staff, 2016.

_____. *Joint Publication 3-09: Joint Fire Support*. Washington, D.C.: Chairman of the Joint Chiefs of Staff, 2014.

_____. *Joint Publication 3-09: Joint Fire Support*. Washington, D.C.: Chairman of the Joint Chiefs of Staff, 2019.

_____. *Joint Publication 3-09.3: Joint Close Air Support*. Washington, D.C.: Chairman of the Joint Chiefs of Staff, 2014.

_____. *Joint Publication 3-09.3: Joint Close Air Support*. Washington, D.C.: Chairman of the Joint Chiefs of Staff, 2019.

_____. *Joint Publication 3-30: Joint Air Operations*. Washington, D.C.: Chairman of the Joint Chiefs of Staff, 2019.

_____. *Joint Publication 3-31: Joint Land Operations*. Washington, D.C.: Chairman of the Joint Chiefs of Staff, 2019.

_____. *Joint Publication 3-50: Joint Personnel Recovery*. Washington, D.C.: Chairman of the Joint Chiefs of Staff, 2015.

_____. *Joint Publication 3-52: Joint Airspace Control*. Washington, D.C.: Chairman of the Joint Chiefs of Staff, 2019.

Chief of the Air Corps. *Training Regulations No. 440–15: Air Corps, Employment of the Air Forces of the Army*. Washington, D.C.: War Department, 1935.

Chief of the Air Service. *Training Regulations No. 440–15: Air Service, Fundamental Principles for the Employment of the Air Service*. Washington, D.C.: War Department, 1926.

Department of the Air Force. *TAC Manual 3-3 Volume III: Mission Employment Tactics, Fighter Fundamentals, A-10*. Langley Air Force Base: Headquarters Tactical Air Command, 1989.

Office of the Secretary of Defense Joint Warfighters Joint Test and Evaluation. *Commander's Handbook for Joint Time-Sensitive Targeting*. Washington, D.C.: National Defense University Institute for National Strategic Studies, 2002. https://apps.dtic.mil/dtic/tr/fulltext/u2/a403414.pdf.

Seventh Air Force. *Seven Air Force Pamphlet 55-1: Seventh Air Force in Country Tactical Air Operations Handbook*. San Francisco: Headquarters Seventh Air Force, 1968.

US Air Force. *Air Force Manual No. 1–3: Air Doctrine, Theater Air Operations*. Washington, D.C.: Department of the Air Force, 1953.

_____. *AFM 2–1, Tactical Air Operations: Counter Air, Close Air Support, and Air Interdiction*. Washington, D.C.: Headquarters US Air Force, 1965.

_____. *Air Force Manual No. 1–1: Aerospace Doctrine, United States Air Force Basic Doctrine.* Washington, D.C.: Headquarters US Air Force, 1971.

_____. *Air Force Doctrine Document 1–1: Basic Aerospace Doctrine of the United States Air Force.* 2 vols. Washington, D.C.: Headquarters US Air Force, 1992.

_____. *Air Force Doctrine Document 1: Basic Air Force Doctrine.* Maxwell Air Force Base: Curtis E. Lemay Center for Doctrine, 2003.

_____. *CFETP 13LXX, Parts I and II: AFSC 13 LXX Air Liaison Officer Career Field Education and Training Plan.* Washington, D.C.: Headquarters, US Air Force, 2013.

_____. *United States Air Force Doctrine.* 3 vols. and 29 annexes. Maxwell Air Force Base: Curtis E. Lemay Center for Doctrine, 2015–2019.

_____. *Air Force Instruction 1-2A-10C Volume 1: Flying Operations, A-10 Aircrew Training.* Joint Base Langley-Eustis, Virginia: Headquarters, Air Combat Command, 2019.

_____. *Air Force Instruction 11-2F-16 Volume 1: Flying Operations, F-16 Aircrew Training.* Joint Base Langley-Eustis, Virginia: Headquarters Air Combat Command, 2019.

_____. *Air Force Manual 11–202 Volume 1: Flying Operations, Aircrew Training.* Washington, D.C.: Headquarters Air Force, 2019.

_____. *Air Force Tactic Techniques and Procedures 3–3. A-10: Combat Aircraft Fundamentals. Unclassified.* Nellis Air Force Base, NV: 561st Joint Tactics Squadron, 2018.

_____. *Air Force Tactic Techniques and Procedures 3–3. F-16: Combat Aircraft Fundamentals. Unclassified.* Nellis Air Force Base, NV: 561st Joint Tactics Squadron, 2020.

US Army. *Field Manual 100-5: Operations.* Washington, D.C.: Headquarters, Department of The Army, 1976.

_____. *Field Manual 100-5: Operations.* Washington, D.C.: Headquarters, Department of The Army, 1982.

_____. *Field Manual 100-5: Operations.* Washington, D.C.: Headquarters, Department of The Army, 1986.

US Army and US Air Force. *TRADOC Training Text 17-50-3: Joint Air Attack Team.* Fort Rucker, AL: Commander USAAVNC, 1983.

US Army, Marine Corps, Navy, and Air Force. *Dynamic Targeting: Multi-Service Tactics, Techniques, and Procedures for Dynamic Targeting.* Fort Leavenworth, KS: Training and Doctrine Command, 2015.

_____. *JFire: Multi-Service Tactics, Techniques, and Procedures for Joint Application of Firepower*. Fort Leavenworth, KS: Training and Doctrine Command, 2019.

US Joint Forces Command, Joint Warfighting Center. *Commander's Handbook for Joint Time-Sensitive Targeting*. Washington, D.C.: National Defense University Institute for National Defense Strategic Studies, 2002.

US War Department. *Army Air Force Field Manual 1-10: Tactics and Technique of Air Attack*. Washington, D.C.: United States Government Printing Office, 1940.

_____. *FM 31-35: Basic Field Manual, Aviation in Support of Ground Forces*. Washington, D.C.: United States Government Printing Office, 1942.

_____. *Army Air Force Field Manual 1-15: Tactics and Technique of Air Fighting*. Washington, D.C.: United States Government Printing Office, 1942.

_____. *War Department Field Manual FM 100-20: Command and Employment of Air Power*. Washington, D.C.: United States Government Printing Office, 1943.

Service Histories and Historical Studies

Air Force Historical Research Agency, Numbered Air Force Historical Studies.

No. 10. *Organization of the Army Air Arm, 1935–1945 (Revised)*, by Chase C. Mooney and Edward C. Williamson (1956). Iris Num. 467602.

No. 17. *Air Action in the Papuan Campaign, 21 July 1942 to 23 January 1943*, by Richard L. Watson (1944). IRIS Num 467609.

No. 24. *Command of Observation Aviation: A Study in Control of Tactical Airpower*, by Robert F. Futrell (1952). IRIS Num 903385.

No. 30. *Ninth Air Force in the Western Desert Campaign to 23 January 1943*, by Harry C. Coles (1945). IRIS Num 467621.

No. 32. *Ninth Air Force in the ETO [European Theater of Operations], 16 October 1943 to 16 April 1944*, by John F. Ramsey (1945). IRIS Num 467623.

No. 34. *Army Air Forces in the War against Japan, 1941–1942*, by E. Kathleen Williams (1945). IRIS Num 467625.

No. 35. *Guadalcanal and the Origins of the Thirteenth Air Force*, by Kramer J. Rohfleisch (1945). IRIS Num 467626.

No. 36. *Ninth Air Force, April to November 1944*, by Robert H. George (1945). 375 pages. IRIS NUM 467627.

No. 37. *Participation of the Ninth and Twelfth Air Forces in the Sicilian Campaign*, by Harry L. Coles (1945). IRIS Num 467628.

No. 38. *Operational History of the Seventh Air Force, 6 November 1943 to 31 July 1944*, by James C. Olson (1945). IRIS Num 467629.

No. 41. *Operational History of the Seventh Air Force, 7 December 1941 to 6 November 1943*, by James C. Olson (1945). IRIS Num 467632.

No. 43. *The Fifth Air Force in the Conquest of the Bismarck Archipelago, November 1943 to March 1944*, by Harris G. Warren (1946). IRIS Num 467634.

No. 44. *Evolution of the Liaison-Type Airplane, 1917–1944*, by Irving B. Holley Jr. (1946). IRIS Num 467635.

No. 70. *Tactical Operations of the Eighth Air Force, 6 June 1944–8 May 1945*, Juliette Hennessy (1952). IRIS Num 467659.

No. 71. *United States Air Force Operations in the Korean Conflict, 25 June–1 November 1950*, by Robert F. Futrell (1951). IRIS Num 467660.

No. 72. *United States Air Force Operations in the Korean Conflict, 1 November 1950–30 June 1952*, by Robert F. Futrell (1955). IRIS Num 467661.

No. 80. *Air Force Participation in Joint Army-Air Force Training Exercises, 1947–1950*, by Ralph D. Bald Jr. (1952). IRIS Num 467669.

No. 82. *Combat Squadrons of the Air Force, World War II*, edited by Maurer Maurer (1969). IRIS Num pt. 1. 916794, pt. 2. 916794.

No. 86. *Close Air Support in the War against Japan*, by Joe G. Taylor (1955). IRIS Num 916797.

No. 88. *The Employment of Strategic Bombers in a Tactical Role, 1941–1951*, by Robert W. Ackerman (1953). IRIS Num 467673.

No. 89. *The Development of Air Doctrine in the Army Air Arm, 1917–1941*, by Thomas H. Greer (1953). IRIS Num 467674.

No. 91. *Biographical Data on Air Force General Officers, 1917–1952*, by Robert P. Fogerty (1953). 2 volumes: Volume 1—A thru K; Volume 2—L thru Z. IRIS Num pt. 1. 467676, pt. 2. 467677.

No. 98. *The United States Army Air Arm, April 1861 to April 1917*, by Juliette A. Hennessy (1958). IRIS Num 467683.

No. 100. *History of the Air Corps Tactical School, 1920–1940*, by Robert T. Finney (1955). IRIS Num 916808.

No. 101. *The AAF in the South Pacific to October 1942*, by Kramer J. Rohfleisch (1944). IRIS Num 467685.

No. 105. *Air Phase of the North African Invasion, November 1942*, by Thomas J. Mayock (1944). IRIS Num 467689.

No. 108. *The AAF in the Middle East: A Study of the Origins of the Ninth Air Force*, by Edith Rogers (1945). IRIS Num 467692.

No. 111. *Army Air Action in the Philippines and Netherlands East Indies, 1941–1942*, by Richard L. Watson Jr. (1945). IRIS Num 467695.

No. 113. *The Fifth Air Force in the Huon Peninsula Campaign, January to October 1943*, by Richard L. Watson Jr. (1946). IRIS Num 467697.

No. 114. (U) *The Twelfth Air Force in the North African Winter Campaign, 11 November 1942 to the Reorganization of 18 February 1943*, by Thomas J. Mayock (1946). IRIS Num 467698.

No. 115. (U) *Air Phase of the Italian Campaign to 1 January 1944*, by Albert F. Simpson (1946). IRIS Num 467699.

No. 116. (U) *The Fifth Air Force in the Huon Peninsula Campaign, October 1943 to February 1944*, by Richard L. Watson Jr. (1947). IRIS Num 467700.

No. 127. (U) *United States Air Force Operations in the Korean Conflict, 1 July 1952–27 July 1953*, by Robert F. Futrell (1956). IRIS Num 467711.

No. 129. (U) *Air Forces Participation in Joint Army-Air Force Training Exercises, 1951–1954*, by Ralph D. Bald Jr. (1957). IRIS Num 467712.

No. 131. (U) *The United States Air Forces in Korea, 1950–1953*, by Robert F. Futrell (1961). IRIS Num 1012389.

No. 138. (U) *A History of the United States Air Force, 1907–1957*, edited by Alfred Goldberg et al (1957). IRIS Num 1012390.

No. 139. (U) *Ideas, Concepts, Doctrine: A History of Basic Thinking in the United States Air Force, 1907–1964*, by Robert F. Futrell (1971; revised 1974). Vol. 1 IRIS Num 1098990. Vol. 2 IRIS Num 1098991.

No. 140. (U) *Air Force Combat Units of World War II*, edited by Maurer Maurer (1963). IRIS Num 916846.

No. 141. (U) *The World War I Diary of Col Frank P. Lahm*, edited by Albert F. Simpson (1970). IRIS Num 1011074.

No. 144. (S) *The United States Air Force in Southeast Asia: The Advisory Years to 1965*, Robert F. Futrell with the assistance of Martin Blumenson (1981). IRIS Num 01007441 (on classified server).

No. 147. (U) *The US Air Service in World War I*, edited by Maurer Maurer, 4 volumes (1978–1979). IRIS Num Vol. 1 1039709, IRIS Num Vol. 2 1039710, IRIS Num Vol. 3 1039711, IRIS Num Vol. 4 1039712.

Air Force Historical Support Division.

Koontz, Christopher. *Valley of Fire: Air Power at the Battle of COP Keating.* Washington, D.C.: Air Force Historical Support Division, 2018. Secret.

US Air Force Statistical Digests. *USAF Statistics Since 1945,* by the Comptroller of the Air Force. https://www.afhistory.af.mil/USAF-STATISTICS/.

Other Air Force Historical Studies.

Carter, Kit C., and Robert Mueller. *The Army Air Forces in World War II: Combat Chronology 1941–1945.* 1973. Reprint, Washington, D.C.: Center for Air Force History, 1990.

Craven, Wesley, and James Cate, eds. *The Army Air Forces in World War II.* 7 vols. 1948. Reprint, Washington, D.C.: Office of Air Force History, 1983.

Davis, Richard G. *Air Staff Historical Study. The 31 Initiatives: A Study in Air Force-Army Cooperation.* Washington, D.C.: Office of Air Force History, 1987.

Griset, Rick. *56th Fighter Wing and Luke Field/Air Force Base Heritage Pamphlet 1940–2015.* Luke Air Force Base: 56th Fighter Wing History Office, 2015. https://www.luke.af.mil/Portals/58/Documents/Luke%20Heritage%20Pamphlet%201941-2015%20Expanded.pdf?ver=2015-11-10-190808-463.

Jacques, David R., and Dennis Strouble. *A-10 Thunderbolt II (Warthog) Systems Engineering Case Study.* Wright-Patterson Air Force Base: Air Force Center for Systems Engineering, Air Force Institute of Technology, 2010. https://apps.dtic.mil/dtic/tr/fulltext/u2/a530838.pdf.

Keaney, Thomas A., and Eliot A. Cohen. *Gulf War Air Power Survey, Unclassified.* 6 vols. Washington, D.C.: Department of the Air Force, 1993. IRIS Num 876065-876087.

———. *Gulf War Air Power Survey.* Secret. 6 vols. Washington, D.C.: Department of the Air Force, 1993. IRIS Num 876055-876064, 876089, 876091-876093, 876096, 876098, 876100-876102.

Haulman, Daniel L. *US Unmanned Aerial Vehicles in Combat, 1991–2003.* Washington, D.C.: Office of Air Force History, 2003. https://www.afhra.af.mil/Portals/16/documents/Studies/AFD-070912-042.pdf.

———. *Aberrations of Air War: Operations Enduring Freedom and Iraqi Freedom.* Maxwell Air Force Base: Air Force Historical Research Agency, 2015. https://www.afhra.af.mil/Portals/16/documents/Airmen-at-War/Haulman-AberrationsOfAirWar.pdf?ver=2016-08-22-131404-273.

Headquarters, Army Air Forces, Office of Assistant Chief of Air Staff, Intelligence. *Wings at War Series, No. 1. The AAF in the Invasion of Southern France, an Interim Report.* 1944. Reprint, Washington, D.C.: Center for Air Force History, 1992.

_____. *Wings at War Series, No. 5. Air-Ground Teamwork on the Western Front: The Role of XIX Tactical Air Command during August 1944 an Interim Report.* 1944. Reprint, Washington, D.C.: Center for Air Force History, 1992.

_____. *Wings at War Series, No. 6. The AAF in Northwest Africa: An Account of the Twelfth Air Force in the Northwest African Landings and the Battle for Tunisia, an Interim Report.* 1944. Reprint, Washington, D.C.: Center for Air Force History, 1992.

Owen, Robert C. *Deliberate Force: A Case Study in Effective Air Campaigning. Final Report of the Air University Balkans Air Campaign Study.* Maxwell Air Force Base: Air University Press, 2000.

Racenstein, Charles A. *Reference Series. Air Force Combat Wings: Lineage and Honors Histories 1947–1977.* Washington, D.C.: Office of Air Force History, 1984.

Reed, William B., Thomas C. Quinlan, Chester K. Shore, Ell E. Geissler, and Robert S. Gerdy, eds. *Condensed Analysis of the Ninth Air Force in the European Theater of Operations.* 1946. Reprint, Washington, D.C.: Office of Air Force History, 1984.

United States Air Force. *Operation Anaconda: An Air Power Perspective (Unclassified)* by Headquarters Air Force AF/XOL. February 7, 2005. https://media.defense.gov/2014/Sep/17/2001329845/-1/-1/0/anaconda%20ADA495248.pdf.

_____. *Operation Anaconda: An Air Power Perspective (Secret)* by Headquarters Air Force AF/XOL. February 7, 2005. USAFCENT History Office Classified Archive.

Watson, George M. *The Office of the Secretary of the Air Force 1947–1965.* Washington, D.C.: Center for Air Force History, 1993.

Wolf, Richard I. *The United States Air Force Basic Documents on Roles and Missions.* Washington, D.C.: Office of Air Force History, 1987.

US Army Historical Studies.

Briscoe, Charles H. "Fighting through the 'Fog of War': The Battle of An Najaf, 8–9 January 2007—Part I." *Veritas* 4, no. 2 (2008). https://www.soc.mil/ARSOF_History/articles/v4n2_fog_of_war_page_1.html.

_____. "Fighting through the 'Fog of War': The Battle of An Najaf, 8–9 January 2007—Part II." *Veritas* 4, no. 3 (2008). https://www.soc.mil/ARSOF_History/articles/v4n3_fog_of_war_page_1.html.

Cosmas, Graham A. *CMH Pub 91-6-1. United States Army in Vietnam: MACV The Joint Command in the Years of Escalation, 1962–1967.* Washington, D.C.: Center of Military History United States Army, 2006.

Stamps, T. Dodson, and Vincent J. Esposito. *A Short Military History of World War I: Atlas.* West Point: United States Military Academy, 1950.

_____. *A Military History of World War II: Atlas.* (West Point: United States Military Academy, 1953.

US Army. *CMH Pub 70-30-1. Whirlwind War: The United States Army in Operations Desert Shield and Desert Storm.* Washington, D.C.: Center of Military History, 1995.

_____. *CMH Pub 70-117-1. War in the Persian Gulf: Operations Desert Shield and Desert Storm August 1990–March 1991.* Washington, D.C.: Center of Military History, 2010.

United States Military Academy and the United States Air Force Academy. "Kuwait and Vicinity, 1991—The Liberation of Kuwait, 24–26 February." *The First Gulf War.* https://westpoint.edu/sites/default/files/inline-images/academics/academic_departments/history/Gulf%20War%20and%20Iraq/GulfWarFirst03.pdf.

US Joint Chiefs of Staff Historical Studies.

Office of the Chairman of the Joint Chiefs of Staff. *History of the Joint Chiefs of Staff: The Joint Chiefs of Staff and National Policy.* 12 vols. Washington, D.C.: Office of Joint History, 1996–2015.

US Marine Corps Historical Studies

Quilter II, Charles J. *US Marines in the Persian Gulf, 1990–1991: With the I Marine Expeditionary Force in Desert Shield and Desert Storm.* Washington, D.C.: History and Museums Division, Head Quarters US Marine Corps, 1993.

Memoirs

Amend, Dale. *A Duck Looking for Hunters.* Los Lunas, NM: Sage Mesa Publications, 2009.

Bell, Don, and The Tiger Forward Air Controllers. *The Tiger Forward Air Controllers: A Dance with the Devil.* Denver: Outskirts Press, 2014.

Bohm, Norris. *Out of the Pack: 8th TFW Wolf Forward Air Controllers.* Self-published, Akron: 48 Hour Books, 2014.

Broughton, Jack. *Thud Ridge.* New York: Popular Library, 1969.

Brulle, Robert V. *Angels Zero: Close Air Support in Europe*. Washington, D.C.: Smithsonian Books, 2000.

Clancy, Tom, and Chuck Horner. *Every Man a Tiger*. New York: G.P. Putnam's, 1999.

Cleveland, W. M. *Mosquitoes in Korea*. Portsmouth: Peter R. Randall Publisher, 1991.

Cobleigh, Ed. *War for the Hell of It*. New York: Berkley Caliber, 2005.

Cosentino, Neil. *Letters from the Cockpit*. Self-published, CreateSpace, 2017.

Detwiler, Ross C. *The Great Muckrock and Rosie*. Bloomington: Abbott Press, 2013.

Duehring, Craig W. *The Lair of the Raven*. Self-published, Craig W. Duehring, 2014.

Edgerton, Clyde. *Solo: My Adventures in the Air*. Chapel Hill: Algonquin Books of Chapel Hill, 2005.

Foulois, Benjamin D., and C. V. Glines, *From the Wright Brothers to the Astronauts: The Memoirs of Major General Benjamin D. Foulois, with Colonel C.V. Glines, USAF*. New York: McGraw, 1968.

Gilmore, William H. *Dear Jason and Jodi: A Memoir*. Morgan Hill, CA: Powerup, 2017.

Gross, Chuck. *Rattler One-Seven: A Vietnam Helicopter Pilot's War Story*. Denton: University of North Texas Press, 2004.

Haave, Christopher E., and Phil M. Haun. *A-10s Over Kosovo: The Victory of Airpower over a Fielded Army as Told by the Airmen Who Fought in Operation Allied Force*. Maxwell Air Force Base: Air University Press, 2003.

Harrison, Marshall. *A Lonely Kind of War: Forward Air Controller, Vietnam*. Novato: Presidio, 1997.

Howard, Barry J. *Wow! And They Even Paid Me, by Colonel Lucky*. Self-published, North Charleston, SC: CreateSpace, 2013.

Jackson, Mike, and Tara Dixon-Engel. *Naked in Da Nang: A Forward Air Controller in Vietnam*. St. Paul: Zenith Press, 2004.

Johnson, Howard C., and Ian A. O'Conner. *Scrappy: Memoir of a US Fighter Pilot in Korea and Vietnam*. Jefferson, NC: McFarland, 2007.

LeMay, Curtis E., and MacKinlay Kantor. *Mission with LeMay: My Story*. Garden City, N.Y.: Doubleday, 1965.

McCarthy, Mike. *Phantom Reflections*. Westport, Conn.: Praeger Security International, 2007.

Newman, Rick, and Don Shepperd. *Bury Us Upside Down: The Misty Pilots and the Secret Battle for the Ho Chi Minh Trail*. New York: Ballantine Books, 2006.

Olds, Robin, and Christina Olds. *Fighter Pilot*. New York: St. Martin's Press, 2010.

Pittard, Dana J. H., and Wes J. Bryant. *Hunting the Caliphate*. New York: Post Hill Press, 2019.

Plumb, Joseph Charles, and Glen DeWerff. *I'm No Hero: A POW Story as Told to Glen DeWerff*. Self-published, Tremendous Life Books, 1995.

Pocock, Charles. *Viper-7: Forward Air Controlling in South Vietnam in 1966*. Saline, MI: McNaughton and Gunn, 2000.

Pocock, Charles. *Forward Air Controller History Book: A Partial History of Participation by USAF Forward Air Controllers in the Southeast Asia Wars 1961–1975*. CD ROM. Self-published, The Forward Air Controllers Association, 2004.

Richey, Paul, and Diana Richey. *Fighter Pilot*. London: Cassell, 2001.

Robbins, Christopher. *The Ravens: The Men Who Flew in America's Secret War in Laos*. New York: Crown, 1987.

Roper, Jim. *Quoth the Raven*. Baltimore: America House, 2001.

Schultz, Alfred W., with Kirk Neff. *Janey: A Little Plane in a Big War*. Middletown, CT: Southfarm Press, 1998.

Schwarzkopf, Norman H., with Peter Petre. *It Doesn't Take a Hero*. New York: Linda Gray, 1992.

Thenault, Georges. *The Story of the LaFayette Escadrille, Told by Its Commander Captain Georges Thenault*. Translated by Walter Duranty. Boston: Small, Maynard, 1921.

Tibbets, Paul W. *Return of the Enola Gay*. Columbus: Mid Coast Marketing, 1998.

Secondary Sources

Anderegg, C. R. *Sierra Hotel: Flying Air Force Fighters in the Decade After Vietnam*. Washington, D.C.: Air Force History and Museums Program, 2001.

Andrews, William F. *Airpower against an Army: Challenge and Response in CENTAF's Duel with the Republican Guard*. Maxwell Air Force Base: Air University Press, 1998.

Askonas, Jon, and Colby Howard. "Fuzzy Thinking About Drones." *War on the Rocks*, January 26, 2018. https://warontherocks.com/2018/01/fuzzy-thinking-drones/.

Atkinson, Rick. *An Army at Dawn: The War in North Africa, 1942–1943*. New York: Henry Holt and Co, 2002.

_____. *The Day of the Battle: The War in Sicily and Italy 1943–1944.* London: Little, Brown, 2007.

Ballard, John R., David W. Lamm, and John K. Wood. *From Kabul to Baghdad and Back: The US at War in Afghanistan and Iraq.* Annapolis: Naval Institute Press, 2012.

Barnett, Jeffery R. *Future War: An Assessment of Aerospace Campaigns in 2010.* Maxwell Air Force Base: Air University Press, 1996.

Becthold, B. Michael. "A Question of Success: Tactical Air Doctrine and Practice in North Africa, 1942–43." *The Journal of Military History* 68, no. 3. (July 2004): 821–851. https://www.jstor.org/stable/i367783.

Belote, Howard D. "Counterinsurgency Airpower: Air-Ground Integration for the Long War." Air and Space Power Journal 20, no. 3. (Fall 2006): 55–68.

Benitez, Mike. "How Afghanistan Distorted Close Air Support and Why it Matters." *War on the Rocks,* June 29, 2016. https://warontherocks. com/2016/06/how-afghanistan-distorted-close-air-support-and-why-it-matters/.

Benitez, Mike. "21st-Century Forward Air Control: The Roots to Rebuild." *War on the Rocks.* March 1, 2017. https://warontherocks. com/2017/03/21st-century-forward-air-control-the-roots-to-rebuild/.

_____. "Heed the Grail Knight: Can the Air Force Choose Wisely?" *War on the Rocks,* November 19, 2018. https://warontherocks.com/2018/11/heed-the-grail-knight-can-the-air-force-choose-wisely/.

Berger, Carl, ed. *The United States Air Force in Southeast Asia, 1961–1973.* Washington, D.C.: Office of Air Force History, 1977.

Bilodeau, Peter M. *2035 Air Dominance Requirements for State-On-State Conflict.* Maxwell Air Force Base: Air University Press, 2012. https://media. defense.gov/2017/Dec/04/2001852030/-1/-1/0/MP_0060_BILODEAU_AIR_DOMINANCE.PDF.

Bingham, Price T. "Air Power in Desert Storm and the Need for Doctrinal Change." *Airpower Journal* 5, no. 4 *(Winter 1991): 33–46.*

Boothby, R., Jerry P. Jarmasz, and K. C. Wulterkens. *Forward Air Controller: Task Analysis and Development of Team Training Measures for Close Air Support.* Guelph, Ontario: Humansystems Incorporated, 2007. https://apps.dtic.mil/dtic/tr/fulltext/u2/a477167.pdf.

Bowden, Mark. *Black Hawk Down: A Story of Modern War.* New York: Signet, 2001.

Boyne, Walter J. "Foulois." *Air Force Magazine.* June 20, 2008. https://www. airforcemag.com/article/0203foulois/.

_____. *The Influence of Air Power Upon History*. New York: Pelican Publishing Company, 2003.

Bradbeer, Thomas G. *Battle for Air Supremacy Over the Somme: 1 June–30 November 1916*. Fort Leavenworth: US Army Command and General Staff College, 2004.

Brands, Hal. *Making the Unipolar Moment: U.S. Foreign Policy and the Rise of the Post-Cold War Order*. Ithaca: Cornell University Press, 2016.

Breau, Jesse D. *Forward Air Controller (Airborne): Training to Win Large Scale Combat Operations*. Thesis. School of Advanced Military Studies, Fort Leavenworth, 2019. Jesse Breau personal archive.

Buckley, John. *Air Power in the Age of Total War*. Bloomington: Indiana University Press, 1999.

Budiansky, Stephen. *Air Power: The Men, Machines, and Ideas That Revolutionized War, From Kitty Hawk to Iraq*. New York: Penguin Books, 2004.

Butler, Amy. "USAF Eyes New Era of Close Air Support: US Air Force's campaign to reinvent CAS." *Aviation Week and Space Technology*, March 30, 2015. http://aviationweek.com/defense/usaf-eyes-new-era-close-air-support.

Butz Jr., J. S. "Forward Air Controllers in Vietnam—They Call the Shots." *Air Force and Space Digest*, 49, no. 5 (May 1966): 60–66.

Campbell, Douglas N. *The Warthog and the Close Air Support Debate*. Annapolis: Naval Institute Press, 2003.

Cameron, Rebecca Hancock. *Training to Fly: Military Flight Training 1907–1945*. Washington, D.C.: Air Force History and Museums Program, 1999.

Cannon, Hardy D. *Box Seat Over Hell: The True Story of America's Liaison Pilots and Their Light Planes in World War Two*. 1984. Reprint, San Antonio: Alamo Liaison Squadron, 2007.

Carter, John R., and Air University. *Airpower and the Cult of the Offensive*. Maxwell Air Force Base: Air University Press. 1998.

Celeski, Joseph D. *Special Air Warfare and the Secret War in Laos: Air Commandos 1964–1975*. Maxwell Air Force Base: Air University Press, 2019.

Chadsey, David S. *Rebuilding the Joint Airborne Forward Air Controller: Analyzing Joint Air Tasking Doctrine's Ability to Facilitate Effective Air-Ground Integration*. Thesis, US Army Command and General Staff College, Fort Leavenworth, 2000.

Chadsey, David S., Jason Feuring, and Clement W. Rittenhouse. "Improving Close Air Support: An Army-Air Force Collaborative Approach." *Small*

Wars Journal, accessed January 24, 2019. https://smallwarsjournal.com/jrnl/art/improving-close-air-support-army-air-force-collaborative-approach.

Chandler, Charles Forest, and Frank P. Lahm. *How Our Army Grew Wings: Airmen and Aircraft before 1914.* New York: The Ronald Press Company, 1943.

Chandler, Michael J. *Gen Otto P Weyland, USAF: Close Air Support in the Korean War.* Thesis. School of Advanced Air and Space Studies, Maxwell Air Force Base, 2007. https://apps.dtic.mil/dtic/tr/fulltext/u2/a470482.pdf.

Chitwood, Adam. "The False Promise of the OA-X." *War on the Rocks,* November 24, 2017. https://warontherocks.com/2017/11/false-promise-oa-x/.

Chollet, Derek, and James Goldgeier. *America between the Wars: From 11/9 to 9/11. The Misunderstood Years Between the Fall of the Berlin Wall and the Start of the War on Terror.* New York: Public Affairs, 2008.

Chun, Clayton K. S. *Aerospace Power in the Twenty-First Century: A Basic Primer.* Maxwell Air Force Base: Air University Press, 2001.

Churchill, Jan. *Hit My Smoke: Forward Air Controllers in Southeast Asia.* Manhattan: Sunflower University Press, 1997.

Churchill, Jan. *Classified Secret: Controlling Airstrikes in the Clandestine War in Laos.* Manhattan: Sunflower University Press, 2000.

Clevenger, Daniel R. *"Battle of Khafji" Air Power Effectiveness in the Desert, Volume 1.* Washington, D.C.: Air Force Studies and Analyses Agency, 1996.

Clodfelter, Mark. *The Limits of Air Power: The American Bombing of North Vietnam.* New York: The Free Press, 1989.

_____. "Of Demons, Storms, and Thunder: A Preliminary Look at Vietnam's Impact on the Persian Gulf Air Campaign." *Airpower Journal* 5, no. 4 (Winter 1991): 17–32.

Cody, James R. *AWPD-42 to Instant Thunder: Consistent, Evolutionary Thought or Revolutionary Change.* Thesis. School of Advanced Airpower Studies, Maxwell Air Force Base, 1996.

Cohen, Raphael S. *Air Force Strategic Planning: Past, Present, and Future.* Santa Monica, CA: RAND, 2017. https://www.rand.org/pubs/research_reports/RR1765.html.

Cohen, Warren I. *The New Cambridge History of Foreign Relations, Vol. 4: Challenges to American Supremacy, 1945 to the Present.* New York: Cambridge University Press, 2013.

Coll, Steve. *Ghost Wars: The Secret History of the CIA, Afghanistan and Bin Laden, from the Soviet Invasion to September 10, 2001*. New York: Penguin, 2004.

Cooling, Benjamin Franklin, ed. *Case Studies in the Achievement of Air Superiority*. Washington, D.C.: Air Force History and Museums Program, 1994.

Coram, Robert. *Boyd: The Fighter Pilot Who Changed the Art of War*. Boston: Little Brown, 2002.

Corbett, Julian S. *Principles of Maritime Strategy*. 1911. Reprint. New York: Dover Publications, 2004.

Correll, John T. *The Air Force in Vietnam*. Arlington: Aerospace Education Foundation, 2004.

Corum, James S., and Wray R. Johnson. *Airpower in Small Wars*. Lawrence: University of Kansas Press, 2003.

Courtwright, David T. *Sky as Frontier: Adventure, Aviation, and Empire*. College Station: Texas A&M University Press, 2004.

Cox, Gary C. *Beyond the Battle Line: US Air Attack Theory and Doctrine, 1919–1941*. Maxwell Air Force Base, AL: Air University Press, 1996.

Crane, Conrad C. *American Airpower Strategy in Korea 1950–1953*. Lawrence: University Press of Kansas, 2000.

_____. "Raiding the Beggar's Pantry: The Search for Airpower Strategy in the Korean War." *The Journal of Military History* 63, no. 4 (October 1999): 885–920.

Cumings, Bruce. *The Korean War: A History*. New York: Modern Library, 2010.

Daalder, Ivo H., and Michael E. O'Hanlon. *Winning Ugly: NATO's War to Save Kosovo*. Washington, D.C: Brookings Institution Press, 2000.

De Serversky, Alexander P. *Victory through Air Power*. New York: Simon and Schuster, 1942.

Deaile, Melvin G. "The SAC Mentality." *Air and Space Power Journal* 29, no. 2. (March–April 2015): 48–73.

Dean, David J. *The Air Force Role in Low-Intensity Conflict*. Maxwell Air Force Base: Air University Press, 1986.

Deptula, David A. *Effects Based Operations: Change in the Nature of Warfare*. Arlington: Aerospace Education Foundation, 2001.

Digman, Kevin L. *Unmanned Aircraft Systems in a Forward Air Controller (Airborne) Role*. Thesis. Air Command and Staff College, Maxwell Air Force Base, 2009. https://apps.dtic.mil/dtic/tr/fulltext/u2/a539662.pdf.

Dorr, Robert F., and Thomas D. Jones. *Hell Hawks: The Untold Story of the American Fliers Who Savaged Hitler's Wehrmacht.* Minneapolis: Zenith Press, 2008.

Dotson, Robert S. "Tactical Air Power and Environmental Imperatives." *Air University Review,* 28, no. 5 (July–August 1977): 27–35. https://apps.dtic. mil/dtic/tr/fulltext/u2/a539662.pdf.

Doughty, Robert A. *The Evolution of US Army Tactical Doctrine, 1946– 76.* Fort Leavenworth: US Army Command and General Staff College, 1979.

Douhet, Giulio. *The Command of the Air.* Translated by Dino Ferrari. 1942. Reprint, New York: Arno Press, 1972.

Drake, Ricky James. *The Rules of Defeat: The Impact of Aerial Rules of Engagement on USAF Operations in North Vietnam, 1965–1968.* Maxwell Air Force Base, AL: Air University, 1992.

Dudney, Robert S. "The Gulf War II Air Campaign, by the Numbers." *Air Force Magazine* 86, no. 7 (July 2003): 36–42. https://www.airforcemag. com/PDF/MagazineArchive/Magazine%20Documents/2003/July%20 2003/0703Numbers.pdf.

Dunlap, Charles J. *Shortchanging the Joint Fight? An Airman's Assessment of FM 3–24 and the Case for Developing Truly Joint COIN Doctrine.* Maxwell Air Force Base, AL: Air University Press, 2008.

Echevarria II., Antulio J. *An American Way of War or Way of Battle?* Strategic Studies Institute: US Army War College, 2004. https://ssi. armywarcollege.edu/pubs/display.cfm?pubID=662

_____. *Toward an American Way of War.* Carlisle, PA: Strategic Studies Institute, US Army War College, 2004.

Ehlers Jr., Robert S. *The Mediterranean Air War: Airpower and Allied Victory in World War II.* Lawrence: University Press of Kansas, 2015.

Egginton, Jack B. *Ground Maneuver and Air Interdiction: A Matter of Mutual Support at the Operational Level of War.* Thesis. School of Advanced Airpower Studies, Maxwell Air Force Base, 1994. https:// permanent.access.gpo.gov/websites/dodandmilitaryejournals/www. maxwell.af.mil/au/aul/aupress/SAAS_Theses/SAASS_Out/Eggington/ eggington.pdf.

Ellis, John W. "The Airborne Forward Air Controller: Future Needs and Opportunities." *Air University Review* 30, no. 2 (January–February 1979): 38–47. https://www.airuniversity.af.edu/Portals/10/ASPJ/journals/ 1979_Vol30_No1-6/1979_Vol30_No2.pdf.

Engel, Jeffrey A. *When the World Seemed New: George H.W. Bush and the End of the Cold War.* Boston: Houghton Mifflin Harcourt, 2017.

Estilow, R. A. *US Military Force and Operations Other Than War: Necessary Questions to Avoid Strategic Failure.* Maxwell Air Force Base: Air University, 1996. https://media.defense.gov/2017/Dec/04/2001851769/-1/-1/0/MP_0003_ESTILOW_OPERATIONS_OTHER_THAN_WAR.PDF.

Evangelista, Matthew and Henry Shue. *The American Way of Bombing: Changing Ethical and Legal Norms, from Flying Fortresses to Drones.* Ithaca: Cornell University Press, 2014.

Farley, Robert. "AirLand Battle: The Army's Cold War Plan to Crush Russia (That Ended Up Crushing Iraq)." *The National Interest,* August 1, 2018. https://nationalinterest.org/print/blog/buzz/airland-battle-armys-cold-war-plan-crush-russia-ended-crushing-iraq-27477.

Farley, Robert. *Grounded: The Case for Abolishing the United States Air Force.* Lexington: University Press of Kentucky, 2014.

Farmer, J., and J. M. Strumwasser. *The Evolution of the Airborne Forward Air Controller: An Analysis of Mosquito Operations in Korea.* Santa Monica, CA: RAND, 1967.

Felker, Edward J. *Airpower, Chaos, and Infrastructure: Lords of the Rings.* Maxwell Air Force Base: Air University Press, 1998. https://media.defense.gov/2017/Dec/04/2001851785/-1/-1/0/MP_0014_FELKER_AIRPOWER_CHAOS_INFRASTRUCTURE.PDF.

Fehrenbach, T. R. *This Kind of War: A Study in Unpreparedness.* New York: Macmillan, 1963.

Frandsen, Bert. "America's First Air-Land Battle." *Air and Space Power Journal* 17, no. 4. (Winter 2003): 31–39. https://www.airuniversity.af.edu/Portals/10/ASPJ/journals/Volume-17_Issue-1-4/win03.pdf.

Frisbee, John L. ed. *Makers of the United States Air Force.* Washington, Office of Air Force History, 1987.

Frostic, Fred. *Air Campaign against the Iraqi Army in the Kuwaiti Theater of Operations.* Santa Monica, CA: RAND, 1994. https://www.rand.org/pubs/monograph_reports/MR357.html.

Fyfe, John M. *The Evolution of Time Sensitive Targeting: Operation Iraqi Freedom Results and Lessons.* Maxwell Air Force Base: Airpower Research Institute, 2005.

Gallogly, Patrick Coffey. *The Evolution of Integrated Close Air Support: World War II, Korea, and the Future of Air-Ground Combined Arms Synergy.* Thesis. School of Advanced Air and Space Studies, Maxwell Air Force Base, 2011. https://apps.dtic.mil/dtic/tr/fulltext/u2/1019694.pdf.

Gardner, Lloyd C. *The Long Road to Baghdad: A History of U.S. Foreign Policy from the 1970s to the Present.* New York: The New Press, 2008.

Gentile, Gian, Michael E. Linick, and Michael Shurkin. *The Evolution of US Military Policy from the Constitution to the Present.* Santa Monica, CA: RAND, 2017.

Givens, Robert P. *Turning the Vertical Flank: Airpower as a Maneuver Force in the Theater Campaign.* Maxwell Air Force Base: Air University Press, 2002.

Goldberg, Alfred, and Donald Smith. *Army Air Force Relations: The Close Air Support Issue.* Santa Monica, CA: RAND, 1971. https://www.rand.org/pubs/reports/R0906.html.

Gooderson, Ian. *Air Power at the Front: Allied Close Air Support in Europe 1943–1945.* London: Frank Cass, 1998.

Goodson, Larry. *Afghanistan's Endless War: State Failure, Regional Politics and the Rise of the Taliban.* Seattle: University of Washington Press, 2001.

Grant, Rebecca. "Marine Air in the Mainstream." *Air Force Magazine.* June 8, 2008. https://www.airforcemag.com/article/0604marine/.

————. "The Great Escape." *Air Force Magazine,* June 18, 2008. https://www.airforcemag.com/article/0303escape/.

Grant, Rebecca. "The Second Offset." *Air Force Magazine.* July 2016. 32–36. https://www.airforcemag.com/article/the-second-offset/.

Grau, Lester, ed. *The Bear Went over the Mountain: Soviet Combat Tactics in Afghanistan.* Washington, D.C.: National Defense University Press, 1996.

Gray, Colin S. *Weapons Don't Make War: Policy, Strategy, and Military Technology.* Lawrence: University Press of Kansas, 1993.

————. *The Airpower Advantage in Future Warfare: The Need for Strategy.* Maxwell Air Force Base: Airpower Research Institute, 2007.

————. *Understanding Airpower, Bonfire of the Fallacies.* Maxwell Air Force Base: Airpower Research Institute, 2008.

————. *Airpower for Strategic Effect.* Maxwell Air Force Base: Air University Press, 2012.

Gray, Peter. *Air Warfare: History, Theory, and Practice.* London: Bloomsbury Academic. 2015.

Greer, Thomas H. *The Development of Air Doctrine in the Army Air Arm, 1917–1941.* Washington, D.C.: Office of Air Force History, 1985.

Grier, Peter. "The First Offset." *Air Force Magazine* 99, no. 6. (June 2016): 56–60.

Griffith, Charles. *The Quest: Haywood Hansell and American Strategic Bombing in World War II.* Maxwell Air Force Base, Air University Press, 1999.

Groom, Winston. 2013. *The Aviators: Eddie Rickenbacker, Jimmy Doolittle, Charles Lindbergh and the Epic Age of Flight.* Washington, D.C.: National Geographic.

Grosscup, Beau. *Strategic Terror: The Politics and Ethics of Aerial Bombardment.* London: Zed Books, 2006.

Guttman, John. *The USAS 1st Pursuit Group.* New York: Osprey Publishing. 2008.

Guzinger, Mark. *Shaping America's Future Military toward a New Force Planning Construct.* Washington, D.C.: Center for Strategic and Budgetary Assessments, 2013.

Hagen, Jeff, David Blancett, Michael Bohnert, Shuo-Ju Chou, Amado Cordova, Thomas Hamilton, Alexander C. Hou, Sherrill Lingel, Colin Ludwig, Christopher Lynch, Muharrem Mane, Nicholas A. O'Donoughue, Daniel M. Norton, Ravi Rajan, and William Stanley. *Needs, Effectiveness, and Gap Assessment for Key A-10C Missions: An Overview of Findings.* Santa Monica, CA: RAND, 2016.

Hall, David Ian. *Learning How to Fight Together: The British Experience with Joint Air-Land Warfare.* Maxwell Air Force Base: Air Force Research Institute, 2009.

Hall, James Norman, and Charles Bernard Nordhoff. *The Lafayette Flying Corps during the First World War.* 2 Vols. US: Leonaur Ltd., 2014.

Hall, Dewayne P. *Integrating Joint Operations Beyond the FSCL: Is Current Doctrine Adequate?* Maxwell Air Force Base: Air University, 1997. https://media.defense.gov/2017/Dec/04/2001851788/-1/-1/0/MP_0012_HALL_INTEGRATING_JOINT_OPERATIONS.PDF.

Hallion, Richard P. *Strike from the Sky: The History of Battlefield Air Attack, 1911–1945.* Washington, D.C.: Smithsonian Institute Press, 1989.

Hammond, Grant. *The Mind of War: John Boyd and American Security.* Washington, D.C., Smithsonian Books, 2004.

Hanson, Victor Davis. *Carnage and Culture: Landmark Battles in the Rise of Western Power.* New York: Double Day, 2001.

Haulman, Daniel L. "The United States Air Force and Bosnia, 1992–1995." *Air Power History* 60, no. 3 (Summer 2015): 24–31. https://www.afhistory.org/wp-content/uploads/Fall_2013_Issue_All.pdf.

_____. "The US Air Force in the Air War over Serbia, 1999." *Air Power History* 62, no. 2 (Summer 2015): 6–21. https://www.afhistory.org/wp-content/uploads/Air_Power_History_Summer-2015.pdf.

Haun, Phil M. *Airpower Versus a Fielded Force: Misty Forward Air Controllers of Vietnam and A-10 Forward Air Controllers of Kosovo, a Comparative Analysis*. Maxwell AFB, AL: Maxwell Air Force Base: Air University Press, 2004.

_____. "Direct Attack—A Counterland Mission." *Air and Space Power Journal* 17, no. 2 (Summer 2003): 9–16. https://www.airuniversity.af.edu/Portals/10/ASPJ/journals/Volume-17_Issue-1-4/sum03.pdf.

_____. *Lectures of the Air Corps Tactical School and the American Strategic Bombing in World War II*. Lexington: University of Kentucky Press, 2019

_____. "Misty Forward Air Controllers of the Vietnam War." *Air Power History* 50, no.4 (Winter 2003): 38–45. https://www.afhistory.org/wp-content/uploads/2003_winter.pdf.

Haun, Phil M., and Colin Jackson. "Breaker of Armies: Air Power in the Easter Offensive and the Myth of Linebacker I and II in the Vietnam War." *International Security* 40, no. 3 (Winter 2015–2016): 139–178.

Haux, Hailey. "Sequestration threatens America's air power advantage." US Air Force, News. March 23, 2015. https://www.af.mil/News/Article-Display/Article/581217/sequestration-threatens-americas-air power-advantage/.

Head, William P. "The Battle for Ra's Al-Khafji and the Effects of Air Power January 29–Februrary 1, 1991, Part 2" *Air Power History*, 60, no. 2 (Summer 2013): 23–33. https://www.afhistory.org/wp-content/uploads/Summer_2013_Issue.pdf.

Head, William P. "The Battles of Al-Fallujah: Urban Warfare and the Growth of Air Power" *Air Power History*, 60, no. 4 (Winter 2013): 32–51. https://www.afhistory.org/wp-content/uploads/Winter_2013_All_v2.pdf.

_____. "Dirty Little Secret in the Land of a Million Elephants: Barrel Roll and the Lost War." *Air Power History*, 64, no. 4 (Winter 2017): 7–28.

_____. *War from Above the Clouds: B-52 Operations during the Second Indochina War and the Effects of the Air War on Theory and Doctrine*. Maxwell Air Force Base: Air University Press, 2002. https://media.defense.gov/2017/May/05/2001742911/-1/-1/0/FP_0007_HEAD_WAR_FROM_ABOVE_CLOUDS.PDF.

Hebert, Adam J. "The Balkan Air War." *Air Force Magazine*. February 23, 2009. https://www.airforcemag.com/article/0309balkan/.

Hendrix, Jerry. "How the Air Force Lost Its Way, and Most of its Bombers, too." *National Review*. January 28, 2019. https://www.nationalreview.com/wp-content/uploads/2019/01/xml_20190128_hendrix.html

Hennessy, Juliette A. "Men and Planes of World War I and a History of the Lafayette Escadrille." *Air Power History* 64, no. 2 (Summer 2017): 43.

Henriksen, Dag, ed. *Airpower in Afghanistan 2005–10: The Air Commanders' Perspectives.* Maxwell Air Force Base: Air University Press, 2014.

Herring, George C. *From Colony to Superpower: U.S. Foreign Relations since 1776.* The Oxford History of the United States. New York: Oxford University Press, 2008.

Hess, William N. *Pacific Sweep: The 5th and 13th Fighter Commands in World War II.* New York: Double Day, 1974.

Heuser, Beatrice. *The Evolution of Strategy, and The Strategy Makers: Thoughts on War and Society from Machiavelli to Clausewitz.* Santa Barbara: Praeger Security International, 2010.

Hewitt, Mark A. *Time Sensitive Targeting: Overcoming the Intelligence Gap in Interagency Operations.* Thesis. Naval War College, Newport, RI. 2003. https://www.hsdl.org/?view&did=450696.

Hobson, Christopher Michael. *Vietnam Air Losses.* Hinkley, England: Midland Publishing, 2001.

Hodgson, Jim. "World War I in Texas, Aviation Arrives." Presentation at the Texas State Historical Association Annual Convention, San Marcos, TX, March 8–10, 2018.

Hoeffding, Oleg. *Bombing North Vietnam: An Appraisal of Economic and Political Effects.* Santa Monica, CA: RAND, 1966

Holley Jr., I. B. *Ideas and Weapons.* Washington, D.C.: Office of Air Force History, 1983.

―――. *Technology and Military Doctrine: Essays on a Challenging Relationship.* Maxwell Air Force Base: Air University Press, 2004.

Hooyer, David A. *Forward Air Control Today: Will it Work in Europe?* Fort Leavenworth: US Army Command and General Staff College, 1979. https://apps.dtic.mil/dtic/tr/fulltext/u2/a077287.pdf.

Horwood, Ian. *Interservice Rivalry and Airpower in the Vietnam War.* Fort Leavenworth: Combat Studies Institute Press, 2006. https://apps.dtic.mil/dtic/tr/fulltext/u2/a460480.pdf.

Hudson, James J. *Hostile Skies: A Combat History of the American Air Service in World War I.* Syracuse: Syracuse University Press. 1968.

Hughes, Thomas Alexander. "Anglo-American Tactical Air operations in World War II." *Air and Space Power Journal,* 18, no. 4 (Winter 2004): 34–45.

―――. *Overlord: General Pete Quesada and the Triumph of Tactical Air Power in World War II.* New York: The Free Press, 1995.

Hume, David B. *Integration of Weaponized Unmanned Aircraft into the Air-to-Ground System*. Maxwell Air Force Base: Air University Press, 2007. https://media.defense.gov/2017/Dec/04/2001851902/-1/-1/0/MP_0041_ HUME_WEAPONIZED_UNMANNED_AIRCRAFT.PDF.

Hunt, Michael H. *The American Ascendancy: How the United States Gained and Wielded Global Dominance*. Chapel Hill: University of North Carolina Press, 2007.

Hura, Myron, Gary W. McLeod, Eric V. Larson, James Schneider, Dan Gonzales, Daniel M. Norton, Jody Jacobs, Kevin M. O'Connell, William Little, Richard Mesic, and Lewis Jamison. *Interoperability: A Continuing Challenge in Coalition Air Operations*. Santa Monica, CA: RAND, 2000.

Hurley, Alfred F. *Billy Mitchell: Crusader for Air Power*. Bloomington: Indiana University Press, 1975.

Hurst, James W. *Pancho Villa and Black Jack Pershing: The Punitive Expedition in Mexico*. Westport: Praeger, 2008.

Hurst, Jules. "The Developing Fight for Tactical Air Control." *War on the Rocks*. March 28, 2019. https://warontherocks.com/2019/03/the-developing-fight-for-tactical-air-control/.

_____. "Small Unmanned Aerial Systems and Tactical Air Control." *Air and Space Power Journal* 33, no. 1, (Spring 2019): 19–33.

Hynes, Samuel. *The Unsubstantial Air: American Fliers in the First World War*. Kindle Ed. New York: Farrar, Straus and Giroux, 2014.

Ignatieff, Michael. "Issue: The Virtual Commander," Annals of Diplomacy, *The New Yorker* (August 2, 1999).

Janus, Allan. "Lion Cubs? Yeah, We've Got Lion Cubs, Too." *National Air and Space Museum Archives Division, Online*. December 27, 2010. https:// airandspace.si.edu/stories/editorial/lion-cubs-yeah-weve-got-lion-cubs-too.

Jayne, Randy. "The Last Prop Figher: Sandys, Hobos, Fireflies, Zorros, and Spads." *Air and Space Power Journal* 31, no. 2 (Summer 2017): 82–90. https://www.airuniversity.af.edu/Portals/10/ASPJ/journals/Volume-31_ Issue-2/V-Jayne.pdf.

Jeremiah, David E. "What's Ahead for the Armed Forces." *Joint Forces Quarterly* 1, no. 1 (Summer 1993): 25–35. https://ndupress.ndu.edu/ portals/68/Documents/jfq/jfq-1.pdf.

Johnson, David E. *Fast Tanks and Heavy Bombers: Innovation in the US Army 1917–1945*. Ithaca: Cornell University Press, 1998.

Johnson, Michael H. *Cleared to Engage: Improving Joint Close Air Support Effectiveness*. Maxwell Air Force Base, Air University Press, 2008. https://apps.dtic.mil/dtic/tr/fulltext/u2/a602566.pdf.

Jones, David. *Perceptions of Airpower and Implications for the Leavenworth Schools: Interwar Student Papers*. US Army Combined Arms Center Fort Leavenworth: US Army Command and General Staff College Press, 2014.

Jones, Seth. *In the Graveyard of Empires: America's War in Afghanistan.* New York: W. W. Norton, 2010.

Kaaoush, Kamal J. "The Best Aircraft for Close Air Support in the Twenty-First Century." *Air and Space Power Journal* 30, no. 3 (Fall 2016): 39–53.

Kamps, Charles Tustin. "The JCS 94-Target List: A Vietnam Myth that Still Distorts Military Thought." *Aerospace Power Journal* 15, no. 1 (Spring 2001): 67–80. https://apps.dtic.mil/dtic/tr/fulltext/u2/a524103.pdf.

Kemper, Todd G. *Aviation Urban Operations: Are We Training Like We Fight?* Maxwell Air Force Base: Air University Press, 2004. https://media.defense.gov/2017/Dec/04/2001851894/-1/-1/0/MP_0033_KEMPER_AVIATION_URBAN_OPERATIONS.PDF.

Kessner, Thomas. *The Flight of the Century: Charles Lindbergh and the Rise of American Aviation.* Oxford: Oxford University Press, 2010.

King, Philip H. "CAS innovations: Digital data burst and ATHS." *Marine Corps Gazette* 80, no. 5 (May 1996): 60–62.

King, Brett A., "Coercive Airpower in the Precision Age: The Effects of Precision Guided Munitions on Air Campaign Duration" PhD diss., University of Nebraska, Lincoln, 2014. http://digitalcommons.unl.edu/poliscitheses/32.

Knox, Raymond O. *The Terminal Strike Controller: The Weak Link in Close Air Support*. Thesis. School of Advanced Military Studies, Fort Leavenworth, 1988. https://apps.dtic.mil/dtic/tr/fulltext/u2/a208256.pdf.

Krause, Merrick E. "Airpower in Modern War." *Air and Space Power Journal,* 29, no. 3. (May–June 2015): 42–56. https://www.airuniversity.af.edu/Portals/10/ASPJ/journals/Volume-29_Issue-3/V-Krause.pdf.

Krause, Peter John Paul. "The Last Good Chance: A Reassessment of US Operations at Tora Bora." *Security Studies* 17, no. 4. (Winter 2008): 644–684. http://web.mit.edu/SSP/people/krause/PKrause%20_The%20Last%20Good%20Chance_.pdf.

Kuehl, Daniel T., and Charles E. Miller. "Roles, Missions, and Functions: Terms of Debate." *Joint Forces Quarterly* 94 (Third Quarter, July 1994): 103–105. https://apps.dtic.mil/dtic/tr/fulltext/u2/a528570.pdf.

Kugler, Richard L., Michael Branick, and Hans Binnendijk. *Operation Anaconda: Lessons for Joint Operations*. Washington, D.C.: National

Defense University, 2009. https://apps.dtic.mil/dtic/tr/fulltext/u2/
a496469.pdf.

Lamb, Michael W. *Operation Allied Force: Golden Nuggets for Future
Campaigns.* Maxwell Air Force Base: Air University Press, 2002.
https://media.defense.gov/2017/Dec/04/2001851840/-1/-1/0/MP_0027_
LAMB_OPERATION_ALLIED_FORCE.PDF.

Lambeth, Benjamin S. "Air Land Reversal." *Air Force Magazine* 97, no. 2.
(February 2014): 60–64. https://www.airforcemag.com/PDF/Magazine
Archive/Documents/2014/February%202014/0214reversal.pdf

_____. *Air Power Against Terror: America's Conduct of Operation Enduring
Freedom.* Santa Monica, CA: RAND, 2001.

Lambeth, Benjamin S. *NATO's Air War for Kosovo.* Santa Monica, CA:
RAND, 2001.

Larson, Eric V., David T. Prletsky, and Kristin Leuschner. *Defense Planning
in a Decade of Change: Lessons from the Base Force, Bottom-Up Review,
and Quadrennial Defense Review.* Santa Monica, CA: RAND, 2001.

Laslie, Brian D. *The Air Force Way of War: US Tactics and Training after
Vietnam.* Lexington: University Press of Kentucky, 2015.

Laughbaum, R. Kent. *Synchronizing Airpower and Firepower in the Deep
Battle.* Maxwell Air Force Base: Air University Press, 1999.

Leary, Willian M. "Supporting the Secret War: CIA Air Operations in Laos,
1955–1974." *Studies in Intelligence.* (Winter 1999–2000). https://www.
cia.gov/library/center-for-the-study-of-intelligence/csi-publications/
csi-studies/studies/winter99-00/art7.html.

Leffler, Melvyn P. *A Preponderance of Power: National Security, the Truman
Administration, and the Cold War*. Stanford: Stanford University Press,
1992.

Leonhard, Robert. *The Art of Maneuver: Maneuver-Warfare Theory and
AirLand Battle.* New York: Ballantine Books, 1991.

Lester, Gary Robert. *Mosquitoes to Wolves: The Evolution of the Airborne
Forward Air Controller.* Maxwell Air Force Base: Air University Press,
1997.

Levine, Isaac D. *Mitchell Pioneer of Air Power.* New York: Duell, Sloan and
Pearce, 1943.

Lewis, Adrian. *The American Culture of War: A History of US Military Force
from World War II to Operation Iraqi Freedom.* New York: Routledge,
2007.

Lewis, Kevin N. *National Security Spending and Budget Trends Since World
War II.* RAND: Santa Monica, CA, 1990.

_____. *The US Air Force Budget and Posture over Time.* RAND: Santa Monica, CA, 1990.

Li, Xiaobing. *China's Battle for Korea: The 1951 Spring Offensive.* Bloomington: Indiana University Press, 2014.

Lind, J. R., K. Harris, and S. G. Spring. *Fast-Val: A Study of Close Air Support (Briefing Summarizing the Comparisons of Model with Combat Results and Illustrating the Influence of Supporting Arms on Fire-fight Outcomes).* Santa Monica, CA: RAND, 1971.

Linn, Brian McAllister. *Elvis's Army: Cold War GIs and the Atomic Battlefield.* Cambridge: Harvard University Press, 2016.

Loosbrock, John F., ed. "The Air War in Vietnam." Special Issue, *Air Force and Space Digest* 49, no. 3 (March 1966).

Lorell, Mark A. *The US Combat Aircraft Industry, 1909–2000: Structure, Competition, Innovation.* Santa Monica, CA: RAND. 2003.

Manchester, William. *American Caesar: Douglas MacArthur, 1880–1964.* (Boston: Little, Brown, 1978).

Mane, Nicholas A. O'Donoughue, Daniel M. Norton, Ravi Rajan, and William Stanley. *Needs, Effectiveness, and Gap Assessment for Key A-10C Missions: An Overview of Findings.* Santa Monica, CA: RAND, 2016. https://www.rand.org/pubs/research_reports/RR1724z1.html.

Mann, Edward C., III. *Thunder and Lightning: Desert Storm and the Airpower Debates.* Maxwell AFB, AL: Air University Press, 1995.

Mark, Eduard. "A New Look at Operation STRANGLE." *Military Affairs* 52, no. 4 (1988): 176–184. https://www.jstor.org/stable/pdf/1988449.pdf.

Mark, Eduard Maximilian. *Aerial Interdiction: Air Power and the Land Battle in Three American Wars.* Washington, D.C.: Center for Air Force History, 1994.

Matisek, Jahara, and Jon McPhilamy. "Why Airpower Needs Landpower." *Modern War Institute,* November 5, 2018. https://mwi.usma.edu/airpower-needs-landpower/.

Matsumura, John, John Gordon IV, and Randall Steeb. *Defining an Approach for Future Close Air Support Capability.* Santa Monica, CA: RAND, 2017.

McCaffrey, Terrance J. *What Happened to Battlefield Air Interdiction? Army and Air Force Battlefield Doctrine Development from Pre—Desert Storm to 2001.* Maxwell Air Force Base: Air University Press, 2004.

McCarthy, Michael C. *Air-to-Ground Battle for Italy.* Maxwell Air Force Base: Air University Press, 2004.

McClendon, R. Earl. *Autonomy of the Air Arm*. Washington, D.C.: Air Force History and Museums Program, 1996.

McCullough, David. *The Wright Brothers*. New York: Simon and Schuster. 2015

McGarry, Brendan. "Air Force Sees Rising Demand for Joint Terminal Attack Controllers." *Military.com* March 10, 2015. http://www.military.com/daily-news/2015/03/10/air-force-sees-rising-demand-joint-terminal-attack-controllers.html.

Mearsheimer, John J. *Liddell Hart and the Weight of History*. Ithaca: Cornell University Press, 1988.

Meilinger, Phillip A. *Airmen and Air Theory: A Review of the Sources*. Rev. ed. Maxwell Air Force Base: Air University Press, 2001.

_____. *Airpower: Myths and Facts*. Maxwell Air Force Base: Air University Press, 2003.

_____. *Bomber: The Formation and Early Years of Strategic Air Command*. Maxwell AFB: Air University Press, 2012.

_____. "The Problem with Our Air Power Doctrine." *Airpower Journal*, 6, no. 1 (Spring 1992): 24–31.

_____. *Ten Propositions Regarding Air Power*. Washington, D.C.: Air Force Historical Studies Office, 1995. https://apps.dtic.mil/dtic/tr/fulltext/u2/a469807.pdf.

Meilinger, Phillip A., ed. *The Paths of Heaven: The Evolution of Airpower Theory*. Maxwell Air Force Base: Air University Press, 1997.

Mets, David R. *The Air Campaign: John Warden and the Classical Airpower Theorists*. Revised ed. Maxwell Air Force Base: Air University Press, 1998.

_____. *Land-Based Air Power in Third World Crises*. Maxwell Air Force Base: Air University Press, 1986.

_____. *The Long Search for a Surgical Strike: Precision Munitions and the Revolution in Military Affairs*. Maxwell Air Force Base: Air University Press, 2001.

Michalke, Jeff. "Air Commands History Revived in 1960s." *16th SOW History Office*. October 13, 2006. https://www.afsoc.af.mil/DesktopModules/ArticleCS/Print.aspx?PortalId=86&ModuleId=11155&Article=163454.

Michel III, Marshall L. *Clashes: Air Combat over North Vietnam 1965–1972*. Annapolis: Naval Institute Press, 1997.

_____. *The 11 Days of Christmas: America's Last Vietnam Battle*. New York: Encounter Books, 2002.

_____. "The Revolt of the Majors: How the Air Force Changed After Vietnam." PhD diss. Auburn University, 2006.

Miller, Roger G. *A Preliminary to War: The First Aero Squadron and the Mexican Punitive Expedition.* Washington, D.C.: Air Force History and Museums Program. 2003.

————. *Like a Thunderbolt: The Lafayette Escadrille and the Advent of American Pursuit in World War I.* Washington, D.C.: Air Force History and Museums Program. 2007.

Mitchell, William. *Winged Defense: The Development and Possibilities of Modern Air Power— Economic and Military.* London: G. P. Putnam's Sons, 1925.

Momyer, William W., James C. Gaston, and Arthur J. C. Lavalle. *Airpower in Three Wars.* Maxwell Air Force Base: Air University Press, 1978.

Moore, Harold G., and Joseph L. Galloway. *We Were Soldiers Once—and Young: Ia Drang, the Battle That Changed the War in Vietnam.* New York: Harper Perennial, 1993.

Morgan, Prentice G. "The Forward Observer." *Military Affairs*, 23, no. 4 (Winter, 1959–1960): 209–212 http://www.jstor.org/stable/1984606.

Morrow, John H. *The Great War in the Air: Military Aviation from 1909 to 1921.* Washington, D.C.: Smithsonian Institute Press. 1994.

Mortensen, Daniel R. *A Pattern for Joint Operations: World War II Close Air Support North Africa.* Washington, D.C.: Office of Air Force History and US Army Center of Military History, 1987.

————, ed. *Airpower and Ground Armies: Essays on the Evolution of Anglo-American Air Doctrine, 1940–43.* Maxwell Air Force Base: Air University Press, 1998.

Mowbray, James A. "Air Force Doctrine Problems: 1926-Present." *Airpower Journal* 9, no. 4 (Winter 1995): 21–41. https://www.airuniversity.af.edu/Portals/10/ASPJ/journals/Volume-09_Issue-1-Se/1995_Vol9_No4.pdf

Mrozek, Donald J. *Air Power and the Ground War in Vietnam: Ideas and Actions.* Maxwell Air Force Base: Air University Press, 1988.

————. *The US Air Force After Vietnam: Postwar Challenges and Potential for Responses.* Maxwell Air Force Base: Air University Press, 1998.

Mueller, Karl P. *Air Power.* Santa Monica, CA: RAND, 2010. https://www.rand.org/pubs/reprints/RP1412.html

————. *Precision and Purpose: Airpower in the Libyan Civil War.* Santa Monica, CA: RAND, 2015. https://www.rand.org/pubs/research_reports/RR676.html.

Munson, Tom. *Operation Allied Force: Operational Planning and Political Constraints.* Newport: US Naval War College, 2000.

Murray, Williamson. *Strategy for Defeat: The Luftwaffe 1933–1945.* Maxwell Air Force Base: Air University Press, 1983.

Nalty, Bernard C. *Air War over South Vietnam 1968—1975.* Washington, D.C.: Air Force History and Museums Program, 2000.

_____. *The War against Trucks: Aerial Interdiction in Southern Laos 1968–1972.* Washington, D.C.: Air Force History and Museums Program, 2005.

_____. *Winged Shield, Winged Sword: A History of the United States Air Force, Vol. 1, 1907–1950.* Washington, D.C.: Air Force History and Museums Program, 1997.

_____. *Winged Shield, Winged Sword: A History of the United States Air Force. Vol. 2, 1950–1997.* Washington, D.C.: Air Force History and Museums Program, 1997.

Nardulli, Bruce R., Walter L. Perry, Bruce Pirnie, John Gordon IV, and John G. McGinn. *Disjointed War: Military Operations in Kosovo, 1999.* Santa Monica, CA: RAND, 2002.

Neufeld, Jacob, Kenneth Schaffel, and Anne E. Shermer. *Guide to Air Force Historical Literature, 1943–1983.* Washington, D.C: Office of Air Force History, United States Air Force, 1985.

Neufeld, Jacob, and George M. Watson Jr., eds. *Coalition Air Warfare in the Korean War, 1950–1953. Proceedings: Air Force Historical Foundation Symposium Andres AFB, Maryland, May 7–8, 2002.* Washington, D.C.: US Air Force History and Museums Program, 2005.

New, Terry L. "Where to Draw the Line between Air and Land Battle." *Airpower Journal.* 1996, 10, no. 3 (Fall 1996): 34–49. https://www.airuniversity.af.edu/Portals/10/ASPJ/journals/Volume-10_Issue-1-Se/1996_Vol10_No3.pdf.

NPR Staff. "What a Downed Black Hawk in Somalia Taught America." *All Things Considered, NPR.* October 5, 2013. https://www.npr.org/2013/10/05/229561805/what-a-downed-black-hawk-in-somalia-taught-america.

Olsen, John Andreas. *Airpower Applied: US, NATO and Israeli Combat Experience.* Annapolis: Naval Institute Press. 2017.

_____. *Airpower Reborn: The Strategic Concepts of John Warden and John Boyd. History of Military Aviation Series.* Annapolis: Naval Institute Press, 2015.

_____. "Warden Revisited: The Pursuit of Victory Through Air Power." *Air Power History,* 64, no. 4 (Winter 2017): 39–53.

_____, ed. *A History of Air Warfare.* Washington, D.C.: Potomac, 2010.

Overy, Richard. *The Air War 1939–1945.* New York: Stein and Day, 1980.
_____. *Bombers and the Bombed.* New York: Penguin Group, 2014.

Owen, Robert C. *Deliberate Force: A Case Study in Effective Air Campaigning.* Maxwell Air Force Base: Air University Press, 2000.

Pape, Robert A. *Bombing to Win: Air Power and Coercion in War.* Ithaca: Cornell University Press, 1996.

Parton, Kames, ed. *Impact: The Army Air Forces' Confidential Picture History of World War II.* First edition, 8 vols. New York: The Air Force Historical Foundation with James Parton and Company, 1980–1989.

Peeler, David L. *Counterinsurgency Aircraft Procurement Options: Processes, Methods, Alternatives, and Estimates.* Maxwell Air Force Base: Air University Press, 2009. https://media.defense.gov/2017/Dec/04/2001851655/-1/-1/0/WF_0040_PEELER_COUNTERINSURGENCY_AIRCRAFT.PDF.

Perry, Walter L., Richard E. Darilek, Laurinda L. Rohn, and Jerry M. Sollinger, eds. *Operation Iraqi Freedom: Decisive War, Elusive Peace.* Santa Monica, CA: RAND, 2015.

Phinney, Todd R. "Reflections on Operation Unified Protector." *Joint Force Quarterly* 73, (2nd Quarter 2014): 73–86. https://ndupress.ndu.edu/Portals/68/Documents/jfq/jfq-73/jfq-73_86-92_Phinney.pdf?ver=2014-03-26-120652-783.

Pisano, Dominick A., Thomas J. Dietz, Joanne M. Gernstein, and Karl S. Schneide. *Legend, Memory and the Great War in the Air.* Seattle: University of Washington Press. 1992.

Pietrucha, Mike. "The Five-Ring Circus: How Airpower Enthusuasts Forgot About Interdiction." *War on the Rocks,* September 29, 2015. https://warontherocks.com/2015/09/the-five-ring-circus-how-airpower-enthusiasts-forgot-about-Interdiction/.

_____. "Slaying the Unicorn: The Army and Fixed-Wing Attack." *War on the* Rocks, December 9, 2019. https://warontherocks.com/2019/12/slaying-the-unicorn-the-army-and-fixed-wing-attack/.

_____, and Jeremy Renken. "Airpower May Not Win Wars, but It Sure Doesn't Lose Them." *War on the Rocks.* August 19, 2015. https://warontherocks.com/2015/08/airpower-may-not-win-wars-but-it-sure-doesnt-lose-them/.

Pirnie, Bruce R., Alan Vick, Adam Grissom, Karl P. Mueller, and David T. Orletsky. *Beyond Close Air Support: Forging a New Air-Ground Partnership.* Santa Monica, CA: RAND, 2005.

Pivarsky Jr, Carl R. *Airpower in the Context of a Dysfunctional Joint Doctrine.* Maxwell Air Force Base: Air University Press, 1997. https://media.

defense.gov/2017/Dec/04/2001851762/-1/-1/0/MP_0007_PIVARSKY_
DYSFUNCTIONAL_JOINT_DOCTRINE.PDF.

Post, Carl A. "Forward Air Control: A Royal Australian Air Force Innovation." *Air Power History* 52, no. 4. (Winter 2006): 4–11.

Purdham Jr, Aldon E. *America's First Air Battles: Lesson Learned or Lessons
Lost?* Maxwell Air Force Base: Air University Press, 2003.

Purtee, Edward O. *History of the Army Air Service 1907–1926.* Wright-
Patterson Air Force Base, Historical Office, Executive Secretariat, Air
Materiel Command, 1948.

Putney, Diane T. *Airpower Advantage: Planning the Gulf War Air
Campaign 1989–1991.* Washington, D.C.: Air Force History and
Museums Program, 2004.

Quartararo, Joe, Michael Rovenolt, and Randy White. "Libya's Operation
Odyssey Dawn: Command-and-control." *Prisim* 3, no.2. (October 2013):
141–156. https://cco.ndu.edu/PRISM/PRISM-volume-3-no-2/.

Rabinovich, Abraham. *The Yom Kippur War: The Epic Encounter That
Transformed the Middle East.* New York: Schocken, 2005.

Renken, Jeremy L. "Strategic Reform: A Battle of Assumptions." *Air and
Space Power Journal* 29 no. 3 (2015): 73.

Reynolds, Richard T. *Heart of the Storm: The Genesis of the Air Campaign
Against Iraq.* Maxwell Air Force Base: Air University Press, 1995.

Ricks, Thomas E. *Fiasco: The American Military Adventure in Iraq.* New
York: Penguin Books, 2007.

Robertson, Scott. "The Development of Royal Air Force Strategic Bombing
Doctrine between the Wars: A Revolution in Military Affairs?" *Airpower
Journal* (Spring 1998).

Roman, Gregory A. *The Command or Control Dilemma When Technology and
Organizational Orientation Collide.* Maxwell Air Force Base: Air University Press, 1997. https://media.defense.gov/2017/Dec/04/2001851763/-1/-
1/0/MP_0008_ROMAN_COMMAND_OR_CONTROL_DILEMMA.
PDF.

Romjue, John L. *From Active Defense to Airland Battle: The Development
of Army Doctrine, 1973–1982.* Fort Monroe: Historical Office, US Army
Training and Doctrine Command, 1984.

Ruehrmund Jr., James C., and Christopher J. Bowie. *Arsenal of Airpower:
USAF Aircraft Inventory 1950–2009.* Arlington: Mitchell Institute, 2010.

Saffold, Timothy L. *The Role of Airpower in Urban Warfare: An Airman's
Perspective.* Maxwell Air Force Base: Air Command and Staff College,
1998. https://apps.dtic.mil/dtic/tr/fulltext/u2/a405880.pdf.

Sarver, Andrew D. "24th Tactical Air Control Squadron Activates, Focuses on Close Air Support." *Nellis Air Force Base News,* March 16, 2018. https://www.nellis.af.mil/News/Article/1468696/24th-tass-activates-focuses-on-close-air-support/.

Scales Jr., Robert H. *Firepower in Limited War.* Revised Ed. Novato, CA: Presidio Press, 1995.

Schaefer, John J. "Responsive Close Air Support." *Joint Forces Quarterly* 67, (Fourth Quarter 2012): 91–96. https://ndupress.ndu.edu/Portals/68/Documents/jfq/jfq-67/JFQ-67_91-96_Schaefer.pdf.

Schlenoff, Dan. "Aircraft Communication, 1915." *Scientific* American, September 25, 2015. https://blogs.scientificamerican.com/anecdotes-from-the-archive/aircraft-communication-1915/.

Schlight, John. *Help from Above: Air Force Close Air Support of the Army 1946–1973.* Washington, D.C.: Air Force History and Museums Program, 2003.

_____. *The United States Air Force in Southeast Asia: The War in South Vietnam, the Years of the Offensive 1965–1968.* Washington, D.C.: Air Force History and Museums Program, 1999.

Schneider, Jacquelyn, and Julia Macdonald. "Why Troops Don't Trust Drones." *Foreign Affairs*, December 20, 2017. https://www.foreignaffairs.com/articles/united-states/2017-12-20/why-troops-dont-trust-drones?cid=int-fls&pgtype=hpg.

Seroka, Steven G. "In Search of an Identity: Air Force Core Competencies." Thesis, School of Advanced Studies, Air University, Maxwell Air Force Base, 1997. https://apps.dtic.mil/dtic/tr/fulltext/u2/a391795.pdf.

Shand, Michael P. *JTAC and Forward Air Controller(A) Training: How History Illustrate the Path to the Future.* Thesis. US Marine Corps Command and Staff College, Quantico, 2008.

Sherman, William C. *Air Warfare.* 1936. Reprint, Maxwell Air Force Base: Air University Press, 2002.

Sherry, Michael S. *The Rise of American Air Power: The Creation of Armageddon.* New Haven: Yale University Press, 1987.

Shiner, John F. *The Army Air Arm in Transition: General Benjamin D. Foulois and the Air Corps. 1931—1935.* Ann Arbor: Michigan University. 1975.

_____. *Foulois and the US Army Air Corps, 1931—1935.* Washington, D.C.: Office of Air Force History, 1984.

Shipmen, Brian, and Gary Ellison. *L-Bird: The Little Plane that Did.* Brian Shipmen and Gary Ellison Productions, *2010,* 44 min. http://lbirdthemovie.com/.

Shores, Christopher, Giovanni Massimello, and Russell Guest. *A History of the Mediterranean Air War 1940–1945*. 4 Vols. Philadelphia: Casemate, 2012–2018.

Shu Guang Zhang. *Mao's Military Romanticism: China and the Korean War, 1950–1953*. Lawrence: University Press of Kansas, 1995.

Shulimson, Jack. "The Single Manager Air Controversy of 1968." *Marine Corps Gazette* 80, no. 5. (May 1996): 66–71. https://libproxy.library. unt.edu:2165/publicationissue/C011BF6C87FD4DF8PQ/$B/1/ Marine+Corps+Gazette/01996Y05Y01$23May+1996$3b++Vol.+80+$28 5$29/$N?accountid=7113.

Shultz Jr, Richard H., and Robert L. Pfaltzgraff Jr., eds. *The Future of Air Power in the Aftermath of the Gulf War.* Maxwell Air Force Base: Air University Press.

Shwedo, Bradford J. *XIX Tactical Air Command and ULTRA: Patton's Force Enhancers in the 1944 Campaign in France.* Maxwell Air Force Base: Air University Press, 2001.

Skinner, Douglas W. *Airland Battle Doctrine.* Alexandria: Center for Naval Analysis, 1988.

Slife, James C. *Creech Blue: Gen Bill Creech and the Reformation of the Tactical Air Forces, 1978–1984.* Maxwell Air Force Base: Air University Press.

Smart, Bobby W. *An Air Force Ground Attack Control Capability to Support AirLand Battle.* Thesis, Air War College, Maxwell Air Force Base, 1990.

Smithsonian Institution. "Brodie System." *Smithsonian's National Air and Space Museum.* https://airandspace.si.edu/multimedia-gallery/ nasm-9a001183640jpg.

Sneiderman, Barney. *Warriors Seven: Seven American Commanders, Seven Wars, and the Irony of Battle.* Havertown: Savas Beatie. 2006.

Spector, Ronald H. *Eagle against the Sun: The American War with Japan.* New York: Free Press, 1985.

Spires, David. *Air Power for Patton's Army: The XIX Tactical Air Command in the Second World War.* Washington, D.C.: Air Force History and Museums Program, 2002.

St. Clair, Matthew G. *The Twelfth US Air Force: Tactical and Operational Innovations in the Mediterranean Theater of Operations, 1943–1944.* Maxwell Air Force Base: Air University Press, 2007.

Stephens, Alan. *The War in the Air 1914–1994.* Edited, American edition. Maxwell Air Force Base: Air University Press, 2001.

Steadman, Kenneth A. *A Comparative Look at Air-Ground Support Doctrine and Practice in World War II.* Fort Leavenworth KS: Combat Studies Institute, 1982. https://www.armyupress.army.mil/Portals/7/combat-studies-institute/csi-books/comparat.pdf.

Stockfish, J. A. *The 1962 Howze Board and Army Combat Developments.* Santa Monica, CA: RAND, 1994.

Stokesbury, James L. *A Short History of Air Power.* New York: William Morrow and Company, 1986.

Sumner, Ian. *German Air Forces 1914–1918.* New York: Osprey Publishing, 2006.

Sunderland, Riley. *Evolution of Command-and-control Doctrine for Close Air Support.* Washington, D.C.: Office of Air Force History, 1973.

Tanaka, Yuki, and Marilyn Blatt Young. *Bombing Civilians.* New York: New Press, 2009.

Tate, James P. *The Army and its Air Corps: Army Policy toward Aviation, 1919–1941.* Maxwell Air Force Base: Air University Press, 1998.

Taylor, Maxwell D. *The Uncertain Trumpet.* New York: Harper, 1960.

Theisen, Eric E. *Ground-Aided Precision Strike: Heavy Bomber Activity in Operation Enduring Freedom.* Maxwell Air Force Base: Air University Press, 2003. https://media.defense.gov/2017/Dec/04/2001851904/-1/-1/0/MP_0031_THEISEN_GROUND_AIDED_PRECISION_STRIKE.PDF.

Thomas, Bill. "The Last Soldier Buried in the Tomb of the Unknowns Wasn't Unknown." *The Washington Post.* November 8, 2012. https://www.washingtonpost.com/lifestyle/magazine/last-soldier-buried-in-tomb-of-the-unknowns-wasnt-unknown/2012/11/06/5da3e7d6-0bdd-11e2-a%E2%80%A6/.

Thomas, Gary L. *United States Marine Corps Air-Ground Integration in the Pacific Theater.* Maxwell Air Force Base: Air Command and Staff College. https://www.jstor.org/stable/resrep13742

Thompson, James C. *Rolling Thunder: Understanding Policy and Program Failure.* Chapel Hill: The University of North Carolina Press, 1980.

Tilford, Earl H. *Setup: What the Air Force Did in Vietnam and Why.* Maxwell Air Force Base: Air University Press, 1991.

Tirpak, John A. "F-15X In, Light Attack Out in FY20 Budget." *Air Force Magazine,* February 4, 2019. https://www.airforcemag.com/f-15x-in-light-attack-out-in-fy20-budget/.

————. "Kosovo Retrospective." *Air Force Magazine,* July 8, 2008. https://www.airforcemag.com/article/0400kosovo/.

Toll, Ian W. *The Conquering Tide: War in the Pacific Islands, 1942—1944*. New York: W. W. Norton, 2015.

Trest, Warren A. *Air Force Roles and Missions: A History*. Washington, D.C.: Air Force History and Museums Program, 1998.

Van Atta, Richard H. et al *DARPA Technical Accomplishments Volume II: An Historical Review of Selected DARPA Projects*. Alexandria: Institute for Defense Analysis, 1991.

Van Creveld, Martin. *The Age of Airpower*. New York: Public Affairs. 2011.

_____, Kenneth S. Brower, and Steven L. Canby. *Air Power and Maneuver Warfare*. Maxwell Air Force Base: Air University Press, 1994.

Van der Veer Jr., David G. *Air Power: A Decisive Coercive Strategy?* Newport: US Naval War College, 2006.

Venable, Heather. "Balanced Air Power, Not Bombers: How the Air Force Found its Way." *War on the Rocks*, January 29, 2019. https://warontherocks.com/2019/01/balanced-airpower-not-bombers-how-the-air-force-found-its-way/?fbclid=IwAR0JujjvAXtNLP3J1X8_ihAU9pCNt1wAUyqWVpEc8nz-jstxQCNV5e2MZy6I.

_____. "More Than Planes and Pickle Buttons: Updating the Air Force's Core Missions for the 21st Century." *War on The Rocks*, March 28, 2019. https://warontherocks.com/2019/03/more-than-planes-and-pickle-buttons-updating-the-air-forces-core-missions-for-the-21st-century/?utm_source=W%E2%80%A6.

Vick, Alan J. *Force Presentations in US Air Force History and Airpower Narratives*. Santa Monica, CA: RAND, 2018.

_____. *Proclaiming Air Power: Air Force Narratives and American Public Opinion from 2014 to 2017*. Santa Monica, CA: RAND, 2015.

Warden III, John A. *The Air Campaign: Planning for Combat*. Washington, D.C.: National Defense University Press, 1988.

_____. "The Enemy as a System." *Airpower Journal* 9, no. 1 (Spring 1995), 40–55.

Warner, Christopher G. *Implementing Joint Vision 2010: A Revolution in Military Affairs for Strategic Air Campaigns*. Maxwell Air Force Base: Air University Press, 1999.

Watts, Barry D. *The Foundations of US Air Doctrine: The Problem of Friction in War*. Maxwell Air Force Base: Air University Press, 1984.

Wakefield, Ken. *The Fighting Grasshoppers: US Liaison Aircraft Operations in Europe 1942–1945*. Stillwater, MN: Specialty Press, 1990.

Walker, John R. *Bracketing the Enemy: Forward Observers and Combined Arms Effectiveness during the Second World War.* PhD diss., Kent State University, Kent, OH, 2009. https://etd.ohiolink.edu/.

Walsh, Matthew, William W. Taylor, and John A. Ausink. *Independent Review and Assessment of the Air Force Ready Aircrew Program: A Description of the Model Used for Sensitivity Analysis.* Santa Monica, CA: RAND, 2019.

Walton, Andrew R. *The History of the Airborne Forward Air Controller in Vietnam.* Thesis, US Army Command and General Staff College, Fort Leavenworth, KS, 2004.

Wawro, Geoffrey. *A Mad Catastrophe: The Outbreak of World War I and the Collapse of the Habsburg Empire.* New York: Basic Books. 2014.

_____. *Quicksand: America's Pursuit of Power in the Middle East.* New York: Penguin Press, 2010.

_____. *Sons of Freedom: The Forgotten American Soldiers Who Defeated Germany in World War I.* New York: Basic Books, 2018.

Weigley, Russell F. *The American Way of War: A History of United States Military Strategy and Policy.* New York: Macmillan, 1973.

Welsh, Mark A. "Day of the Killer Scouts." *Air Force Magazine,* September 3, 2008. https://www.airforcemag.com/article/0493scouts/?signoff=true.

_____. "CSAF to Airmen: Sequestration creates tough choices." *USAirForce, News.* March 18, 2014. https://www.af.mil/News/Article-Display/Article/474951/csaf-to-airmen-sequestration-creates-tough-choices/.

Westenhoff, Charles M. *Military Airpower: A Revised Digest of Airpower Opinions and Thoughts.* Maxwell Air Force Base: Air University Press, 2007.

Wetterhahn, Ralph. "Ravens of Long Tieng." *Air and Space Magazine,* November 1998. https://www.airspacemag.com/military-aviation/ravens-of-long-tieng-284722/.

Whitcomb, Darrel. "Farm Gate." *Air Force Magazine,* December 2005. https://www.airforcemag.com/article/1205farmgate/.

_____. "PAVE NAIL: There at the Beginning of the Precision Weapons Revolutions." *Air Power History* 58, no. 1 (Spring 2011): 14–27. https://www.afhistory.org/wp-content/uploads/2011_spring.pdf.

Wilbanks, James H. *The Tet Offensive: A Concise History.* New York: Columbia University Press, 2007.

Wills, Craig D. *Airpower, Afghanistan, and the Future of Warfare: An Alternative View.* Maxwell Air Force Base, AL: Air University Press, 2006.

Williams, George K. *Biplanes and Bombsights: British Bombing in World War I.* Maxwell Air Force Base: Air University Press, 1999.

Winnefeld, James A., Preston Niblack, and Dana J. Johnson. *A League of Airmen: US Air Power in the Gulf War.* Santa Monica, CA: RAND, 1994.

Witteried, Peter F. "A Strategy of Flexible Response." *Parameters.* Vol. II, No. 1, 1972. 2–16.

Worden, Mike. *Rise of the Fighter Generals: The Problem of Air Force Leadership 1945–1982.* Maxwell Air Force Base: Air University Press, 1998.

Y'Blood, William T. *Down in the Weeds: Close Air Support in Korea.* Washington, D.C.: Air force History and Museums Program, 2002.

Yurovich, Douglas P. *Operation Allied Force: Air Power in Kosovo. A Study in Coercive Victory.* Carlisle Barracks, PA: US Army War College, 2001.

Zhang, Xiaoming. *Red Wings over the Yalu: China, the Soviet Union, and the Air War in Korea.* College Station: Texas A&M University Press, 2002.

Zimmerman, S. Rebecca, Kimberly Jackson, Natasha Lander, Colin Roberts, Dan Madden, and Rebecca Orrie. *Movement and Maneuver: Culture and the Competition for Influence Among the US Military Services.* Santa Monica, CA: Rand, 2019.

Index

Operation Enduring Freedom, 185–186, 204, 206
Operation Husky, 42, 45–46, 48–49, 78
Operation Iraqi Freedom, 187–188, 190, 207, 215–216
Operation Overlord/D-Day, 5, 55–57, 60
Operation Shingle, 42, 50
Operation Torch, 42–44, 46, 48–49, 63–64, 67, 132
Operation Unified Protector, 204, 207
OSD Office of the Secretary of Defense/Secretary of Defense, 133–134,
        140, 153–154, 156, 176, 187

P

Patton, George, 5, 61, 78
Phu Cat, 119
Pineapple, 51–53, 57, 64, 103, 213

Q

Quesada, Elwood R. (Pete), 5, 34, 55, 57, 59, 69–71, 75

R

Raven, 6, 102, 117–118, 126
rocket(s), 2, 54, 60, 67, 70, 84–85, 97, 99, 104, 111–114, 120–121,
        124–127, 143, 152, 171
Rover Joe, 51–53, 57, 59, 61–62, 64–65, 67, 77, 86, 89, 148, 213
Royal Air Force (RAF)/Royal Flying Corps (RFC), 21, 24, 188
Royal Australian Air Force, 79

S

Sandy, 127–128, 137, 197, 212
Schwarzkopf, Norman, 156–160
Serbia/Serbian, 169, 171–175, 214
Signal Corps, 20–21, 24, 34
Slow FAC, 119, 139, 148, 152, 219, 221
smoke, 2, 29–30, 59, 65–66, 84–85, 97, 104–105, 114, 124, 126, 179
Sniper Pod, 203
South Korea/South Korean, 1, 2, 73–94, 128
South Vietnam, 95–130, 135–136, 162, 215
squadron, 2–3, 5, 20, 22, 24–25, 27, 29, 31–32, 35–36, 47, 57, 60, 62–63,
        68, 70, 77, 80–81, 83–84, 86, 91, 96, 99–103, 106–109, 112,